Geohazards

ASSOCIATION OF GEOSCIENTISTS FOR INTERNATIONAL DEVELOPMENT

AGID Report Series
THE GEOSCIENCES IN INTERNATIONAL DEVELOPMENT

Other Titles in the Series

Geohazards
Natural and man-made

Edited by

G.J.H. McCall
Consultant and Fellow, Liverpool University

D.J.C. Laming
Consultant, Herrington Geoscience, Exeter

and

S.C. Scott
Co-ordinator, Geological Hazards Assessment, Mitigation and Information Unit, Polytechnic South West, Plymouth.

CHAPMAN & HALL
London · New York · Tokyo · Melbourne · Madras

Published by Chapman & Hall, 2–6 Boundary Row, London SE1 8HN

Chapman & Hall, 2–6 Boundary Row, London SE1 8HN, UK

Chapman & Hall, 29 West 35th Street, New York NY10001, USA

Chapman & Hall Japan, Thomson Publishing Japan, Hirakawacho Nemoto Building, 6F, 1-7-11 Hirakawa-cho, Chiyoda-ku, Tokyo 102, Japan

Chapman & Hall Australia, Thomas Nelson Australia, 102 Dodds Street, South Melbourne, Victoria 3205, Australia

Chapman & Hall India, R. Seshadri, 32 Second Main Road, CIT East, Madras 600 035, India

First edition 1992

© 1992 G.J.H. McCall, D.J.C. Laming and S.C. Scott

Typeset by Herrington Geoscience and Exe Valley Dataset Ltd
Printed in Singapore by Fong & Sons Printers Pte Ltd

ISBN 0 412 43920 4 (HB)   0 412 43930 1 (PB)

A catalogue record for this book is available from the British Library

Library of Congress Cataloging-in-Publication data available

Contents

Contributors

N.N. Ambraseys   Department of Civil Engineering, Imperial College, London, UK.

J.L. Anderson   Center for the Study of Active Volcanoes, University of Hawaii at Hilo, Hilo, Hawaii.

R.J. Blong   School of Earth Sciences, Macquarie University, Australia.

D. Brook   Land Stability Branch, Minerals and Land Reclamation Division, Department of the Environment, London, UK.

C.W.A. Browitt   British Geological Survey, Edinburgh, Scotland, UK.

R.W. Decker   Center for the Study of Active Volcanoes, University of Hawaii at Hilo, Hilo, Hawaii.

M.R. Degg   Department of Geography, Chester College, Chester, UK.

Perla J. Delos Reyes   Department of Science and Technology, Philippine Institute of Volcanology and Seismology, Hizon Building, Quezon City, The Philippines.

W.S. Fyfe   Faculty of Science, University of Western Ontario, London, Canada.

D.J. George   Dames & Moore International, Booth House, Church Street, Twickenham, UK.

P.J. Gregory   Department of Soil Science, University of Reading, Reading, UK.

M.L. Hall   Instituto Geofisico, Escuela Politecnica Nacional, Quito, Ecuador.

D.K.C. Jones   Department of Geography, London School of Economics, London, UK.

G.J.H. McCall   Consultant and Fellow of Liverpool University, Liverpool, UK.

Patricia A. Mothes   Instituto Geofisico, Escuela Politecnica Nacional, Quito, Ecuador.

S. Nortcliff   Department of Soil Science, University of Reading, Reading, UK.

D.W. Redmayne   British Geological Survey, Edinburgh, Scotland, UK.

J.M. Reynolds   Department of Geological Sciences, Polytechnic South West, Plymouth, UK.

S.C. Scott   Geological Hazards Assessment, Mitigation and Information Unit, Department of Geological Sciences, Polytechnic South West, Plymouth, UK.

D.A.V. Stow   Department of Geology, The University, Southampton, UK.

Wang Sijing   Institute of Geology, Academia Sinica, Beijing, China.

H.Th. Verstappen   International Institute for Aerospace Survey and Earth Sciences (ITC), Enschede, The Netherlands.

Alice B. Walker   British Geological Survey, Edinburgh, Scotland, UK.

Zhao Xitao   Institute of Geology, Academia Sinica, Beijing, China.

Acknowledgements

The editors would like to express their sincere thanks to all who helped in the production of this book, especially Kit Patrick of Herrington Geoscience, Mick Lear of M.J.L. Graphics, and all the reviewers of the individual contributions.

Preface

Dorrik A.V. Stow

Editor in Chief, Association of Geoscientists for International Development (AGID)

AGID is particularly pleased to see published this latest report in its Geosciences in International Development Series, as a significant contribution to the onset of the UN Decade of National Disaster Reduction, and as a mark of AGID's growing concern over the potential and actual effects of geohazards throughout the developing world.

The problem of geohazards is increasing, not because the rate of earth processes is accelerating, nor because the voice of the media appears to be paying more and more attention to natural drama and disaster, but primarily because the burgeoning world population requires the cultivation of land more prone to hazard. More people and property are thus exposed to the risk of catastrophic disaster than ever before, and the death toll inevitably rises.

As we go to press, reports of major volcanic eruptions in the Philippines and in SW Japan are flashing across our television screens. One persistent problem, perceived by several authors in this volume, is the question of *warning time*: if the authorities issue an alarm and evacuate people too early, the credibility of the warning is lost; fearful of looting or other damage, they return to protect their homes—who would not do the same? But, it was exactly this scenario that led to the awful disaster of Armero in Colombia where 25,000 people perished in a sea of mud; many of those lives could have been saved if proper warning procedures had been adopted.

Also as we go to press, the terrible results of the cyclone in Bangladesh are being revealed. The people and the Government of that country are only too well aware of the ever-present threat from severe climatic conditions and river flooding. But, a poor country with over 100 million people has many priorities demanding attention and finance, and it is wrong to criticise the inadequate level of preparedness for a disaster that has a 10 or 20 year return period; but it is right to ask whether climatic change is exacerbating the cyclonic effects, and that some moral and financial support should be given by the developed world towards helping countries such as Bangladesh to tackle the problems of disasters.

Almost certainly as this volume is read, there will be other disasters—earthquakes, volcanoes, landslides, hurricanes, floods—that are wreaking havoc, destroying livelihood and lives in some corner of the globe.

As geoscientists there are perhaps three concerns that should be uppermost in our minds as we join an international effort to combat the adverse effects of natural hazards. The first must be to improve our scientific understanding of the nature and causes of such hazards and to work towards more reliable prediction of their occurrence and magnitude.

The second is to communicate this knowledge in an effective and timely manner to politicians, planners and decision makers who are involved in the instigation of policies and programmes to mitigate against disaster, as well as in the issuing of public warnings and the implementation of evacuation procedures. If the existing communication gap between scientists and politicians is allowed to persist then the application of the results of geohazards research by many thousands of scientists the world over will be severely impaired.

The third concern and another important theme in the papers in this volume is involvement of local people. This is vital for effective disaster mitigation, whether in a developing or a developed country; whether arranging for a tree nursery to be established in Ethiopia or an earthquake-resistant building code in Algeria, a solution imposed from above has less chance of being effective than one that takes hold at the level of ordinary people. Programmes of public education about natural hazards, participation in disaster preparedness and early warning schemes, and learning to live as safely as possible in a disaster-prone region, are all areas where good communication between scientists and the public is essential.

AGID is a network of geoscientists throughout the world concerned with the application of the geosciences to grassroots development. AGID has taken a particular interest in Natural Hazards in recent years; the application of geoscientific knowledge in this field can significantly improve the quality of life in many parts of the developing world. The collection of papers in this volume illustrate clearly this role and, we hope, will provide both data and ideas from which further advances can be made.

Dorrik A.V. Stow

1 Natural and man-made hazards: their increasing importance in the end-20th century world

G.J.H. McCall

Introduction

This volume is based on a meeting devoted to the theme "Geo-Hazards—Natural and Man-Made" held at the Geological Society of London on 18th October 1989.

Sponsored by three specialist groups of the Society— the Joint Association of Geoscientists for International Development, the Joint Association for Geophysics and the Engineering Group—it was timed to coincide with the commencement of the United Nations International Decade for Natural Disaster Reduction (IDNDR) and aimed especially at countries of the Third World. These countries are particularly vulnerable to natural hazards —"small countries are more vulnerable to hazard impacts than large countries, and small *poor* countries the most vulnerable" (Johnson and Blong, 1985). Equally, these countries are highly vulnerable to man-made hazards— for example, soil loss due to poor agricultural practices, constructional failure due to poor engineering practices, and pollution due to poor water-supply management.

The theme of Geohazards is one of extreme topicality, especially with the media—there is a popular belief that we are moving into an age of disaster, and this may, to a degree, be the truth. Whereas geohazards have always been with us, there are numerous other factors increasing the incidence and scale of hazard and disaster:

(a) increased population concentrations;

(b) increased technological development;

(c) overintensive agriculture, increased industrialisation;

(d) excessive use of the internal combustion engine and other noxious fume-emitters, combined with wasteful transport systems;

(e) bad technological practices in construction, water management and waste disposal;

(f) excessive emphasis on commercial development; and

(g) increased scientific tinkering with Nature without concern for possible long-term effects or disasters.

Also, we are only just now beginning to comprehend the scientific implications of many processes: for example, natural processes such as radon emanation from the ground; man-made processes such as acid rain and destruction of the ozone layer by chlorofluorocarbons; and controversial natural and man-made interactions such as the "greenhouse effect".

Media treatment tends to be sensational, shallow and unquestioning, but behind the "hype" there is a growing series of major environmental problems that require expert scientific research and careful evaluation.

That the theme is indeed topical was emphasised by the occurrence of the Loma Prieta earthquake near San Francisco on the night before the meeting of 18th October (Muir Wood, 1989).

To the original contributions presented at the meeting, we have added a number of further papers, mainly from those unable through constraints of distance and travel expense to be present at the meeting in person.

Hazards—natural and man-made

The UN, in initiating the International Decade for Natural Disaster Reduction, has incorporated the term "Natural" in its definition and terms of reference (Anon,

1989a). For this volume, we have adopted a wider definition, not because we see "Natural" and "Man-Made" hazards as separate entities, but rather because we see them as so intricately linked that they must be considered together. Also we are conscious that, whereas natural geohazards are at present the major cause of loss of life and property, the effects of man-made geohazards are increasing, particularly the "quiet", pervasive, slow-onset hazards, and may eventually replace the natural hazards as the main cause.

A **hazard** relates to human life and property: strictly speaking, there is no hazard unless humans, their possessions and their activities are involved. Doornkamp (1989) defined a geohazard as "a hazard of geological, hydrological or geomorphological nature which poses a threat to man and his activities". Perhaps the definition should be even wider—a geohazard being "one that involves the interaction of man and any natural process of the Planet": this brings in soil science, geophysics, meteorology, glaciology and climatology, and involves the solid Earth, the oceans and the atmosphere, even cosmic processes.

The line between natural and man-made geohazards is finely drawn and may be blurred. Doornkamp (1989) remarked "the concept of a natural hazard is an ambiguous one, for many catastrophic events within the environment are man-induced, or at least made worse by the intervention of man. This reduces their 'naturalness' but not their impact, and gives added weight to the need for hazard mapping . . ." **Natural hazard mapping** is critical in guiding Man's siting of various constructions, settlements and activities and preventing man-made hazards, as well as reducing vulnerability to natural hazards. The two types of geohazard are intricately linked.

Just as Man may influence and trigger off major, essentially natural hazards, there are other geohazards which are more obviously "man-made": even so, most of these involve geological processes. Examples of this are the man-made pollution in the Red Sea (Antonius, 1989) and oil spills from tankers on the Panama Coast (Jackson and others, 1989), both of which disastrously damaged coral growth and other marine life, destroying the marine environment over large areas. Many hazards associated with construction—for example, those described at the 18th October 1989 meeting by T.I. Longworth, the slow-moving La Butte landslide in Mauritius, and by G.J. Hearn, the hazards to roads in the Himalayas (Laming, 1989; McCall and Laming, 1989)—involve triggering of natural forces, though related to poor engineering practices or insufficient prior investigation. Mining also has introduced a category of geohazards, making land unfit for reuse by other development due to dangers of subsidence, again triggering natural forces.

Intensive and pervasive geohazards

It is the spectacular, *rapid-onset*, *intensive* hazards (Johnson and Blong, 1985; Blong and Johnson, 1986) that catch the media headlines (volcanic eruption, earthquakes, tsunamis, floods, lahars) but, in the long term, it may well be the *slow-onset*, *pervasive* geohazards that cause the greatest loss of life, causing starvation and disease to large populations (for example, the loss of soil, a major global hazard discussed by Nortcliff and Gregory, this volume, and Fyfe—in this volume and also in Fyfe, 1989a). Other such hazards include those involving rising groundwater under cities (George, this volume); subsidence of cities related to groundwater abstraction (Nutalaya, 1989); and coastal hazards related to rising sea level (Wang Sijing and Zhao Xitao, this volume).

Geohazards can only be successfully approached with a fully interdisciplinary treatment, and they require as well an international or even global approach. Scientists have for too long found it convenient to keep within the boundaries of their disciplines and subdisciplines, but the pressing demands of the environment are at last breaking these barriers down—for example, a global problem such as the "greenhouse effect" can only be solved by co-operation between numerous scientific disciplines. National barriers in science are likewise breaking down.

There is also a need to consider geohazards from a sociological viewpoint as well as from a scientific one, for the *perception* and *response* of the local populace and all tiers of government are as important as scientific and technological understanding (Blong and Johnson, 1986; papers in this volume by Brook, Blong, Delos Reyes, Hall, and Mothes).

The subjects covered in this volume include papers on both the well-known hazards such as earthquakes and volcanic eruptions, and on the lesser-known glacial hazards, soil loss and rising groundwater. There are also papers on what can be done towards reducing their impact, from sophisticated analysis of remotely sensed images to volunteer monitoring schemes by residents of volcanic and earthquake-prone areas.

Thus we may seem to have covered a very wide range of geohazards, certainly worthy of a volume dedicated to the subject. Yet it is really only the "tip of the iceberg". It does not, for instance, cover three of the most important subjects at the present time—radon emission, acid rain and global warming (the "greenhouse effect"). The first two relate mostly to developed countries, though galloping urbanisation and industrialisation of the Third World will undoubtedly extend the problem of acid rain to those countries in time. The "greenhouse effect" is a subject that needs a conference of its own at the very least—in fact a number of meetings are to be devoted to this theme in Britain alone this year! It is a complex geohazard, probably involving man-made acceleration of natural change.

Carbon dioxide is the most abundant of the volcanic gases and is abundant also in many hydrocarbon-bearing "natural gases". There is a tendency to suppose that the global intensity of volcanic and other natural gas emissions at the surface of the Earth has remained stable and will remain so through time: the reality is otherwise—the history of the Earth is one of continuing fluctuations in this natural process, as van Andel (1989) recently pointed out, and the gas output from within the planet almost certainly undergoes secular variations. We are far from understanding the possible cosmic influences and the role of oceanic processes in carbon dioxide and other gas transfer to the atmosphere (such factors as upwelling and nutrient supply and consumption by plankton populations: Wilson, 1989). Though we must do all we can to reduce Man's contribution to the "greenhouse effect", we may yet be overestimating that contribution as against the natural output of gases. If so, the "greenhouse effect" may well be irreversible by Man, as it is largely a product of earth processes.

Then there are the whole range of problems posed by Man's harnessing of the nuclear processes. The problems of the disposal of high-level nuclear waste remain unsolved and even disposal of low-level waste is subject to vigorous local social protest. There is the problem of the safety of the installations themselves: and this involves the geosciences. A nuclear power station had to be closed down in the USSR following the Spitak Earthquake in Armenia, and the construction of another was halted (Muir Wood, 1989). Earthquake risk zoning in the USSR was found to be inadequate—such as regions of apparent moderate risk where an occasional damaging earthquake might occur at century intervals; which require much more accurate risk assessment than high incidence regions, where one would never site a nuclear power station and risk a naturally-triggered Chernobyl-scale disaster. Partly because of the conjunction of a devastating earthquake and the nuclear installation construction, the importance of the improved earthquake risk evaluation has been highlighted (Ambraseys, Degg, this volume).

The contribution by Fyfe concerns the global hazards posed by the Earth's escalating population. This theme has been discussed recently by the Duke of Edinburgh; it was raised by the present author in the context of the need for land-use planning, the conservation of resources and the protection of the natural environment (McCall, 1991) and, again, by Fyfe (1989a) at Strasbourg. Whereas Fyfe related our new general awareness of the fragility of Man's hold on the biosphere, the shallow global "shell" in which he can survive, to awareness of Man's ability to destroy himself by nuclear catastrophe; perhaps instead the awakening came from seeing our

water-covered planet, splendid and blue with the eyes of the lunar astronaut, and realising that our habitat is unique in the solar system—destroy it, and we are destroyed?

Space exploration, particularly the unmanned missions, has given us an awareness of the precariousness of Man's hold on Planet Earth and his cosmic isolation as never before. The type of hazard covered by Nortcliff and Gregory (this volume), and mentioned again by Fyfe, has quite frightening statistics—i.e., the annual incremental global loss of soil. Soil is as important to Man's continuance on the Earth as the air we breathe and the water we drink. Equally frightening are the statistics of population growth (and the looming future of hundreds of megacities, mainly in the Third World, mentioned by Degg, this volume). The statistics of nitrate, phosphate and other fertiliser pollutions of the soil and groundwater given by Fyfe (1989a, 1989b, and this volume) are just as daunting: also the impossibility of disposing of Man's toxic and non-toxic waste materials if the global population escalates much further (Fyfe, this volume). Paul Ehrlich (Anon, 1989b) stated that the human population has already exceeded the Earth's long-term carrying capacity: "With current and foreseeable technologies even today's 5.2 billion people can only be supported by a continual depletion of humanity's one-time inheritance from the planet—deep, rich agricultural soils, fossil groundwater and the diversity of non-human species. Food supplies will get steadily worse in Africa and Latin America, and with the prospect of climatic change [i.e. the 'greenhouse effect'], the picture becomes much bleaker".

The global hazard posed by Ehrlich and Fyfe may be summed up by posing the question—"Is not the Earth even now, in a sense, overgrazed by Man?"

This is the one which over-rides all other geohazards —the hazard of overpopulation and excessive human activity, adverse to the environment. Politicians turn a deaf ear to this hazard, for they appear to be unable or unwilling to think in the long term. To protect the environment, politically drastic and unpopular measures will be needed, not, as at present, rejecting necessary but drastic measures because "the public would not like it!". Politicians must also start thinking globally and for the long term. To quote van Andel (1989) again, "The greening of government without necessarily requiring painful action against the environmental mischief committed around us every day" is a meaningless facade. For example, we should not regard agricultural land as a destroyable resource simply because there is a short-term glut of foodstuffs. We should conserve it, for it is a finite resource; once destroyed, it cannot be remade by natural processes for hundreds or thousands of years. We should also be planning for the day when fossil fuels become exhausted, introducing less wasteful and environmentally damaging transport systems.

The main challenge

It is the insidious, pervasive long-term geohazards that threaten all mankind, much more than the intensive, rapid-onset natural hazards. Nevertheless, there is much that we can do, and should be doing, to counter the locally destructive geohazards which are the subject of most of the contributions to this book (and the setting up of the unit at Plymouth described by Scott, this volume, is a welcome recognition of this fact). From the point of view of natural disasters, the International Decade for Natural Disaster Reduction will produce many contributions to scientific research, but this output will be largely wasted if we cannot overcome the *perception/response* problems (see Delos Reyes, this volume) in the countries most likely to be affected by hazards.

Problems of finance, lack of perception or inertia tend to prevent any action being taken—and it is no use educating if those educated take no action. Likewise, the perception/response problem concerns engineers and developers, who must be persuaded to consider *safety* rather than *cost reduction* as paramount.

The real challenge of the IDNDR may not be the scientific advances made in the geohazard field, but rather the degree of success achieved in obtaining perception and response in the vulnerable countries and by the engineers and developers in, or working in, those countries.

References

Anon., 1989a. Natural hazards—Tokyo declaration of the International Decade for Natural Disaster Reduction. *AGID News*, vol 58, p 14.

Anon., 1989b. Experts meet, discuss global environment. *Geotimes*, vol 34(8), p 24–26.

Antonius, A., 1989. Coral pathology and seawater pollution in the eastern Red Sea. *Abstract Volume*, 1989 Annual Meeting of the International Society for Reef Studies, Marseille, 14–19 decembre 1989, p 25.

Blong, R.J., and R.W. Johnson, 1986. Geological hazards in the Southwest Pacific and Southeast Asian region: identification, assessment and impact. *BMR Journal of Australian Geology and Geophysics*, vol 10, p 1–15.

Doornkamp, J.C., 1989. Hazards. In: G.J.H. McCall and B.R. Marker (eds), *Earth Science Mapping for Planning, Development and Conservation*, Graham and Trotman, London, p 157–173.

Fyfe, W.S., 1989a. Earth: the year 1,001,989. Acceptance lecture, Arthur Holmes Medal, Vth Meeting of European Union of Geosciences. *Terra Nova*, vol 1 (4), p 315–317.

Fyfe, W.S., 1989b. Soil and global change. *Episodes*, vol 12 (4), p 249–254.

Jackson, J.B.C., J.D. Cubit, B.D. Keller, V. Battista, K. Burns, H.M. Caffey, R.M. Caldwell, S.D. Garrity, C.D. Getter, C. Gonsalez, H.H. Guzman, K.W. Kaufmann, A.H. Knap, S.C. Levings, M.J. Marshall, R. Steger, R.C. Thompson and E. Weil, 1989. Effects of a major oil spill on Panamanian Coastal Marine Communities. *Science*, vol 243, p 37–44.

Johnson, R.W., and R.J. Blong, 1985. Vulnerability and the identification and assessment of geological hazards in the Southwest Pacific and Southeast Asia. In: M.B. Katz and E.J. Langevad (eds), *Geosciences for Development*, AGID Report Series no 9, p 217–225.

Laming, D.J.C., 1989. Conference Report: Geo-Hazards. *AGID News*, no 60, December 1989, p 18–19.

McCall, G.J.H., 1991. The "Alternative" Earth Science Mapping for Planning and Development. In: D.A.V. Stow and D.J.C. Laming (eds), *Geoscience in Development*, AGID Report Series No 14, Balkema Publishers, Rotterdam.

McCall, G.J.H., and D.J.C. Laming, 1989. Conference Report: GeoHazards, natural and man-made. *Journal of the Geological Society of London*, vol 147, p 879–881.

Muir Wood, R., 1989. After Armenia. *Terra Nova*, vol 1 (2), p 209–212.

Nutalaya, P., 1989. Flooding and land subsidence in Asia. *Episodes*, vol 12 (4), p 239–248.

Van Andel, T.H., 1989. Global change: do only the present and future count? *Terra Nova*, vol 1 (3), p 236–237.

Wilson, T.R.S., 1989. Climatic change: possible influence of ocean upwelling and nutrient concentration. *Terra Nova*, vol 1 (2), p 172–176.

Part One

Volcanic Hazards

2 Volcano risk mitigation through training

J.L. Anderson and R.W. Decker

Abstract An educational program in volcano monitoring is currently being offered at the Center for the Study of Active Volcanoes (CSAV) at the University of Hawaii at Hilo. The Center is a cooperative effort of the Geological Survey's Hawaiian Volcano Observatory and the University. A main objective of the program is the mitigation of volcanic risk worldwide, with special emphasis on the Circum-Pacific area.

The volcano monitoring program is designed to assist developing nations in attaining self-sufficiency in the area of applied volcanology. The training emphasizes volcano monitoring methods, both data collection and interpretation, currently in use by the US Geological Survey. In addition, the training program addresses the assessment of volcanic hazards and the interrelationship of scientists, governing officials, and the news media during volcanic crises. The program is intended to decrease volcanic risk in developing nations by providing local scientists with the necessary skills to collect base-line geologic, geophysical, and geochemical data on dangerous volcanoes for which such data is currently unavailable or inadequate. It is anticipated that the CSAV training program will improve international cooperation in the area of volcano monitoring and help make possible quick and effective international response to emergency situations.

Introduction

Active volcanoes have been and will continue to be a significant danger to mankind, as an estimated 10% of the world's population totaling approximately 360 million people now live in close proximity to them (Peterson, 1986). On the average, 50 volcanoes erupt each year (Simkin and others, 1981) with the majority of this activity, approximately 80%, in the Circum-Pacific area. Volcanoes worldwide kill about 1000 people and cause US$100 million in property damage each year. Of course, in some years no one is killed by volcanoes, but twice in this century single eruptions have brought death to more than 20,000 people.

Volcano monitoring methods

Volcano monitoring is an effective means of providing early warning of impending eruptions. Precursory seismicity and deformation, for example, are two important physical clues widely used to reliably forecast activity at volcanoes. Other methods that show promise include geophysical techniques such as geoelectric, geomagnetic, microgravity, radar, and thermal radiation. Also, geochemical monitoring methods are currently being studied to determine whether or not there are consistently detectable variations in composition, quantity, or rate of emission of volcanic gases prior to an eruption. A comprehensive approach would include all of these methods. However, the most reliable immediate results are currently obtainable through monitoring of seismicity and deformation.

In addition to these instrumental approaches, it is important to characterize the eruptive history of a given volcano. This involves basic geologic mapping that includes characterization of the eruptive products, dating of various eruptive episodes, and analysis of the structure of the volcano and surrounding areas. This information contributes to understanding the statistical frequency, anticipated explosivity, type, and potential areal extent of eruptive events. It also provides insight into the probability of gravity-driven sector collapse and resulting long-runout debris avalanches.

The various instrumental monitoring methods that are either in use or under development are frequently "high tech" and relatively expensive. This means that only a limited number of volcanoes can be fully instrumented and that such instrumentation may not be affordable in developing nations. Robert Tilling has correctly pointed out that an "urgent need exists to develop reliable but widely applicable 'low tech' methods, [that are] more affordable and easier to use" (Tilling, 1989).

In addition, personnel need to be trained in the use of such methods so that a greater number of active or potentially active volcanoes can be monitored. Developing nations, for example, should be able to have the capability of collecting their own baseline seismic, deformation, and geologic data. Such information could greatly increase the amount of time available to evacuate areas at risk and could increase the effectiveness of international response efforts.

Hawaii, a natural laboratory

Appropriately focused training programs in volcano monitoring are currently not included in typical geology curricula. A significant limitation is finding relevant field localities located close to educational institutions where realistic training exercises can be conducted. However, the volcanoes of Hawaii are ideally suited for such efforts: they are among the most consistently active on earth and have the additional advantage of being relatively approachable.

The Hawaiian Islands are centrally located with respect to the Circum-Pacific area. Kilauea Volcano is one of the world's best-documented volcanoes with nearly 80 years of detailed records accumulated at the United States Geological Survey's Hawaiian Volcano Observatory (HVO) and two centuries of historical accounts. It has long been a focus of international attention in the area of volcanology, and has consequently been the subject of a wealth of contributions to the scientific literature.

The methods developed to forecast volcanic eruptions in Hawaii have been applied successfully to volcanoes in other tectonic settings in many parts of the world. The consistent eruptive history and approachability of the volcano has made it an effective proving ground for developing volcano monitoring methods.

The University of Hawaii has recently established a baccalaureate degree program in geology with associated equipment, laboratories, and classrooms at its campus in Hilo, 40 km northeast of Kilauea Volcano. This facility is ideally located to provide educational programs relevant to volcano monitoring training.

Center for the Study of Active Volcanoes

In 1989, the University Board of Regents and the State Legislature formally established and funded the Center

for the Study of Active Volcanoes (CSAV) to be head-quartered in the building housing the Department of Geology at the Hilo Campus. The Center, as chartered, is a cooperative effort of the Hawaiian Volcano Observatory (HVO) and the University of Hawaii.

The purpose of the CSAV is to help save lives and prevent destruction of property by increasing the number of persons capable of monitoring active volcanoes and by promoting awareness of volcanic hazards.

The establishment of the Center at Hilo was made possible in large part because no capital investment was necessary to fund construction of buildings, a major obstacle to previous international efforts to establish such a Center. The physical facilities needed for the CSAV program already existed; classroom, laboratory, and office space was already present in support of the Department of Geology. The Hawaiian Volcano Observatory, located at Kilauea Volcano, serves as a base of operations for field activities. Research analytical facilities and additional teaching faculty are available at the University of Hawaii at Manoa, located on the island of Oahu, 400 km to the northwest.

The two organizations, HVO and CSAV, are funded by the Federal Government and State of Hawaii respectively. Each has its own organizational structure. However, the two are linked by the Charter for the Center wherein the Scientist-in-Charge of HVO is a permanent member of the Advisory Board and, in that capacity, interacts with the Director of the Center from the University of Hawaii at Hilo. HVO plays an essential advisory role in developing the scientific/technical elements of the program; the University of Hawaii at Hilo, on the other hand, carries out its implementation. This cooperative venture gives the Volcano Observatory greatly expanded educational capabilities while, at the same time, it gives the University the opportunity to offer courses of study that are truly unique in terms of applicability and relevance to the area of volcanic risk mitigation.

The CSAV program in volcano monitoring

A summer curriculum in volcano monitoring has been designed to be offered annually for students from developing nations and to operate independently from the normal summer session of the University. Class size is kept small to maximize the amount of attention given to each student. A pilot course held in 1987 involved four students from Indonesia, and this number was judged to be an approximate limit for the first formal course held in 1990. Those who complete the curriculum earn a certificate in volcano monitoring.

The course, as currently structured, consists of four subject modules including:

1. physical volcanology and hazards,
2. deformation monitoring,
3. seismological monitoring, and
4. geochemical monitoring.

However, the amount of attention devoted to each is variable and is dependent upon the background and needs of the particular group of students in attendance. The primary need of one group might be, for example, in the area of seismology. The modules themselves are designed to emphasize hands-on experience with as much as 90% of the time spent in the field or laboratory. Discussions are conducted as frequently as possible at field localities.

Most participants in the program will have had some scientific training. However, there is a clear need for developing the capability of providing training for persons with little or no formal scientific experience. This is particularly true of crisis-response situations: these are currently being dealt with flexibly on a case-by-case basis and the curriculum is tailored to the needs of a particular group. Background enquiries are made for each participant and emphasis is placed on the specific volcanoes or types of volcano present in his or her particular country.

The physical volcanology and hazards portion of the curriculum focuses on essential basics, with emphasis on case histories relevant to volcano hazards assessment. Each case history is taught by a scientist with first-hand experience. Examples include Nevado del Ruiz (Colombia), Pozzuoli and Mount Vesuvius (Italy), Mount St Helens (USA), Kilauea and Mauna Loa (Hawaii, USA), and others. Basic principles include mapping techniques and the fundamentals of how volcanoes work.

Kilauea and Mauna Loa Volcanoes provide excellent opportunities for assessing actual hazards to communities on the Island of Hawaii. A lava flow from Mauna Loa, for example, nearly reached the town of Hilo in 1984. Hilo is the second largest city in the State of Hawaii and is the only other significant seaport outside of Honolulu. The US Geological Survey has reported that the summit of Mauna Loa is now 75% reinflated since the 1984 eruption and that the probability of a new eruption is increasing during the 1990s; indeed, the smaller com-

Figure 1 Leveling survey along the east rift zone of Kilauea Volcano (photograph courtesy of J.D. Griggs, US Geological Survey).

Figure 2 Electronic distance measurements being conducted at Kilauea Volcano using a short-range geodimeter (photograph courtesy of J.D. Griggs, US Geological Survey).

munity of Kalapana, located southwest of Hilo, was almost totally destroyed during April and May of 1990 by lava flows from Kilauea Volcano. Other communities such as Kailua Kona are also in jeopardy from potential eruptions of either Mauna Loa or Hualalai Volcanoes. Hualalai, located about 35 km from the town of Kailua Kona, last erupted in 1801.

Monitoring methods
Deformation monitoring includes methods designed to detect horizontal and vertical ground motion on volcanoes. This includes precise leveling surveys (Figure 1), dry tilt, and electronic distance measurements (Figure 2). Students take part in extensive traverses along established lines which are used to monitor Kilauea and Mauna Loa Volcanoes, and become experienced at reducing and interpreting the data. Students are also introduced to surveying methods employed on steep-sided volcanoes such as trigonometric tilt; the steeper slopes of the older volcanoes of the island are excellent sites for training of this type. The older volcanoes are capped by an alkalic suite of volcanic rocks emplaced on top of earlier tholeiitic

suite rocks, which produces a steeper, more rounded, and irregular summit in contrast to the gentler slopes and summit calderas of the younger more active volcanoes.

The seismological monitoring methods course involves field deployment of both portable seismographs and permanent seismic arrays. It also includes the requisite communication with those instruments; CSAV at Hilo is currently equipped with a permanent seismic array that is capable of continuously monitoring radio-telemetered analog signals from selected seismic stations from the HVO array at Kilauea and Mauna Loa volcanoes. In addition, students are provided with background in computer-based digital data processing to locate earthquake foci and to interpret the time and spatial sequence of earthquake swarms. Both digital and analog data are considered, in order to provide adequate flexibility in terms of what might be feasible in a particular country.

Seismicity associated with volcanoes
Hawaii is an excellent place to teach seismological monitoring methods because of the variety of seismicity associated with both volcanic and tectonic sources. These

events include both long- and short-period seismicity associated with inflation and deflation of the summit areas. They also include seismicity from a variety of depths including events as deep as 60 km. Tremor and earthquake swarms can also be observed. Moderate to large earthquakes have also occurred in the recent past including, for example, earthquakes of M = 7.2 in 1975 (Rojahn and Morrill, 1977), M = 6.6 in 1983 (Buchanan-Banks, 1987), and M = 6.1 in 1989. Great earthquakes can also occur, as indicated by an event of approximately M = 8 in 1868 (Wyss, 1988). The historic record is too short to estimate the recurrence interval of such events.

The geochemical monitoring component of the curriculum deals with methods under development that show promise of contributing to accurate forecasting of volcanic eruptions. It also provides a broad background in the area of volcanic geochemistry and the hazards posed by eruptive gases. Topics include fumarole gas and crater lake chemistry, radon and hydrogen monitoring, gas sampling methodology, and geothermal systems.

Hawaii has been subjected to varied gas emission effects during the current eruption that began in 1983. Acid rain occurs in two distinct settings: sulfuric acid is produced as the result of degassing near the source vents, while hydrochloric acid, on the other hand, is produced as the result of hydrolysis of sea water upon entry of lava flows into the ocean (Gerlach and others, 1989).

Other hazards treated in the course include the build-up of carbon dioxide in summit lakes such as that at Lake Nyos, Cameroon. Also included are fluoride and hydrogen sulfide poisoning.

Other CSAV programs

Another related program of CSAV is the inclusion of students from developing nations in the University of Hawaii at Hilo four-year baccalaureate degree program in Geology. This provides a more extensive educational experience that includes the undergraduate curriculum in Geology, with emphasis in volcanology, as well as the summer certificate program in volcano monitoring.

CSAV is also attempting to promote public awareness of volcanic hazards and risk by offering courses for educators. The objective is to reach a large number of primary and secondary school students through their teachers.

An important function of the Center will be to serve as a source for information about active volcanoes. HVO currently maintains an unparalleled collection of reprints dealing with Hawaiian volcanism, and CSAV is currently in the process of developing a literature database that will complement that collection. The objective is to build a computer-searchable library of literature and imagery pertaining to active volcanoes outside of Hawaii. This will be a cooperative project of the University of Hawaii library, CSAV, and the HVO library.

Summary

The purpose of the CSAV is to help save lives and prevent destruction of property by increasing the number of persons capable of monitoring active volcanoes and by promoting awareness of volcanic hazards.

The initial program designed for participants from developing nations is intended to help encourage collection of base-line data on active or potentially active volcanoes and to address the question of how to communicate that information to appropriate authorities for effective and timely action. The monitoring methods taught are those that have been proven to be effective in forecasting volcanic eruptions. The question of communication is addressed through the thoughtful examination of case histories.

It is anticipated that the course will evolve as the methods themselves evolve and as we learn more from new eruptive events. However, the general approach will always emphasize practical equipment and methods—that is, the equipment should be obtainable, as affordable as possible, maintainable, effective, and not excessively "high tech". Similarly, the methodology should be as understandable as possible for those who will be using it.

Volcanism will continue to pose a serious threat to humanity in the future, and the number of people at risk will increase in parallel to population pressures that compel mankind to live in harm's way. It is, therefore, imperative that early-warning capabilities and planning, to reduce volcanic risk, be increased worldwide. It is the objective of the CSAV program to contribute to that goal.

References

Buchanan-Banks, J.M., 1987. Structural damage and ground failures from the November 16, 1983, Kaoiki Earthquake. In: R.W. Decker and others (eds), *Volcanism in Hawaii*, US Geological Survey Professional Paper 1350, p 1187–1195.

Gerlach, T.M., J.L. Krumhansl, R.O. Fournier, and J. Kjargaard, 1989. Acid rain from the heating and evaporation of seawater by molten lava: a new volcanic hazard. *EOS, Transactions of the American Geophysical Union*, vol 70(43), p 1421–1422.

Peterson, D.W., 1986. Volcanoes: tectonic setting and impact on society. In: *Studies in Geophysics: Active Tectonics, Panel on Active Tectonics*, National Academy Press, Washington, DC, p 231–246.

Rojahn, C., and B.J. Morrill, 1977. The island of Hawaii earthquakes of November 29, 1975: strong-motion data and damage reconnaissance report. *Seismological Society of America Bulletin*, vol 67, p 493–515.

Simkin, T., L. Siebert, L. McClelland, D. Bridge, C. Newhall, and J.H. Latter, 1981. *Volcanoes of the World: A Regional Directory, Gazetteer, and Chronology of Volcanism During the last 10,000 Years.* Hutchinson & Ross, Stroudsburg, Pennsylvania, 240 p.

Tilling, R.I. (ed.), 1989. *Volcanic Hazards: Short Course in Geology: Volume 1.* American Geophysical Union, 123 p.

Wyss, M., 1988. A proposed source model for the great Kaui, Hawaii, earthquake of 1868. *Seismological Society of America Bulletin,* vol 68, p 1450–1462.

3 Volunteer Observers Program: a tool for monitoring volcanic and seismic events in the Philippines

Perla J. Delos Reyes

Abstract   The Philippine Archipelago is one of the most tectonically active and volcanic regions of the world and, as such, frequent occurrences of volcanic eruptions and destructive earthquakes are to be expected. One of the main objectives in the creation of the Philippine Institute of Volcanology and Seismology (PHIVOLCS) was the mitigation of the effects of volcanic eruptions and earthquakes through prediction and prompt warning.

The Philippines lacks the technology and manpower needed for its program on volcanic and earthquake prediction. Therefore PHIVOLCS devised a Volunteer Observer Program to enlist the help of residents in areas likely to be affected by volcanic and seismic events, and to increase its sources of precursory signs such as crater glow, rumbling sounds, anomalous animal behavior, steaming activity, sulfurous odor, and drying vegetation. Questionnaires on observable manifestations of volcanic or seismic events were distributed to observers in communities in areas presently being monitored for volcanic or seismic activity by PHIVOLCS; these are collected according to set schedules, though unusual phenomena are reported immediately.

The ultimate aim of the program is to encourage local residents to participate actively and, in so doing, awaken the public's awareness by increasing their knowledge of the dangers and risks posed by volcanic eruptions and earthquake occurrences. Preliminary results have shown that this program is indeed useful and could be adapted to other developing countries. Development of the scheme in geologically active areas could eventually lead to the proper implementation of an effective prediction, warning, and disaster-preparedness program.

Introduction

The Philippines, situated in the Circum-Pacific "Belt of Fire", is an active volcanic and earthquake-prone country. Its global location in regard to the distribution of earthquake occurrences confirms it as being one of the most tectonically active regions in the world. Apart from being seismically active, the country is home to 21 active volcanoes (Figure 1), second only in number to Indonesia (Simkin and others, 1981). Values of the Volcanic Explosivity Index (VEI) (Figure 2) of 2 to 4 have been documented in areas as far north as Smith Volcano in Luzon and down to Ragang Volcano in Mindanao (Figure 1). These values are moderate to large on the scale of Newhall and Self, 1982 (Figure 2), which extends to VEI = 7 for the largest historic eruption. To date, the highest VEI recorded (VEI = 4) in the Philippines was the 1814 eruption of Mayon Volcano in Albay Province, southeastern Luzon (Figure 1), when 1200 persons were killed, due mainly to mudflows during and after the eruptive phase.

One of the most destructive eruptions in recent times was the 30 January 1911 eruption of Taal Volcano in Batangas, south of Manila (Figure 1) which claimed a total of 1335 victims and resulted in an undetermined amount of damage to property. Another devastating event was the 4 December 1951 eruption of Hibok-Hibok Volcano on Caminguin Island, just north of Mindanao, which exacted a death toll of 500 people, many of whom were mummified by hot blasts of gases known as *nuées ardentes* (glowing clouds). During this event houses burst into flames, animals roasted to death, and trees and plants were charred due to the intense heat.

From 1599 to 1988, earthquake records show that the

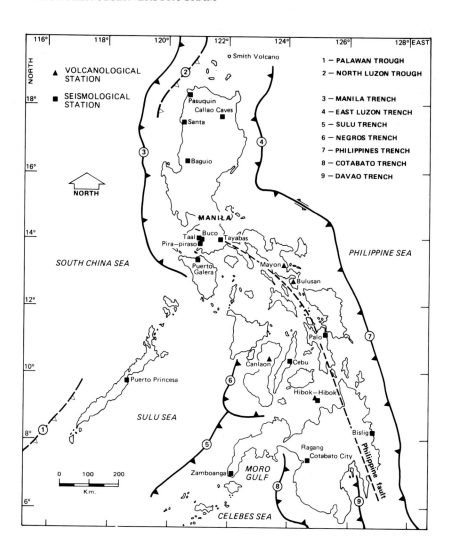

Figure 1 Volcanological and seismological stations in the Philippines established by PHIVOLCS, in relation to major volcanoes and tectonic elements.

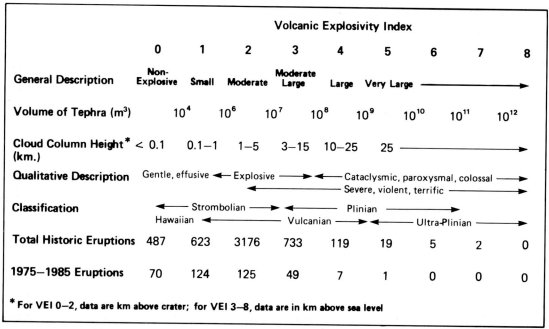

	Volcanic Explosivity Index								
	0	**1**	**2**	**3**	**4**	**5**	**6**	**7**	**8**
General Description	Non-Explosive	Small	Moderate	Moderate Large	Large	Very Large ———————————————→			
Volume of Tephra (m³)	10^4	10^6	10^7	10^8	10^9	10^{10}	10^{11}	10^{12}	
Cloud Column Height* (km.)	< 0.1	0.1–1	1–5	3–15	10–25	25 ———————————————→			
Qualitative Description	Gentle, effusive ←— Explosive ——→ ←— Cataclysmic, paroxysmal, colossal ——→								
	←——————— Severe, violent, terrific ———————→								
Classification	←— Strombolian ——→ ←— Plinian ——→								
	Hawaiian ←— Vulcanian ——→ ←— Ultra-Plinian ——→								
Total Historic Eruptions	487	623	3176	733	119	19	5	2	0
1975–1985 Eruptions	70	124	125	49	7	1	0	0	0

* For VEI 0–2, data are km above crater; for VEI 3–8, data are in km above sea level

Figure 2 Volcanic Explosivity Index (from Newhall and Self, 1982).

Table 1 Data on destructive earthquakes in the Philippines, 1589–1988

Date	Lat (N)	Long (E)	Focal depth (km)	Magnitude (Ms)	Intensity (Mod. Mercalli)	Deaths	Damage
1599 Jun 21	14.6	121.0	—	—	VII	—	Severe
1619 Nov 30	18.2	121.6	—	—	X	Undetermined	Extreme
1743 Jan 12	14.0	121.6	—	—	X	6–7	Extreme
1787 Jul 13	10.7	122.6	—	—	X	15+	Severe
1796 Nov 05	16.1	120.3	—	—	X	0	Severe
1852 Sep 16	14.0	120.4	—	—	IX	3–4	Severe
1863 Jun 03	14.6	121.4	—	—	X	298+	Extreme
1869 Aug 16	12.2	123.7	—	—	IX	—	Moderate
1869 Oct 01	14.8	120.8	—	—	IX	—	Moderate
1873 Nov 14	13.1	123.0	—	—	VII	—	Limited
1880 Jul 18	16.0	121.8	—	—	X	4	Moderate
1885 Jul 23	8.4	123.6	—	—	X	—	Limited
1889 May 26	13.6	121.2	—	—	VII	—	Limited
1892 Mar 16	16.1	120.4	—	—	IX	—	Severe
1893 Jun 21	6.9	125.8	—	—	X	—	Moderate
1897 Sep 21	7.1	122.1	—	8.7	IX	—	Moderate
1897 Oct 18	12.4	125.0	—	8.1	IX	—	Moderate
1901 Sep 10	14.0	122.2	—	—	VII	—	Moderate
1902 Aug 21	8.1	124.2	—	—	X	12	Moderate
1902 Aug 22	10.8	122.6	—	—	IX	—	Severe
1903 Dec 28	7.0	127.0	—	7.8	—	—	Moderate
1907 Apr 18	14.0	123.0	—	7.6	—	2	Severe
1907 May 25	18.0	120.0	—	—	VIII	392	Severe
1907 Nov 24	13.3	123.4	—	—	X	—	Severe
1909 Mar 18	8.0	127.0	—	—	VIII	—	Moderate
1911 Jul 12	9.0	126.0	—	7.7	X	—	Moderate
1913 Mar 14	4.5	126.5	—	7.9	IX	—	Limited
1917 Jan 31	5.6	124.8	—	—	IX	7	Moderate
1918 Aug 15	5.5	123.0	—	8.5	X	102	Severe
1921 Nov 11	8.0	127.0	—	7.5	—	600	Severe
1924 Apr 14	6.5	126.5	—	8.3	IX	—	Limited
1924 Aug 30	8.5	126.5	—	7.3	IX	—	Limited
1925 Nov 13	13.0	125.0	—	7.3	VIII	Undetermined	Limited
1929 Jun 13	8.5	127.0	—	7.2	X	—	Limited
1931 Mar 19	18.3	120.2	—	6.9	VIII	—	Moderate
1937 Aug 20	14.2	122.1	—	7.5	VIII	1	Severe
1948 Jan 24	10.5	122.0	—	8.3	IX	72	Severe
1949 Dec 29	18.0	121.0	—	7.2	—	1	Moderate
1954 Jul 02	13.0	123.9	—	6.8	—	22	Moderate
1955 Mar 31	8.0	124.0	55	7.5	X	465	Extreme
1968 Aug 01	16.5	22.3	36	7.3	IX	207	Severe
1969 Jan 30	4.5	125.5	—	7.2	—	—	—
1970 Jan 10	6.5	125.0	Normal	7.5	RFVI	—	Limited
1970 Feb 05	12.5	122.1	11	6.6	—	3	Moderate
1970 Mar 30	6.5	125.5	—	7.0	—	—	Limited
1970 Apr 07	15.8	121.7	—	7.3	IX	14	Moderate
1970 Apr 12	16.5	122.0	—	7.0	—	—	Limited
1973 Mar 17	13.3	122.7	33	7.5	XI	15	Limited
1976 Aug 16	7.3	123.6	—	7.9	X	6500	Extreme
1977 Mar 18	16.7	122.3	—	7.0	VIII	1	Moderate
1978 Jun 14	8.29	122.4	Shallow	6.7	—	—	—
1978 Dec 12	7.3	123.4	Shallow	6.5	—	—	—
1979 Aug 26	19.0	122.2	Shallow	6.9	—	—	Limited
1980 Jan 02	5.96	126.3	43	6.9	—	—	—
1981 Nov 22	18.7	120.6	—	6.7	VIII	—	Limited
1982 Jan 11	13.7	124.3	26	7.0	—	—	Limited
1982 Jan 24	13.8	124.5	Shallow	6.5	—	—	Limited
1983 Aug 17	18.3	120.9	Shallow	6.5	VIII	24	Severe
1984 Mar 05	8.1	123.8	651	6.7	—	—	—
1984 Nov 20	5.2	125.3	202	7.1	—	—	Limited
1985 Mar 18	7.7	123.6	Shallow	6.5	—	1	Moderate
1985 Apr 24	16.5	120.8	Shallow	6.0	—	6	Limited
1987 Apr 25	15.9	120.3	120	5.6	—	—	Limited
1987 Apr 25	16.0	120.5	123	6.3	—	—	Limited
1987 May 23	8.1	125.5	Shallow	5.4	—	1	Limited
1987 Jun 18	17.2	121.3	Shallow	5.9	V	8	Limited
1988 Feb 24	13.4	124.6	Shallow	7.0	—	—	—
1988 Jun 19	12.3	121.1	16	6.3	—	1	Limited

15

Philippines has been frequently shaken by events of different magnitudes and intensities. About 70 destructive earthquakes have occurred during this period (Tables 1 and 2). Considered to be the most destructive and severe was the 16 August 1976 earthquake, the epicenter of which was found to be located in the Moro Gulf off western Mindanao (Figure 1); damage was extreme, with more than 6500 people killed, 261 injured, and the cost of property damage running into many millions of dollars.

Role of PHIVOLCS in geologic hazard prediction and mitigation

In order to mitigate hazards and disasters arising from volcanic eruptions, the Commission on Volcanology (COMVOL) was created in June 1952 by virtue of Republic Act 766, and is presently known as the Philippine Institute of Volcanology and Seismology (PHIVOLCS) after being reorganized in 1984. The Commission was initiated after the disaster wrought by the 1951 eruption of Hibok-Hibok Volcano and since then, as the Institute, its functions have increased in scope and now include seismological observation and earthquake prediction.

At present, there are five active volcanoes in the Philippines being monitored by PHIVOLCS. Observatories have been established in the vicinity of Mayon Volcano in Albay, Canlaon Volcano in Negros Oriental, Bulusan Volcano in Sorsogon, Taal Volcano in Batangas, and Hibok-Hibok Volcano in Caminguin Island (Figure 1); these are considered the most active in densely populated areas of the country.

Monitoring equipment is housed in the volcano and seismic observatories, particularly seismographs, water-tube tiltmeters, rain gauges and telescopes. Additionally, ground deformation studies are conducted semi-annually to determine the behavior of each volcano, including electronic distance measurement (EDM) surveys, precise leveling surveys, and dry tilt, geophysical, and fissure measurements.

Considering that there are 21 active volcanoes which need to be monitored constantly by the Institute, the present facilities as well as its present allocated budget are very inadequate. Moreover, there are only 12 operational seismological observatories in the country (Figure 1), although an ideal number would be at least 100. These stations are continuously monitoring earthquake occurrences nationwide through the use of strong-motion seismographs, which form the backbone of all seismological observations in the Philippines.

Table 2 Destructive volcanic eruptions in the Philippines, 1911–1984

Date	Volcano	Latitude (N)	Longitude (E)	Casualties	Damage
1911 Jan 30	Taal	14.0	121.0	Dead: 1336 Injured: 19	Devastated area: 360 km^2 Destruction of arable land; accompanied by lahar
1928 Jan	Mayon	13.3	123.7	Number of casualties undetermined	Destruction of arable land; accompanied by lahar
1933 Dec 25	Bulusan	12.8	124.1	—	Destruction of arable land
1938 Jun 05	Mayon	13.3	123.7	Number of casualties undetermined	Destruction of arable land; accompanied by lahar
1947 Jan 07	Mayon	13.3	123.7	Number of casualties undetermined	Ash ejection buried barrios of Masarawag and Guinobatan ankle-deep in ash
1948 Aug	Hibok-Hibok	9.2	120.7	—	Destruction of arable land
1951 Dec 04	Hibok-Hibok	9.2	120.7	Dead: 500	Trees and agricultural crops destroyed, houses burned
1965 Sep 28	Taal	14.0	121.0	Dead: 235	Destruction within 50 km; accompanied by tsunami
1968 Apr 21	Mayon	13.3	123.7	Dead: 1	Thousands evacuated; partial damage to towns of Camalig and Guinobatan; accompanied by lahar
1969 Mar 21	Didicas	19.1	122.2	Dead: 3	No reported damage
1976 Sep 03	Taal	14.0	121.0	—	Destruction of arable land
1978 May 03	Mayon	13.3	123.7	—	About 23,000 people evacuated
1984 Sep 09	Mayon	13.3	123.7	Injured: 4	About 70,000 people evacuated, thousands rendered homeless; damage to crops, infrastructure, livestock, and houses totaling more than US$240,000

Related studies on disaster preparedness and mitigation

Measures are now being undertaken to promote PHIVOLCS's efforts to save lives and property through prediction of potentially hazardous events. A grand plan to set up a seismic and geodetic network for earthquake prediction along the trace of the Philippine Fault Zone is now under way. Likewise, a quick response team (QRT) was formed in January 1988 with the objectives of providing efficient response to emergencies and gathering real-time data on suspected developing seismological and volcanological phenomena. Personnel from the Geology, Seismology and Volcano Monitoring Divisions of the Institute are tapped for emergency surveys, wherever and whenever needed.

To enhance the effectiveness of these undertakings, the Volunteer Observers Program (VOP) for volcanic and earthquake events in the Philippines was formulated by PHIVOLCS in 1986.

Creation of the Volunteer Observers Program

After the successful prediction of the Heicheng earthquake in 1975 (western Inner Mongolia, China), studies applying the Chinese earthquake prediction technique were initiated by scientists all over the world. The Chinese method employed state-of-the-art electronic equipment, but the bulk of the predictive success came from the total commitment, involvement and cooperation of the Chinese people towards the undertaking. The accuracy of the method of prediction amazed Japanese and American scientists; in their countries, with sophisticated equipment, prediction to the exact day and time had not been totally successful. The Chinese technique employed local residents to monitor such bizarre manifestations as subterranean noises coming from the ground, fluctuating water levels in wells, and the strange behavior of animals.

Realizing the need for developing a monitoring scheme similar to the Chinese technique, and considering the lack of sophisticated equipment for volcanic and earthquake monitoring available in the Philippines, the VOP was designed by PHIVOLCS and is currently being implemented. The program aims to enhance public awareness, initiate public involvement, and obtain significant information on volcanic and earthquake event documentation from volcano-slope dwellers, as well as residents living in earthquake-prone areas. In so doing, PHIVOLCS can upgrade its monitoring capability by increasing its range of sources for acquisition of volcano and earthquake-related data, and increase the frequency of report-gathering, without too much additional expense. Ultimately, the information being gathered through the VOP will be valuable in identifying precursor events for predicting earthquakes and volcanic eruptions.

With the proper implementation of the program, the VOP could support the Institute's monitoring and prediction scheme. With enough lead time between prediction and actual occurrence of the event, it could then issue warnings to the proper authorities at the earliest possible time, so that appropriate loss-reduction measures could be undertaken. An effective and prompt warning system would enable the residents concerned to prepare for orderly evacuation.

In the case of the 1984 eruption of Mayon Volcano, there were unconfirmed reports that, months before the outburst, crater glow was visible in some sectors of the volcano. If these sightings had been reported to PHIVOLCS as soon as they were observed, the eruption could have been predicted. Fortunately, no casualties were attributed to this eruption or to subsequent lahars.

Methodology

The response of the people in the vicinity of a volcano or earthquake-prone area to precursory signs must be immediate and an attitude of indifference must be totally eliminated. This is essential if casualties are to be kept to a minimum for catastrophic events.

The residents participating in the VOP would assist PHIVOLCS in its task of monitoring seismic and volcanic activities without any monetary compensation on their part. Their reward would be the possible early warning of an impending event which would ultimately affect their lives and livelihood.

The VOP participants are expected to document daily observations by filling out appropriate questionnaires on visual and aural signs which may be noted in their vicinity. They must also report unusual events in their locality which could be related to earthquake or volcanic activity. The compiled data would then be collected by assigned PHIVOLCS personnel on a regular basis, or whenever the need for such information arises. After data-gathering, corresponding evaluations and recommendations by the Commission on Volcano Assessment (COVA) would be made as soon as reports were received by PHIVOLCS.

Selection of volunteer observers

Every year, in different barangays (small districts) throughout the Philippines, PHIVOLCS conducts a series of community and school-based lecture-seminar workshops and exhibits on various aspects of volcanology and seismology and their attendant geologic hazards, especially in areas most susceptible to such hazards.

PHIVOLCS has recommended the inclusion of geologic hazards in the school elementary science curriculum. The aim of such an undertaking is to introduce these subjects very early in the child's mind so that the response would be spontaneous in the event of such hazards occurring. Furthermore, PHIVOLCS also encourages teachers and students all over the Philippines to support and participate in its information and dissemination program throughout the year. Posters and leaflets which are part of the PHIVOLCS information campaign are regularly distributed to the public, so that people may become more familiar with and be more aware of the dangers posed by volcanic eruptions and earthquakes.

From such community gatherings, possible volunteers are being recruited for the program by the personnel conducting the seminars or exhibits. Strategically located areas where significant data could best be gathered are delineated as observation points for monitoring. For convenience, more often than not, the selected volunteer observers reside in those areas so that, at any given time, observations can be conducted without delay.

Ideally, the whole community should actively participate. The barangay captain or any other local official must act as overall coordinator for PHIVOLCS, but widespread community participation is usually not possible. In most cases, the local volunteer observers are recommended by the officials in their communities and recruited by PHIVOLCS personnel; it is assumed that these officials will vouch for their dependability and trustworthiness. These qualities are foremost in the Institute's requirements, to avoid the gathering of incomplete or inaccurate data, and to guarantee that the information acquired can be utilized and relied upon. In areas where monitoring stations exist, the Resident Volcanologist (RV), Seismic Observer (SO), and/or Officer in Charge (OIC) make the final selection from all possible volunteers. In areas where there is no monitoring station, local volunteers are selected by PHIVOLCS personnel conducting field surveys in the area.

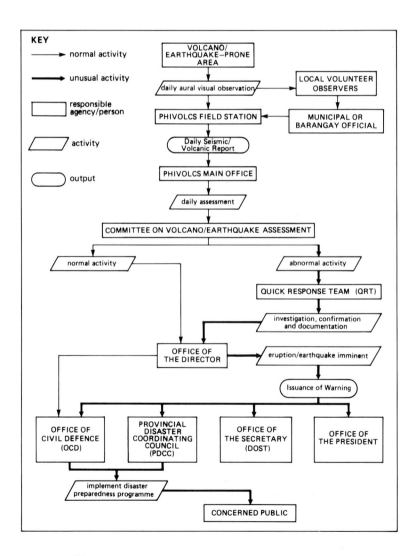

Figure 3 Flowchart of the reporting scheme for volcano and earthquake-related information.

Training program for volunteer observers Local volunteers undergo a training program with a minimum duration of one month under the RV/SO/OIC. The training includes a series of lectures on the historical record of eruptions or earthquake occurrences near their locality, the nature and pattern of all possible observable precursor events, and other related matters. They are provided with a list of parameters to be observed, with emphasis on the need for accuracy in observation and reporting.

Reporting scheme of the gathered observations Accessibility of the areas where observations are to be noted, as well as the importance and significance of the data gathered, play vital roles in the periodicity of the reporting scheme. The RVs, SOs and/or OICs determine the frequency of data collection from each barangay involved in the program.

At present, the collection scheme for VOP data in volcanic areas with no unusual activity is done on a monthly basis, as in the case of Hibok-Hibok Volcano. The local volunteers submit their observations to their barangay official, who in turn submits all information to municipal officials. The gathered reports are either submitted to the Institute or collected by Institute personnel on a regular basis. The flowchart of the reporting scheme for volcano and earthquake-related information is illustrated in Figure 3.

In cases where the local volunteers note any unusual activity in the locality, which may be indicative of impending eruption or earthquake occurrence, it is their responsibility to report the incident to the nearest volcanological or seismological station as soon as possible, unless it coincides with a scheduled data collection.

Further investigation will then be conducted by PHIVOLCS personnel, or a QRT will be dispatched for verification and confirmation immediately on receiving the information. In the event that precursory signs are already appearing, and if the RV present in the vicinity

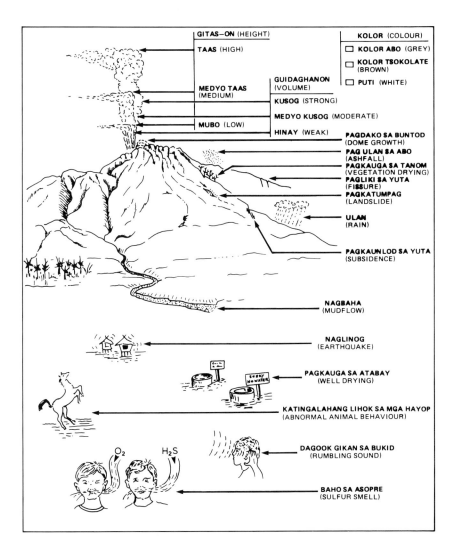

Figure 4 Illustration of the unusual changes that can be observed during imminent or probable volcanic eruption, as issued to local residents around Hibok-Hibok Volcano.

is otherwise involved in performing technical field investigations, it is the responsibility of the SO/OIC to distribute and collect volcano-event questionnaires to residents of the affected areas. This is done to document such events and allow confirmation if there are precursory signs to the event, as perceived by the residents. Results of these questionnaires are immediately forwarded to the central office, or transmitted by radio, telegram or telephone, for proper evaluation and recommendation.

In cases where there are no monitoring stations in the vicinity, the reports will be investigated by a QRT sent to the site. The QRT distributes and collects the VOP questionnaires from the residents for collation and evaluation.

Parameters to be observed

The local volunteers observe precursory signs for volcanic or seismic activity such as crater glow, bluish flames, rumbling sounds, steaming activity, felt earthquakes, etc., details of which are shown in an explanatory sheet

(Figure 4). The results are listed in a Volcano Events Report (Figure 5).

Crater glow The observation of glow in the crater of a volcano is indicative of the presence of magma near its mouth. Such occurrences are best observed at night time, but in some instances an intense glow may be seen even in the early morning hours. It is important to note the time of occurrence, the intensity and duration of the glow, and the position of the moon during the occurrence (which affects how well the glow can be seen), in order to predict how imminent the eruption could be.

Bluish flames Flames resulting from burning heavy gases creeping down the slope of the volcano may signify that an eruption is imminent.

Rumbling sounds These may suggest magma movement beneath the ground and may resemble the beat of

---

```
                                                           [ ] MORNING
  _____          _____                           [ ] AFTERNOON
      DATE                TIME                              [ ] EVENING

      PLACE OF OBSERVATION: _____

  I. STEAMING ACTIVITY
      1. INTENSITY        [ ] STRONG     [ ] MODERATE  [ ] WEAK
      2. HEIGHT           [ ] HIGH       [ ] MEDIUM    [ ] LOW
      3. COLOR            [ ] GRAY       [ ] BROWN     [ ] WHITE
      4. DIRECTION        [ ] NORTH      [ ] SOUTH     [ ] EAST    [ ] WEST

  II. UNUSUAL CHANGES IN THE GROUND
      1. EARTHQUAKE                      [ ] YES                  [ ] NO
      2. RUMBLING SOUNDS                 [ ] YES                  [ ] NO
      3. GROUND        [ ] SUBSIDENCE         [ ] LIQUEFACTION    [ ] FISSURING
      4. SULFUR SMELL                    [ ] YES                  [ ] NO
      5. DRYING OF                       [ ] YES                  [ ] NO
         VEGETATION
      6. UNUSUAL ANIMAL BEHAVIOR         [ ] YES                  [ ] NO
            KIND OF ANIMAL: _____

  III. UNUSUAL CHANGES IN LAKE WATER
  A. LAKE                  [ ] TAAL LAKE          [ ] MAIN CRATER LAKE
      1. BUBBLING                        [ ] YES                  [ ] NO
      2. INCREASE IN TEMPERATURE         [ ] YES                  [ ] NO
      3. INCREASE IN WATER LEVEL         [ ] YES                  [ ] NO
      4. CHANGE IN COLOR                 [ ] YES                  [ ] NO

  B. FISHES
      1. UNUSUAL BEHAVIOR                [ ] YES                  [ ] NO
      2. DEATH                           [ ] YES                  [ ] NO

  C. SULFUR SMELL                        [ ] YES                  [ ] NO

  IV. OTHER OBSERVATIONS/REMARKS
      _____
      _____
      _____
      _____

      NAME OF OBSERVER: _____

      NAME OF BARANGAY: _____
```

Figure 5 Sample of a volcano event questionnaire for Taal Volcano.

a drum. A significant eruption which was preceded by rumbling sounds was the 1853 eruption of Mayon Volcano; earthquakes were felt coincident with these rumbling sounds. Frequent observation of this phenomenon by local volunteers may also be an indication of an abnormality of a volcano's condition.

Steaming activity This is the emission of steam from the crater or any other source such as cracks and fissures. The intensity of the emission should be noted and then rated by the volunteer as weak, moderate, or strong. The volume should be related to the crater area of the volcano: a volume ranging from one tenth to three tenths is classified as weak steaming, from four to six tenths is moderate, and from seven to nine tenths is strong. Complete filling of the crater is classified as voluminous. Other observations to be noted are color, duration, height, and drift of the steam, the point of observation, and the source of the steam. Some descriptive remarks may be added such as whether the emission is profuse, or has a certain degree of force.

Felt earthquakes Earthquakes are manifestations of the sudden motion or trembling in the earth due to abrupt release of slowly accumulated strain resulting from the movement of crustal plates or, in the case of volcanoes, from the force exerted by rising magma. Volcanic earthquakes often occur in a sequence, called a swarm, with no discernable main shock. In contrast, tectonic earthquakes are often preceded, by a few days to weeks, by several smaller shocks (foreshocks) and closely followed by a number of aftershocks, the foreshocks and aftershocks being usually much smaller than the main shock, and are normally readily distinguishable.

Observation of earthquakes is relevant to prediction of volcanic activity, but the onset of a tectonic earthquake not directly linked to a volcanic area can also be monitored by volunteer observers. The observers need to be able to report the strength of any earthquake felt in the vicinity and, for this purpose, the Institute has distributed translated versions of the adapted Rossi-Forel Intensity Scale which indicates the amount of surface damage caused by the earthquake. From intensity reports, the epicentral magnitude of the earthquake can be calculated using the Richter formula which measures the energy released during an earthquake.

Sulfurous odor Hydrogen sulfide may be noted by observers in two forms: as an odor similar to rotten eggs or to that of sandalwood. The observation of the latter type of smell indicates a greater possibility of impending eruption than the former. The presence of this gas, which may indicate renewed activity of the volcano, must be reported by the local volunteers as soon as possible after observation.

Drying of vegetation A volcanic eruption may be preceded by the drying up of vegetation, usually on the upper slopes of the volcano, due to the intense heat of the magma just beneath the crater of the volcano or which may be related to the presence of poisonous gases. Mostly affected are grasses and small plants, which usually turn brown in color.

Fluctuating water-well levels A sudden decrease in the level of wells, hot springs and lakes in the vicinity of a volcano may indicate impending activity, and any changes must be reported by the volunteers as soon as they are noted.

Cracks and fissures Surface fractures in rock along which there is a distinct separation may form as a result of movement of the earth's crust possibly coincident with an impending eruption or earthquake, or as a precursor to it.

Landslides This form of mass movement involves the downward sliding of earth or rock, or both, loosened by subterranean movement resulting from renewed activity beneath a crater or by the presence of water in the affected area.

Creep The slow, more-or-less continuous downslope movement of rock or soil under gravity may be more marked as a result of destabilizing influences such as earthquakes, which act to weaken the materials to a point where they slide, slump, flow, or fall. Widespread occurrence of this phenomenon may indicate that the area is seismically active and that a major earthquake is about to occur.

Ground subsidence The sudden or gradual downward settling of the ground surface may be due to natural geologic processes such as earthquakes, or to the withdrawal of fluid lava from beneath a solid crust. The movement is not restricted in rate, magnitude, or area.

Ground uplift The sudden or gradual uplift of an area to form a dome or arch results from positive movements and, as with ground subsidence, can be due to tectonic or magmatic processes.

Changes in water Unusual changes in water bodies such as streams, hot springs, and lakes, such as increased cloudiness, turbulence, changes in level, bubbling, and bitterness in the water are some of the possible signs of an impending earthquake.

Unusual animal behavior Accounts of previous eruptions have shown that wild animals which may inhabit the upper slopes of a volcano may flock to the lower

DATE OF OCCURRENCE:_____ TIME OF OCCURRENCE:_____
ESTIMATED DURATION:_____ seconds
PLACE OF OBSERVATION:_____

I. EFFECTS OF THE QUAKE:	[ ] OBSERVED	[ ] REPORTED BY OTHERS
1. STRUCTURE AFFECTED	[ ] HOUSE [ ] BUNGALOW [ ] BUILDING	[ ] NO. OF FLRS. [ ] FLR. ON DURING THE SHOCK
2. STRUCTURE MADE OF	[ ] WOOD [ ] CONCRETE [ ] SEMICONCRETE [ ] OTHERS: Specify_____	
3. GROUND UNDER STRUCTURE	[ ] COMPACT [ ] LOOSE	[ ] ROCK [ ] SOIL
4. GROUND DATA	[ ] CRACKS [ ] LANDSLIDES [ ] TILTING	[ ] FISSURES [ ] SUBSIDENCE
II. MOTION OF THE SHOCK:	[ ] LIGHT [ ] MODERATE [ ] HEAVY	[ ] JOLT [ ] VIBRATION [ ] SWAYING
III. OBSERVER'S POSITION	[ ] INDOOR	[ ] OUTDOOR
1. PHYSICAL ACTIVITY OF OBSERVER DURING THE EARTHQUAKE AWAKENED	[ ] WALKING [ ] STANDING [ ] SITTING [ ] YES	[ ] LYING DOWN [ ] RIDING A VEHICLE [ ] SLEEPING: [ ] NO
2. DAMAGE OBSERVED	[ ] NONE [ ] SLIGHT [ ] CONSIDERABLE	[ ] SEVERE [ ] COLLAPSE [ ] LEANING

3. PHYSICAL EFFECTS

[ ] WALLS CRACKED
[ ] TILES/BRICKS FALLEN

[ ] HANGING OBJECTS SWINGING:	[ ] SLIGHTLY [ ] MODERATELY [ ] VIOLENTLY	
[ ] SHIFTING, FALL OF OBJECTS:	[ ] SMALL	[ ] LARGE
[ ] HEAVY FURNITURES: Specify _____	[ ] MOVED	[ ] OVERTURNED
[ ] VEHICLES SHAKEN:	[ ] MOVING [ ] STATIONARY [ ] VIOLENTLY	[ ] SLIGHTLY [ ] MODERATELY
[ ] TREES AND BUSHES SHAKEN:	[ ] SLIGHTLY [ ] MODERATELY	
[ ] OFFSET OF RIVER CHANNELS		
[ ] INFRASTRUCTURE DAMAGE:	[ ] WATER PIPES BROKEN [ ] ELECTRIC/TEL. LINES CUT OFF [ ] RAILWAY TRACKS BENT [ ] CRACKS/WARPS IN PAVED ROADS [ ] DAMS, TUNNELS, BRIDGES: [ ] CRACKS [ ] COLLAPSE	
[ ] ARTESIAN WELLS:	[ ] RAN DRY [ ] BECAME MURKY [ ] CHANGE IN ODOR OF WATER [ ] OTHERS: Specify _____	
[ ] UNUSUAL ANIMAL BEHAVIOR: Specify kind _____	[ ] BEFORE [ ] DURING [ ] AFTER THE SHOCK	
4. SOUNDS	[ ] BOOMING [ ] CRACKLING [ ] RUMBLING	[ ] BEFORE [ ] DURING [ ] AFTER THE SHOCK

IV. OTHER OBSERVATIONS/REMARKS

_____
_____
_____
_____

NAME OF OBSERVER: _____
BARANGAY OF OBSERVER: _____

Figure 6 Sample of an earthquake event questionnaire.

slopes before an eruption. During past eruptions of Taal volcano, wild pigs were seen by residents roaming its lower slopes hours before the volcano erupted. Even the strange behavior of domesticated animals and livestock in the vicinity of a volcano or earthquake-prone area beforehand can be used as a predictive precursor.

Implementation of the Program

During the 1984 eruption of Mayon Volcano, no clear visual manifestations were noted by PHIVOLCS observers. Later investigation and interviews with local residents immediately after the first surge of eruption and during its later quiescence revealed that precursory events actually had occurred, as in previous eruptions of Mayon, but unfortunately these were only visible in certain sectors which were not manned by PHIVOLCS personnel. These events had been seen by people days, weeks, and even months before the eruption, but they did not realize the significance of the sightings. Crater glow, abnormal animal behavior, drying of vegetation, and rumbling sounds were all experienced, but even if the people knew the importance of the signs, they still did not know where and how to report them.

To correct such a situation, the VOP concept was devised during the early part of 1986. By the end of the year, the program was already implemented on Mayon and Hibok-Hibok Volcanoes. At the present time, the program is in operation on each of the five active volcanoes mentioned above; and it is projected that, five years after implementation, there will be local volunteer observers in most barangays near active volcanoes and selected earthquake-prone areas. A more ambitious project would be to develop local volunteers in every barangay in every province throughout the Philippines, who would be trained and equipped by the Government.

Individual volcano-event questionnaires have been devised for the five active volcanoes monitored by PHIVOLCS (Figure 5). Each questionnaire was adapted to conform with the known precursory signs for each volcano, and the forms were translated into dialects familiar to the residents in the key areas. Earthquake-event questionnaires were also devised to encompass all observable phenomena associated with seismic activity (Figure 6).

The application of a community-based scheme is slowly but surely taking shape. At present, PHIVOLCS receives regular monthly reports from volunteer observers in the vicinity of Hibok-Hibok Volcano; these reports largely confirm the Institute's assessment that the volcano's condition is normal. Another concrete example of the applicability of the program is the active involvement and participation of some residents in the vicinity of Taal and Mayon Volcanoes.

During the seismic swarm of Taal Volcano on 30 October 1987, the PHIVOLCS station personnel were unable to observe any unusual activity other than the recording of unfelt earthquakes by the seismographs of Buco and Pira-Piraso Observatories. These tectonic events began to increase in frequency on that date and, by 1 November 1987, the OIC of Pira-Piraso Observatory started to receive reports from residents of the occurrence of perceptible earthquakes in some parts of Taal Island and on the mainland. There were also reports of rumbling sounds and the presence of a sulfurous odor. The QRT immediately responded and examined the evidence for the reported phenomena. On-site investigation by the team revealed that, although Taal Volcano was showing signs of restlessness, there was no indication of an impending eruption, and so no eruption warning was issued.

Another encouraging feedback was the reception given the program by Mayon residents when it was launched in the latter part of 1986: about 80% of the total population of barangays nearest the crater of Mayon Volcano were eager to participate. Due to the limited resources of the Instutute, it was only possible for a few residents to be given volcano monitoring kits, which consisted basically of a handbook for daily observation of the volcano's condition.

One of the main problems being tackled by the Institute is how to isolate erroneous accounts. In the Taal seismic crisis, reports of the rumbling sounds were confirmed, but those of sulfurous odor were not. One of the possible reasons for the discrepancy is that some observations were not reported directly to the PHIVOLCS observers, but are passed on by word of mouth from one person to another, whereby it is possible that the original observation may have become distorted or the information could have been misquoted. To eliminate this kind of error, each observer has questionnaires which he/she can fill out the moment he/she notes any unusual event. This is then immediately forwarded to the Institute and forms the basis for determining whether further investigations are necessary. PHIVOLCS must also scrutinize accounts of some observers who have experienced similar events in the past and who tend to exaggerate their current observations.

Further improvements

Based on the results of the partial implementation of the program described above, PHIVOLCS is adopting further improvements through the following measures:

1. Operation of the volunteer observer program is to be part of the routine responsibility of the RV, SO, OIC, and QRT.

2. The collection of questionnaires is required at regular intervals, either at the end of each month or quarter, or at time of emergency.

3. The Committee on Volcano Assessment (COVA) will create guidelines for evaluating these questionnaires.

4. This committee will also evaluate the questionnaires after each collection interval or emergency.

5. After successful implementation in the vicinity of all monitored volcanoes and earthquake-prone areas, any recommendation for emergency action will be referred to all local governments for their active participation and cooperation.

Conclusions

The action taken by Taal residents in informing PHIVOLCS about their observations emphasizes the need, if the program is to be fully developed, to counter-act the attitude of indifference of some people in the affected areas to the hazards posed by volcanic eruptions and large-magnitude earthquakes. Through the raising of consciousness of these people, they may be tapped continually by the Institute for much-needed assistance in volcano and seismic modelling.

In the drive to disseminate the program by PHIVOLCS, emphasis is given to increasing the aware- ness of residents of the effects of these phenomena on their lives and livelihoods. In most cases, such aware- ness has greatly influenced them to participate in this endeavor.

Proper implementation of the program in the Philippines could mean that the people can take pre- cautionary and safety measures in case such visual and aural precursors lead to an impending earthquake or volcanic eruption. Panic and confusion could thus be minimized and smooth evacuation could be managed with minimum property loss and few or no fatalities.

It is clear even at this early stage that the VOP method is effective, and that it could well be adapted to other countries with similar problems such as Indonesia, Papua New Guinea, Pacific Islands and much of Latin America.

References

Newhall, C.G., and S. Self, 1982. The Volcano Explosivity Index (VEI): an estimate of explosive magnitude for historical volcanism. *Journal of Geophysical Research (Oceans and Atmospheres)*, vol 87, p 1231–1238.

Simpkin, T., L. Siebert, L. McClelland, D. Bridge, C. Newhall, and J. H. Latter, 1981. *Volcanoes of the World: A Regional Directory, Gazetteer, and Chronology of Volcanism During the Last 10,000 Years*. Hutchinson & Ross, Stroudsberg, Pennsylvania, 240 p.

4 Monitoring and warning of volcanic eruptions by remote sensing

D.A. Rothery

Abstract Most of the world's volcanoes are in developing countries, which characteristically have poor internal communications and only rudimentary geological and geophysical survey programmes. Data from satellites can be used to monitor volcanic activity in these regions, and in some cases to give prior warning of eruptions. Three aspects of volcanism can be measured in this way: 1) movement of eruption plumes, which is relevant to fall-out hazards and aircraft safety; 2) changes in thermal radiation from the ground, which can sometimes give warning of eruption hazards; 3) output of gases such as sulphur dioxide, which can be a good indicator of the likelihood of eruption. Future satellite systems may be able to offer a comprehensive global volcano monitoring system, with two extreme options: 1) centralized monitoring (with responsibility for continent-sized areas), recording and processing the highest quality data transmitted by satellites; and 2) local monitoring within provincial volcano observatories. The latter option is feasible today, at least for eruption plume monitoring, using a small antenna to receive image signals from weather satellites and a simple microcomputer as a display system. The total cost for adequate equipment may in some cases be no more than £5000 (US$8000) per station.

Introduction

The effects of volcanic eruptions are generally reckoned to rank behind floods and earthquakes in the seriousness of the physical hazards that they represent worldwide (Smith, 1985; Berz, 1988; Peterson, 1988). Nevertheless, many volcanoes are like time-bombs ticking away adjacent to densely populated areas (Blong, 1984; Tilling, 1989). For example, dense rural populations in Indonesia, the Philippines and Central America live beside frequently active volcanoes, and are threatened by the very sources of their soil's fertility (e.g., Clarke, 1991). Some inhabitants of more-developed countries face immediate perils also: for example, the city of Catania in Sicily (Italy) was devastated by an eruption of Mount Etna in 1669 and there is every reason to suppose that a similar eruption will occur again one day.

Volcanic hazards

There are various direct ways in which volcanoes can kill people: by pyroclastic eruptions; by debris avalanches triggered by slope failure; by mudflows (lahars) due to the rupturing of crater lakes, melting of snow and ice, or volcanogenic rainstorms; by emission of poisonous or suffocating gases; and by fast-moving lava flows. Apart from any immediate fatalities, such events often have serious social and economic consequences.

Other risks from volcanoes are perhaps less obvious. Explosions near or beneath the sea can cause tsunamis ("tidal" waves) such as that resulting from the infamous Krakatau eruption of 1883 that claimed over 30,000 lives (Latter, 1981). A more curious case of volcanic hazard occurred in June and July 1982 when Boeing 747 aircraft (jumbo jets) belonging to British Airways and Singapore Airlines each barely escaped a crash due to engine failure after a normal flight-path took them unwittingly through the eruption plume from Mount Galunggung, Java. The ash particles within the plume choked the engines' air-intakes, resulting in a loss of power, fortunately only temporarily (Fox, 1988; Tilling, 1989). There was a similar near-disaster in December 1989 when a KLM Boeing 747 encountered the eruption plume from Mount Redoubt, Alaska (Kienle and others, 1990).

Eruptions that inject very large quantities of ash and

gases such as sulphur dioxide high into the atmosphere have been blamed for abnormal climatic events, which can affect areas far distant from any volcano or indeed the whole globe. The tremendous eruption of Tambora on Sumba, Indonesia, has been held responsible for the disastrous summer of 1816 in the northern hemisphere, when there was widespread crop failure and famine (e.g. Rampino and Self, 1982). Recent work on the probable consequences of cataclysmic eruptions, of the sort that (on the basis of the geological record) may be expected to occur no more than once in every hundred thousand years or so, suggests that these could produce sufficient airborne particles to block out so much solar energy for several months, or even a few years, that photosynthesis would be impossible and the food-chain would collapse, just as in a 'nuclear winter' (Rampino and others, 1988).

Volcano monitoring

If we discovered that the effect of a threatened volcanic eruption would be to prevent photosynthesis over the whole world for the next two years, a disaster that is not completely unthinkable, we could not do much about it; but there are steps that can be taken if we can reliably predict the smaller (and far more common) eruptions of the sort that endanger surrounding populations or their livelihoods (e.g. Peterson, 1988). To this end, various volcano monitoring techniques have been developed, all of which essentially look for changes in some measurable property at a volcano where the likelihood of future activity is suspected.

The list of techniques is quite large: it includes monitoring the shape of the volcano by measuring changes in distances between fixed points (surveying) and changes in the steepness of slopes (using tilt-meters); seismic monitoring to record the swarms of small earthquakes often associated with magma movement; detection of minute changes in the gravity field associated with magma or steam migration; and on-the-spot measurements of the temperature, composition and rate of emission of any products of the volcano such as lava and fumes. Tilt-metering and seismic monitoring are the most simple and reliable techniques and are therefore the most commonly used (e.g. Suryo and Clarke, 1985; Tilling, 1988; Clarke, 1991).

Using satellite images for volcano monitoring

Remote sensing from satellites can now be added to the list of techniques that should be incorporated in a comprehensive volcano monitoring programme (Mouginis-Mark and others, 1989; Kienle and others, 1990). In principle, satellites can record observations of a volcano anywhere in the world, regardless of whether there is any ground-based monitoring equipment in place, and get the data to the volcanologist in his or her laboratory within seconds.

At present, many eruptions occur at volcanoes that are not monitored at all on the ground, sometimes because they were not previously thought to be active and occasionally because they were not even known to be

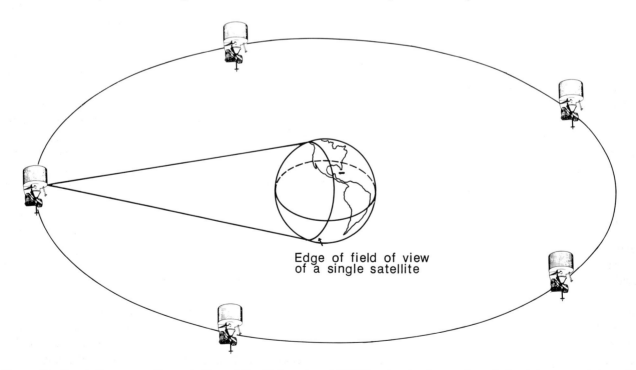

Edge of field of view
of a single satellite

Figure 1 Geostationary weather satellites of the Meteosat and GOES series can image the whole globe, except the polar regions, at half-hourly intervals.

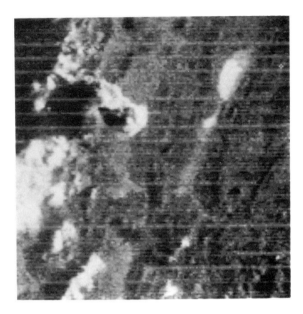 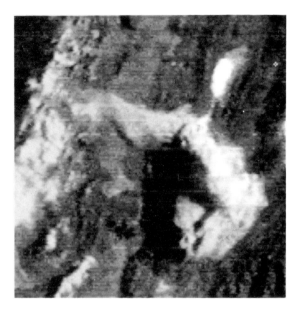

Figure 2 Portions of GOES images acquired half an hour apart on 16 September 1986 showing the development of an eruption plume from Lascar volcano, Chile. Each extract is about 300 km across.

volcanoes (e.g. Mount Lamington in Papua New Guinea, which erupted in 1951). Probably a more common reason for the absence of routine monitoring is because of a lack of trained personnel, equipment and funding, coupled with logistical problems. In such cases, and especially those in which monitoring teams are forced to flee in fear of their lives, satellite images may provide the *only* quantitative data on the state of the volcano before and during its eruption.

There are two main ways in which remote sensing can contribute information on volcanic eruptions. One is the detection and monitoring of eruption plumes, and the other is the detection and measurement of thermal anomalies. I shall review the eruption plume story first, because this is becoming well-established as a routine technique, and then go on to the thermal techniques, which are still in their infancy.

Detection and monitoring of eruption plumes

Being able to spot an eruption plume rapidly has two special advantages. One is that this is often the first evidence of an eruption to reach the central authorities in the country concerned and the scientific community at large. This should enable trained observers and, if necessary, relief teams to be dispatched promptly. The second advantage is that if we know where the plume is, then air traffic controllers can make sure that aircraft avoid it and, by following the dispersal of the plume, the area likely to be affected by fall-out can be mapped.

Weather satellite images

There are no specialized plume-monitoring satellites in orbit, but we do have the next best thing. This is a girdle of weather satellites stationed around the Equator in geostationary orbits at heights of 35,786 km (Figure 1). The members of this family go by various names and acronyms (GOES, Meteosat, GMS) and they have slightly different capabilities. Basically, however, each records an image of most of the hemisphere beneath it as often as once every 30 minutes, making it possible to construct time-sequences such as that in Figure 2. The best spatial resolution available is about 2.5 km (visible channel) and about 5 km (infra-red). Data are transmitted in digital form for reception by "primary data user stations" (PDUS) and in an analogue form (weather facsimile transmissions or WEFAX) that can be received by simpler "secondary data user stations" (SDUS), but which results in a somewhat degraded resolution. Examples of the use of geostationary weather satellite data for detection and monitoring of eruption plumes were given by Matson (1984), Malingreau and Kaswanda (1986), and Sawada (1989).

There are also weather satellites in lower orbits (Figure 3) such as the the United States *NOAA Tiros-N* series of the National Oceanic and Atmospheric Agency (NOAA), which carry an instrument called the Advanced Very High Resolution Radiometer (AVHRR). The resolution of data from the AVHRR is "high" only in meteorological remote-sensing terms; at about 1 km for the best data, it is much coarser than Landsat, for example. However, the 1-km resolution data can be received directly by moderately sophisticated ground stations in High Resolution Picture Transmission (HRPT) format or purchased retrospectively from NOAA as Local Area Coverage (LAC) data if they have been recorded on board for transmission to a ground station in the United States. Lower quality data (about 8 km resolution and less choice of wavelengths) can be received directly in Automatic Picture Transmission (APT) form by simple ground stations.

As there are usually more than one *NOAA Tiros-N* polar orbiters in operation, it is possible to get an AVHRR image of a given area several times per day, which makes these data suitable for use in volcano monitoring (Figure 4). An example of sophisticated use of such data was given by Prata (1989). The Soviet Union operates a comparable series of polar orbiting satellites under the name of *Meteor*.

Images are recorded both in the visible part of the spectrum and in the thermal infra-red, the latter working equally well by day and night; these tell us the temperature of the plume, from which we can estimate its height, once it has become stabilized. When stabilized, the plume as a whole has the same density as the surrounding air, otherwise it would continue rising. The assumption usually made is that, because the ash in the plume is greatly diluted by air, the contribution of the ash to buoyancy forces is negligible, so the plume must be at virtually the same temperature as the normal atmosphere at the same height; this can be deduced from meteorological data such as that provided by radiosonde balloons (e.g. Glaze and others, 1989a).

Data receiving stations

Many numerical aspects of remote sensing require the images to be in digital form, as this preserves all the details recorded by the satellite and makes the data suitable for sophisticated image processing techniques. This requires digital data (PDUS or LAC/HRPT) and there can be considerable delays before the volcanologist can get his or her hands on the data, which are typically received and processed only by national or international agencies. Fortunately there is another, quicker, option, which is becoming sufficiently cheap to use that it should now be considered seriously by volcano observatories. As described above, as well as transmitting their pictures digitally, geostationary and polar orbiting weather satellites also broadcast images in analogue form (WEFAX and APT). These pictures can be received, displayed and interpreted using a receiving dish or simple antenna and monitor screen controlled by a microcomputer or personal computer (PC), costing in total less than £1000 (US$1600) for the most primitive arrangement. The bottom end of the range of available facilities, although useful for educational purposes, is probably suitable only for basic monitoring of meteorological clouds: the image quality is not likely to be good enough to recognize any but the largest volcanic eruption plumes, but the radiometric and spatial resolution improves with the sophistication of the receiver, the computer and the display system.

Adequate facilities for basic volcano plume monitoring could probably be put together for an outlay in the region of £5000–20,000 (US$8000–32,000). This level of expenditure puts the equipment within the budgets of even poorly-funded national volcanological institutes. A schematic diagram of the equipment required is shown in Figure 5.

Many such stations can receive both WEFAX (geostationary satellites) and APT (polar orbiting satellites) images. An ordinary camera could be used to photograph the display screen, in order to make a permanent record of important images. Using such facilities, a volcano observatory could monitor all the volcanoes in a region at once: all that is required is someone with basic skills in recognizing the difference, on the pictures received, between normal meteorological clouds and a plume coming from a volcano. The latter is usually identifiable on the basis of its source at a volcano summit and rapidly increasing height during the first hour or so. There may also be distinctions revealed by comparing images recorded at different wavelengths (e.g. Prata, 1989; Kienle and others, 1990).

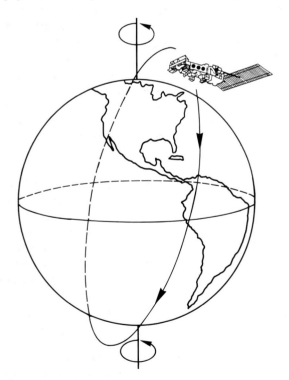

Figure 3 Polar orbiting weather satellites of the *NOAA Tiros-N* series which have an image swath 3000 km wide during each orbit, so a single satellite can provide an image of a given point on the ground twice or more each day.

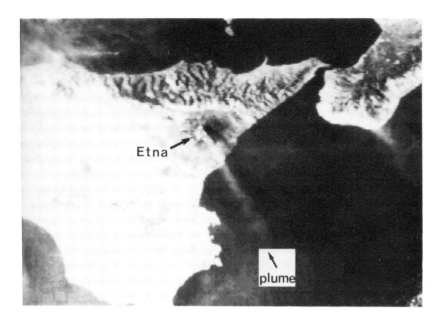

Figure 4 *NOAA Tiros-N* polar orbiter AVHRR image (recorded in HRPT form) showing a plume (arrowed) extending southeast from the summit of Mount Etna, Sicily (11 October 1989).

The cost of buying equipment to receive the top quality LAC or PDUS data has fallen recently, and may bring these facilities, with the improved resolution and capacity to detect smaller phenomena, into the price range affordable by most national volcanological institutes. Facilities are beginning to be marketed by a large number of companies for prices as low as £25,000–£100,000 (about US$40,000–160,000).

Monitoring gas plumes

Remotely sensed data can also be used to detect and measure the concentration of gases emitted during volcanic activity. So far, detection has been limited to sulphur dioxide (e.g. Kreuger, 1983; Matson, 1984) using the Total Ozone Mapping Spectrometer (TOMS) carried by the satellite *Nimbus-7*. This instrument compares the absorption in narrow spectral bands in the ultraviolet to calculate the atmospheric concentration of gases. It is at its most reliable for major eruptions that inject sulphur dioxide into the stratosphere, but can detect plumes below the tropopause as well (L. Walters, personal communication, 1989). At the time of writing, TOMS data can be obtained directly from the NASA Goddard Space Flight Center and are free of charge to *bona fide* scientists.

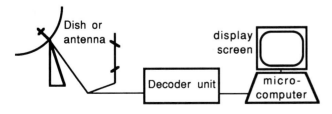

Figure 5 Schematic diagram to show the equipment necessary to receive analogue image data from satellites.

Eruption early warning from satellites

Plume observations, while they are good for spotting eruptions in progress, do not often give much advance warning of the onset of major activity because, by the time it has produced a major ash plume, a volcano may already have caused a large amount of death and destruction. This is where the second remote sensing approach, looking for anomalous thermal activity, may come into its own in the years to come, although it is limited to data recorded when the target volcano is not obscured by clouds or fumes. Nevertheless, immense plumes may be produced by volcanoes in continuous activity without destructive effects, such as Mount Marum, Ambrym Island, in the New Hebrides archipelago (Vanuatu) (Stephenson and others, 1967).

Thermal monitoring

All matter emits radiation, the distribution of which with respect to wavelength is controlled by its temperature. For most environmental monitoring purposes, people look for thermal radiation in what is called the thermal infra-red part of the electromagnetic spectrum, stretching from about 3 μm to about 14 μm in wavelength, because this is where matter at normal atmospheric temperatures radiates most strongly.

There have been attempts to detect rises in ground surface temperature of the order of a few degrees Celsius, heralding lava eruptions, using AVHRR thermal infra-red images. However, these are prone to all kinds of uncertainties due to the low spatial resolution of the images and the differential cooling effects of altitude, slope aspect, wind and moisture (e.g., Bonneville and Kerr, 1987). There is perhaps greater scope for measuring changes in the temperatures of volcanic crater lakes, whose size may

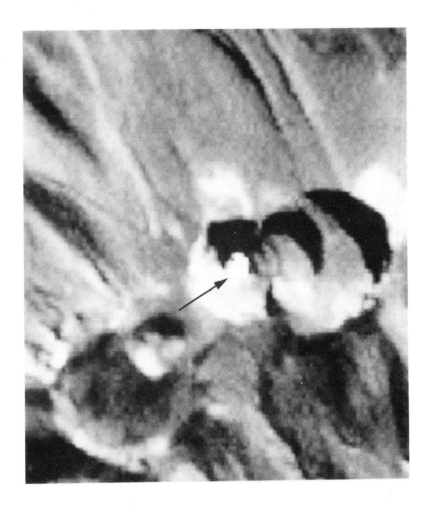

Figure 6 A close up of Lascar volcano (Chile) using Landsat Thematic Mapper band 7 (2.08–2.35 mm) on 16 March 1985. An area radiating thermally is indicated by an arrow. This was a thermal precursor to the eruption of September 1986 (Figure 2), and could have been a lava lake, a hot dome or a fumarole field.

be larger than that of the image pixels.

However, it is simpler to detect thermal anomalies when they are more pronounced. Most large eruptions, and presumably their precursor surface effects, occur within summit craters. These are the least accessible parts of most volcanoes and, especially if there is no seismic or other conventional monitoring, nobody generally knows what is going on inside them.

It has recently been demonstrated that images recorded at shorter wavelengths and higher spatial resolutions than used in conventional thermal infra-red remote sensing can provide evidence of high surface temperatures (Francis and Rothery, 1987; Rothery and others, 1988). This was done using images from the second generation of Landsat satellites, which began operating in 1982. They carry an instrument called the Thematic Mapper (TM) that records images with 30-m pixels in six wavebands, two of which are in the short-wavelength infra-red region (near 1.6 µm and 2.2 µm wavelength). There is a measurable amount of radiation at these wavelengths from surfaces whose temperature exceeds about 200°C. A good example of this is provided by Figure 6 where a region inside one of the summit craters on Lascar volcano, in Chile, was radiating strongly in the 2.2 µm region eighteen months before the eruption shown in Figure 2. In this image, the thermal radiation is very obvious because part of the hot region lies within the shadow of the crater-rim, so clearly we cannot be seeing reflected sunlight. On the same date, Lascar showed a similar, but smaller, thermal anomaly at 1.6 µm wavelength, but there was no sign of this at the next-shortest wavelength (around 0.8 µm wavelength).

If, as is common on volcanoes, the high-temperature thermal anomalies are spatially small compared to the size of the satellite image pixel, the lower temperature limit for detection of an anomaly is higher than the 200°C suggested above. This also increases the upper limit before saturation (overload) of the sensor occurs, so that the measurable temperatures enter the magmatic and high-temperature fumarole range. By comparing the radiance measured at 2.2 µm with that measured at 1.6 µm, Rothery and others (1988) were able to work out that the hot ground of the March 1985 Lascar anomaly was in fact made up of two components; most of the area was cold (160°C) and non-radiant, but there were hot patches within this (too small to be spatially resolved on the image) at 850–1000°C covering somewhat less than 0.5% of the surface area. The likeliest explanations for

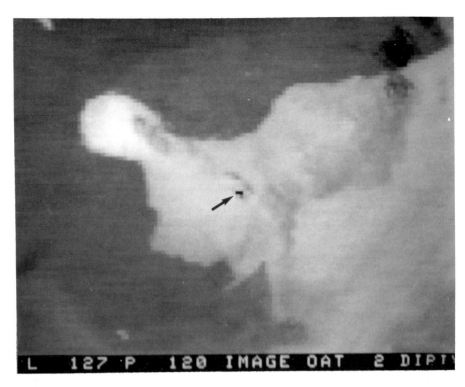

Figure 7 The lava lake (arrowed) on Mount Erebus, Antarctica, detected in a band 3 (3–5 μm) AVHRR LAC image, 13 January 1980. On this image, hot areas are dark and cold areas are bright.

this are hot cracks in a crusted lava lake or dome, or zones of hot rock around fumaroles. We shall never know for sure which is the true explanation, because the evidence was destroyed eighteen months later when Lascar erupted, producing the eruption plume 15 km high imaged in Figure 2. Prior to that eruption, Lascar was not thought to be in an eruptive state, and these Landsat data were the earliest signs we had of the impending activity.

There are several problems to be overcome before short-wavelength infra-red monitoring from satellites is likely to become a widely-used technique for volcano monitoring. At the practical level, when compared with weather satellite data, Landsat data are more expensive to use and can be received only by major receiving stations, with the result that they may take weeks or months to reach the volcanologist. However, future satellite instruments may offer the geologist a better deal (Mouginis-Mark and others, 1989; Oppenheimer and Rothery, 1991). Meantime, the largest and hottest thermal phenomena such as incandescent vents and lava flows are detectable in the infra-red channels recorded by the AVHRR instrument (Figure 7), so that volcano observatories which can receive these data directly are already in a position to do at least some potentially useful, cheap, real-time monitoring of volcanic surface temperatures.

At the scientific level, while it is clear that these data *can* show certain kinds of otherwise unsuspected eruption precursor activity, it has yet to be established just how meaningful the actual temperature measurements are,

and what level of significance to attach to any apparent changes. Further remote-sensing studies are now under-way, backed up by ground-based work, to tackle these problems. A bonus, in addition to the hazard-warning aspect, will be provided if it can be demonstrated, as suggested by Glaze and others (1989b), that remote sensing offers a way of measuring and monitoring the rate of thermal energy release by volcanoes. This will be an important contribution to our understanding of the energetics of volcanic processes.

The future

It is to be hoped that volcanologists and other scientists will continue to be allowed access to cheap satellite data. The potential of these data for providing meaning-ful information about the state of a volcano will be increased with the deployment of the Earth Observing System (EOS) during the decade beginning in the mid-1990s. This will be a joint effort between NASA and NOAA (USA), the European Space Agency (ESA) and Japan, and will carry a wide variety of instruments (see Billingsley and others, 1989, and the accompanying papers). The receiving platforms will carry comparable instruments to those discussed above, with improved thermal sensors. There will also be a variety of imaging spectrometers and atmospheric sounders with the potential to measure concentrations of various gases emanating from volcanoes. Changes in gas composition and flux are known to be linked in many cases to the onset of eruptive activity (e.g. Le Guern, 1983; Stoiber and others, 1983), so EOS will offer an additional means of giving warning of volcanic eruptions by remote sensing techniques.

Summary

At present, major volcanic eruption plumes can be detected and monitored several times a day by remote sensing using quite basic facilities, including low-cost weather satellite data receiving systems. Thermal anomalies can also be detected on certain kinds of weather satellite images. Detailed work and numerical analysis of thermal data requires much more costly high-resolution digital data. Because remote sensing offers a way to monitor all the cloud-free volcanoes in a region from a single laboratory, it should be given serious consideration as a useful tool by volcano observatories.

Acknowledgements

I thank Chris Legg for making available the image data used in Figure 4 and Clive Oppenheimer for providing Figure 7. I am grateful to the Overseas Development Administration for funding the fieldwork and other foreign travel that has stimulated this research.

References

Berz, G, 1988. List of major natural disasters, 1960–1987. *Natural Hazards*, vol 1, p 97–99.

Billingsley, F.C., J. Johnson, E. Greenberg, and M. MacMedan, 1989. Facilitating information transfer in the EOS era. *IEEE Transactions on Geoscience and Remote Sensing*, vol 27, p 117–124.

Blong, R.J., 1984. *Volcanic Hazards*. Academic Press, North Ryde, New South Wales, Australia, 424 pp.

Bonneville, A., and Y. Kerr, 1987. A thermal forerunner of the 28th March 1983 Mt Etna eruption from satellite thermal infrared data. *Journal of Geodynamics*, vol 7, p 1–31.

Clarke, M.C.G., 1991. Volcanic hazards as exemplified by the 1982 eruption of Mount Galunggung, Java. In: D.A.V. Stow and D.J.C. Laming (eds), *Geosciences in Development*, AGID Report No 14, p 167–178.

Fox, T., 1988. Global airways volcano watch is steadily expanding. *International Civil Aviation Organization (ICAO) Bulletin*, April 1988, p 21–23.

Francis, P.W., and D.A. Rothery, 1987. Using the Landsat Thematic Mapper to detect and monitor active volcanoes. *Geology*, vol 15, p 614–617.

Glaze, L., P.W. Francis, S. Self and D.A. Rothery, 1989a. The 16 September 1986 eruption of Lascar volcano, north Chile: satellite investigations. *Bulletin of Volcanology*, vol 51, p 149–160.

Glaze, L., P.W. Francis and D.A. Rothery, 1989b. Multi-temporal radiant thermal energy measurements of active volcanoes: a new satellite technique. *Nature*, vol 338, p 144–146.

Kienle, J., K.G. Dean, H. Garbeil, and W.I. Rose, 1990. Satellite surveillance of volcanic ash plumes, application to aircraft safety. *EOS Transactions of the American Geophysical Union*, vol 71, p 266.

Krueger, A.J., 1983. Sighting of El Chichon sulfur dioxide clouds with the Nimbus-7 Total Ozone Mapping Spectrometer. *Science*, vol 220, p 1377–1380.

Latter, J.H., 1981. Tsunamis of volcanic origin: summary of causes, with particular reference to Krakatoa. *Bulletin Volcanologique*, vol 44, p 467–490.

Le Guern, F., 1983. Magmatic gas monitoring. In: H. Tazieff and J.C. Sabroux (eds), *Forecasting Volcanic Events*, p 293–310, Elsevier Scientific, New York.

Malingreau, J.-P. and Kaswanda, 1986. Monitoring volcanic eruptions in Indonesia using weather satellite data: the Colo eruption of July 28, 1983. *Journal of Volcanology and Geothermal Research*, vol 27, p 179–194.

Matson, M., 1984. The 1982 El Chichon volcano eruptions—a satellite perspective. *Journal of Volcanology and Geothermal Research*, vol 23, p 1–10.

Mouginis-Mark, P.J., D.C. Pieri, P.W. Francis, L. Wilson, S. Self, W.I. Rose and C.A. Wood, 1989. Remote sensing of volcanoes and volcanic terrains. *EOS Transactions of the American Geophysical Union*, vol 70, p 1567–1575.

Oppenheimer, C.M.M., and D.A. Rothery, 1991. Infrared monitoring of volcanoes by satellite. *Journal of the Geological Society of London*, vol 148, p 563–569.

Peterson, D.W., 1988. Volcanic hazards and public response. *Journal of Geophysical Research*, vol 93, p 4161–4170.

Prata, A.J., 1989. Observations of volcanic ash clouds in the 10–12 μm window using AVHRR/2 data. *International Journal of Remote Sensing*, vol 10, p 751–761.

Rampino, M.R., and S. Self, 1982. Historic eruptions of Tambora (1815), Krakatau (1883) and Agung (1963), their stratospheric aerosols and climatic impact. *Quaternary Research*, vol 18, p 127–143.

Rampino, M.R., S. Self and R.B. Strothers, 1988. Volcanic winters. *Annual Review of Earth and Planetary Science*, vol 16, p 73–99.

Rothery, D.A., P.W. Francis and C.A. Wood, 1988. Volcano monitoring using short wavelength infrared data from satellites. *Journal of Geophysical Research*, vol 93, p 7993–8008.

Sawada, Y., 1989. The detection capability of explosive eruptions using GMS imagery, and the behaviour of dispersing eruption clouds. In: J.H. Latter (ed.), *IAVCEI Proceedings in Volcanology* 1, Volcanic Hazards, Springer-Verlag, Berlin and Heidelberg.

Smith, J.V., 1985. Protection of the human race against natural hazards (asteroids, comets, volcanoes, earthquakes). *Geology*, vol 13, p 675–678.

Stephenson, P.J., G.J.H. McCall, R.W. Le Maitre and G.P. Robinson, 1967. *Annual Report of the New Hebrides Geological Survey for 1966*, p 9–15.

Stoiber, R.E., L.L. Malinconico Jr, and S.N. Williams, 1983. Use of the correlation spectrometer at volcanoes. In: H. Tazieff and J.C. Sabroux (eds), *Forecasting Volcanic Events*, p 425–444, Elsevier Scientific, New York.

Suryo, I., and M.C.G. Clarke, 1985. The occurrence and mitigation of volcanic hazards in Indonesia as exemplified at the Mount Merapi, Mount Kelut and Mount Galunggung volcanoes. *Quarterly Journal of Engineering Geology*, vol 18, p 79–98.

Tilling, R.I., 1988. Volcano prediction—lessons from materials science. *Nature*, vol 332, p 108–109.

Tilling, R.I., 1989. Volcanic hazards and their mitigation: progress and problems. *Reviews of Geophysics*, vol 27, p 237–269.

5 Volcanic hazards in Colombia and Indonesia: lahars and related phenomena

H.Th. Verstappen

Abstract   Volcanic mudflows or lahars are a major type of volcanic hazard: they travel at high velocity over large distances, thus often affecting densely populated rural areas and urban settlements on the lower slopes of volcanoes. The mechanisms of lahar development and their various types, such as "hot" lahars associated with pyroclastic flows, "cold" or rain-fed lahars, and lahars caused by emptying of crater lakes or melting of snow and ice covering top areas, are discussed. Subsequently problems of hazard zoning and monitoring/early warning are dealt with. Methods of controlling and/or diverting lahar flows and related engineering structures are assessed. Merapi and Kelud volcanoes, Indonesia, and Nevado del Ruiz volcano, Colombia, serve as examples.

Introduction

Volcanic mudflows, often referred to by the Indonesian name "lahars", are well-known phenomena accompanying the eruptions of many volcanoes around the world. In common with pyroclastic flows, these mixtures of volcanic debris and water mostly travel down volcanic slopes along well-defined tracts or ravines, although sideways spilling, overflowing and spreading may occur, particularly in their lower courses where the ravines are less deeply incised. In many cases they reach considerably further down-slope than pyroclastic flows and thus are a major volcanic hazard, particularly on the middle and lower volcanic slopes which often are densely populated.

Contrary to the more diffuse hazards related to ash-fall, lahar hazard zoning is comparatively easy. However, it may be difficult to enforce the zoning in terms of land-use planning because agriculturists will inevitably be tempted to occupy or reoccupy these fertile areas, particularly in developing countries where no alternative way of making a living may exist. Under such conditions monitoring of volcanic activity on critical ash-covered slopes and, in the case of rain lahars, of rainfall with the aim of establishing an early warning system, becomes of special importance. These methods range from simple devices such as the gong warnings illustrated in Figure 1, to fully automated systems. Since early warnings may not always be fully reliable and/or the warning time critically short, one may in addition contemplate controlling the lahars to some extent by way of engineering structures such as embankments, diversion dams, storage reservoirs, etc (Figures 2 and 3).

It is evident that, for any type of action,

Figure 1   Simple warning device in a village on the southeast slopes of the Semeru volcano, Eastern Java. Four bangs on the gong announce a natural hazard (bahaya alam), in this case mudflows.

Figure 2 Automated warning system at a lahar check dam on the southwest slopes of the Merapi volcano, Central Java. The arrow points to the fastening of the three thin horizontal steel cables that stretch across the checkdam. Note the rise of the valley floor caused by the structure.

(a) hazard zoning/physical planning,
(b) monitoring/early warning and
(c) engineering works/protection,

a thorough knowledge of the mechanisms of volcanic mudflows and their physical characteristics is essential. Exact data in this area are, however, comparatively scarce (a good exception is Irazú volcano, Costa Rica: see Ulate and Corrales, 1966). Flow velocities, for instance, are seldom mentioned: velocities of more than 150 km/hr are on record in upslope areas, but on the lower slopes the pace may be reduced to 5 km/hr. In many cases 20–30 km/hr may be assumed for planning purposes, but exact information is required particularly for estimating the warning time when installing an early-warning system. The ratio of liquids to solids of the mud also has an important effect on the behaviour of the flow such as its velocity, the inclination of its surface in curves, the thickness of the deposits left behind, etc; 50% water may be a good average, but less than 30% suffices to generate

Figure 3 Engineering structure aimed at protecting a village on the southwest slopes Merapi from lahars that occasionally occur in the shallow ravine seen in the distance.

Table 1 Lahar types

1. Primary lahars (directly related to volcanic eruption)
   (a) derived from pyroclastic flows by mixing with water (hot)
   (b) formed by water from a blown-out crater lake (hot/cold)
   (c) caused by melting of snow/ice during eruption (hot/cold)
2. Secondary lahars (developed at some later stage: may be triggered by seismicity)
   (a) resulting from rain onto unconsolidated pyroclastic deposits (cold)
   (b) related to seasonal snow/ice melting (cold)
   (c) initiated by spilling of lake water (cold)

a mudflow while about 80% water has been observed on the occasion of the Nevado del Ruiz eruption, Colombia, in 1985 (Naranjo and others 1986; see also Hall, this volume). If the proportion of solids is high, there is a tendency for the ravines to have their beds raised by each subsequent flow, thus increasing endangering the adjacent land, while if the proportion of liquids is high accumulation in the ravine bed will be limited, allowing it to accommodate subsequent flows easily.

Volcanic mudflows are of various types (Table 1). Many develop as so-called "hot" lahars during or immediately after an eruption, commonly from pyroclastic flows, but they may also be associated with other volcanic phenomena. On the other hand, when the quantity of hot volcanic material is low and the water is mainly admixed with pre-existing volcanic slope debris, a "cold" lahar may develop. This is the case, for instance, in the early phases of lahars related to the emptying of a crater lake or the sudden melting of glacier ice and snow covering an erupting volcano (see Mothes, Reynolds, this volume).

Rain lahars are common on ash-covered slopes when, at some stage after the eruption, heavy rainfall occurs. This may be during the next rainy season or it could be years later. The forecasting problem then amounts to assessing which combination of rainfall intensity and duration will produce dangerous mudflow phenomena in the (known) quantities of deposited ash.

Each of the three examples of volcanic mudflow hazard relates to a specific causative factor: the Nevado del Ruiz lahars in Colombia were so-called "jökulhlaups" generated by melting ice and snow in the top area; those of Merapi volcano, Central Java, Indonesia, were derived from debris of a volcanic plug and related pyroclastic flows and glowing avalanches; those of Kelud volcano, Eastern Java, were due to the presence of a crater lake. They thus cover a wide range of lahar phenomena. At the same time the three cases represent, in this order, an increasing capability of lahar mitigation, including hazard zoning, monitoring/early warning and engineering aspects.

The Nevado del Ruiz lahar disaster of 1985

The devastating mudflows that occurred during the eruption of Nevado del Ruiz volcano, Colombia, of November 13, 1985 were generated by the partial melting of the ice cap covering its top area. They were thus not, like "normal" lahars, liquidized sediment flows with about 50% volume of water and 50% volume of debris (mostly volcanic—see Table 1), but "jökulhlaups", a special type of hyperconcentrated flow, well-known particularly from Iceland, which may have 80% volume of water or even more. The hot ash and volcanic bombs dropping on the glacier ice (Figures 4 and 5) resulted in large volumes of floodwater containing a rather insignificant amount of debris of volcanic origin: the bulk of the solid material was derived by the violent flows from the ravines and valleys.

The ice cap of the Nevado del Ruiz measures about 25 km^2, is on the average 20 m thick and has a volume of 300 million m^3. Since the eruption of 1985 was a comparatively mild (sub)Plinian eruption, only 5–15% of the ice did in fact melt. A more violent eruption could easily have caused total melting or sliding down of the glacier ice thus causing a glacier burst or glacier avalanche. Even so, the partial melting that did occur was sufficient to create the largest mudflow disaster on record, with a death toll of 25,000 people (Figures 6 and 7).

Three major mudflows developed: the one in the Chinchina valley caused destruction and about 1000 casualties in the town of the same name situated to the west of the volcano, at a distance of approximately 35 km. The lahars in the Guali valley affected the towns of Mariquita and Honda in the northeast, the latter at a distance of 90 km.

The most devastating and complex lahars were those on the east flank of the volcano, that wiped out the town of Armero. Lahars formed in two major branch valleys of the Lagunillas and the Azufrado rivers respectively, and reached the town, one after the other, having travelled 60 and 70 km. The presence of a small temporary lake dammed by a previous landslide in the Lagunillas Valley may to some extent have aggravated the situation. When entering the plain the mudflow spread in three lobes, the main central one of which destroyed Armero (Figure 8).

There is ample evidence in the field of prehistoric events of similar nature. Also in the last few centuries the volcano has erupted several times (1595, 1828/29, 1832/33 and 1845). The lahars of the 1845 eruption reached the site where, more recently, the town of Armero rapidly developed; it caused a death toll of 1000 people in the town of Ambalema in the Recio Valley. A minor eruption occurred on September 11, 1985, two months before the main eruption, and generated a lahar 27 km

Figure 4 Helicopter view of the top area of the Nevado del Ruiz volcano shortly after the 1985 eruption. The ice cap and the (hot) ash that has been deposited on top of it are clearly visible.

long in the Azufrado valley; although this did not affect inhabited areas, it gave a clear indication of the hazardous nature of the situation (Rivera, 1986; Sigurdsson and Carey, 1986).

A series of earthquakes starting in November 1984 with their epicentres in the top area of Nevado del Ruiz announced renewed volcanic activity and convinced the authorities that surveillance of the volcano was necessary. Seismic monitoring was operational from July 1985 onward. Many earthquakes were recorded every day and an eruption became likely. However, no continuous lahar observation or early warning system, neither manned with radio/telephone connection nor automated, was installed. The lahar warning time is comparatively long in the case of the Nevado del Ruiz because of the great distance between the crater and the settlements; depending on the path length of the flows and their average velocity (38 km/hr in the case of

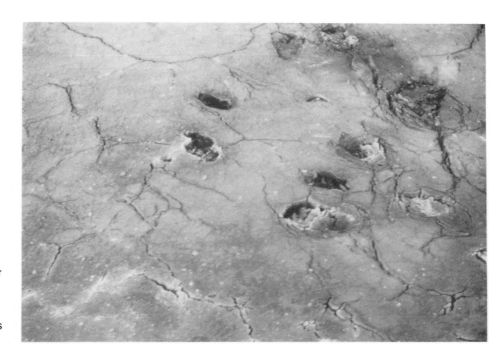

Figure 5 Detailed helicopter view of the ash cover on the Nevado del Ruiz glacier showing deep melting holes caused by hot volcanic bombs (scale approximately 1:3000).

Figure 6 Example of the lahar flows that devastated the slopes of the Nevado del Ruiz volcano in 1985, seen from a helicopter.

Armero) it ranges from 45 minutes to two hours or even more.

Aspects of volcanic hazard zoning also were studied and the resulting hazard zoning map (see Hall, this volume) was published shortly before the eruption. It correctly indicated which areas would be devastated by the lahars a few days later. The map thus clearly showed the importance of hazard zoning in endangered areas for purposes of land-use planning, siting of protective engineering structures, raising awareness among the

population and implementation of emergency operations. It should, however, have been available much earlier, so that its warnings could be acted upon, and be legalized to give it the status of an official document. A more elaborate volcanic hazard zoning map for the area has recently been published and may serve for future use.

A major characteristic of the Nevado del Ruiz lahars was their high fluidity, which had the consequence that the fountaining and spattering flows carried large blocks and filled the valleys up to high levels (25 m has been

Figure 7 Mudmarks on the trees (white arrows) and large deposited blocks (black arrows) resulting from the lahars in Chinchina on the occasion of the Nevado del Ruiz eruption of 1985. Note that no thick layers of lahar material are left.

Figure 8 The devastation in Armero resulting from the 1985 lahar disaster.

observed), but only for a short period. The flow surfaces were strongly inclined at the sharp curves of the valleys, and the mudflows swept over still higher parts of the valley slopes. The water of the deposited mudflows was drained off rather rapidly after the event, however, and ultimately only a comparatively thin layer of lahar deposits was left. The rivers will subsequently incise their courses in these and gradually remove part of the material; considerable storage capacity for future lahars is thus left in the valleys, which is an advantage as compared to less-fluid lahars which tend to build up the valley/ravine bottoms, thus rendering nearby settlements increasingly vulnerable. This is the case with the Merapi volcano, discussed next.

The lahars of Mount Merapi, Central Java

A dome of viscous lava rising in the crater is a main feature of the volcanic activity of Mount Merapi. The dome may be stable for some time and then suddenly grow. Late in 1986, about 1.5 million m³ volume was added within a period of a few months, but no eruption followed. The lava rapidly disintegrated, tumbling down in hot or cold blocks and subsequently disintegrating further. A maximum angle of repose thus was formed in the top area below the dome (Figure 9).

In the case of increased activity, or actual eruption, great quantities of tumbling red-hot lava blocks commonly create rapid "ladu" flows (hot gas-propelled pyroclastic flows forming glowing avalanches, directed down-slope and scorching the land). The process is mainly governed by gravity: since most of the gases have escaped earlier, no gas-propelled glowing avalanche or cloud develops. The material derived from the dome also gives rise to volcanic mudflows (lahars). Most of these are secondary (rain) lahars that occur on the southwest and west slopes during the wet seasons following an eruption (Tazieff, 1983; Zen, 1983).

An effective surveillance system has been established which comprises continuous monitoring and periodic (monthly) observations, visual as well as seismic (six stations), geochemical and geodetic. A long-term investigation programme is also operative, and a lahar monitoring and harnessing programme has been launched. The rainfall on the critical southwestern slopes is monitored every three minutes by X-band radar with a rotating antenna. After filtering out the non-rain-bringing clouds, the rainfall intensity is determined by computer into nine groups. Further, telemetered rain-gauging stations are installed and serve also for calibration; telemetered water-level gauging stations in the major ravines, a lahar wire-sensor and a video-camera complete the monitoring system (VSTC, 1985; 1986). It is not yet operational as an early warning system, because the rainfall intensity and duration needed to trigger lahar flows have yet to be established and, because the short early-warning period due to the high velocity of the flows, this requires an automatic alarm. Attempts are also being made to control the lahar flows to a certain extent by constructing check-dams, consolidation dams, dykes and sediment traps in the major ravines (Figures 2 and 3).

The geomorphological situation of the top area and the southwest slopes is given in Figure 10 with the

Figure 9 Detail of the lava plug in the Merapi top area, Central Java.

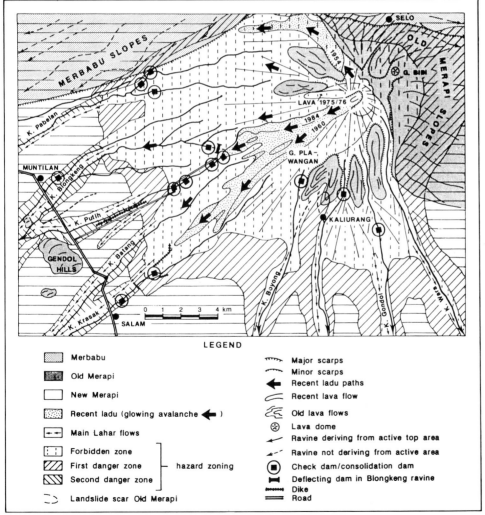

LEGEND

▦	Merabu	⌢⌢⌢	Major scarps
▦	Old Merapi	⌢	Minor scarps
▢	New Merapi	◀	Recent ladu paths
▨	Recent ladu (glowing avalanche ◀)		Recent lava flow
	Main Lahar flows		Old lava flows
▦	Forbidden zone	⊗	Lava dome
▨	First danger zone — hazard zoning		Ravine deriving from active top area
▨	Second danger zone		Ravine not deriving from active area
	Landslide scar Old Merapi	⊙	Check dam/consolidation dam
			Deflecting dam in Blongkeng ravine
			Dike
			Road

Figure 10 Geomorphological situation, hazard zoning and engineering structures of the western slopes of the Merapi volcano, Central Java.

existing hazard zoning (Directorate of Volcanology, 1977) superimposed on it.

The hazard zoning accords very well with the terrain configuration: the limits set by the Merbabu slopes and the protective role of the erosion scarps of the Old Merapi are obvious in the outline of the forbidden zone. A distinction between ravines deriving debris from the top area, and those with headwaters outside the direct influence of volcanic activity, is an important criterion in hazard zoning. The ladu pyroclastic flows and accompanying glowing avalanches probably reach approximately down to the 500 m contour. The first hazard zone only covers a comparatively narrow strip, but includes some rather densely populated mid-slope areas. The second danger zone mainly relates to lahars: locations where the ravines are particularly shallow or narrow and where sudden changes in their general direction occur, are particularly hazardous (Maruyama, 1980; Verstappen, 1982, 1988). One may question whether the at-present-inactive lahars of the southern slope and the active and very dangerous ones of the southwestern slopes should fall in the same category.

Distinction by observation from the air between ladu pyroclastic flows and lahar deposits is feasible as the latter show braided characteristics and are situated lower down the slopes in and near ravines.

Geomorphologically the ladu deposits form a convex "bulge" at the transition of the upper and middle slopes, which tends to spread future ladus over several ravines further downstream. This geomorphological situation invites some discussion as to the optimal policy for lahar control measures: the present control-works aim at accumulating the debris by way of check dams, protected from undercutting by consolidation dams directly downstream, and by sediment traps. This policy clearly favours the build-up of the river beds that, in places, are at dangerously high levels with respect to the adjacent villages. In addition, the lahar hazard has been artificially concentrated in the Putih River by closing off the lower Blengkong River by an earthen dam to protect the Muntilan area. This dam has been raised on several occasions in the past to maintain its effectiveness; increased lahar hazard downstream of the dam along the Putih River is therefore a logical consequence.

One may wonder whether a calculated distribution of the lahar products over several ravines is not preferable to this situation. Part of the material could possibly already be deviated further upstream on largely uninhabited interfluves. As a result of the lesser bed-loads that would result, the ravines would develop the tendency of cutting rather than building up. This would increase their storage capacity for future lahars and also be less dangerous for the villages alongside the channels.

The feasibility of this alternative approach to the lahar problem merits consideration.

The lahars of Kelud Volcano, Eastern Java

The case of Kelud volcano, Eastern Java, Indonesia, is of particular interest for two reasons. First, because the lahar hazard is specifically related to the water of a crater lake that is blown into the air on the occasion of every eruption and, second, because engineering works have been implemented to mitigate the hazard.

The slopes of this relatively low (1731 m) volcano are very irregular as compared to other active Indonesian strato-volcanoes. This is in part due to lava plugs occurring in the top area, but primarily results from the violent, spasmodic erosion processes associated with the emptying of the crater-lake during eruptions before the installation of the control works described further on. The slopes were thus stripped of large quantities of unconsolidated material. This was transported to the Brantas River which, with a very low gradient, loops clockwise around the volcano from the southeast to the north and drains virtually all its slopes. Important natural levees thus were built in its entire middle and lower course. The river therefore lies several metres above the surrounding plains, which gives rise to major irrigation problems. At present, the mudflows being largely under control, further development of the levees is mostly related to accelerated erosion in parts of the catchment area.

The slopes most endangered by mudflows were those in the west (Lahar Gedok, which occured in 1848) and in the southwest (Lahar Badak, etc., 1919) because they corresponded to the lowest part of the crater rim. Prior to 1875 the crater lake measured about 78 million m^3, but in that year a breach was formed in the southwest crater rim after heavy rains and a mudflow 13 km long caused destruction in the Sregat and Blitar districts. Since this disaster was not of direct volcanic origin, it was not included in the eruption record of Kelud volcano. Also the mechanism differed: the lake waters were not flung in the air in all directions, but spilled sideways to the southwest. Thereafter the volume of the crater lake was reduced to approximately 40 million m^3, with some fluctuations depending on changes in the lake bottom configuration caused by volcanic activity, including the formation of lava domes (1920), etc.

The lahars of Kelud volcano are diversified. Those occuring during or directly after an eruption usually start as cold lahars, and turn into hot lahars subsequently when the admixture of hot eruptive material increases. Rain lahars are formed when, at some later stage, heavy rains mobilize the ash, and lahar deposits accumulate on the higher volcanic slopes. Since previous experience (notably the 1901 eruption) had shown the imminent danger for the town of Blitar in the southwest, a diversion dam was constructed in the upper part of the Badak ravine as early as 1905. It proved to be inadequate,

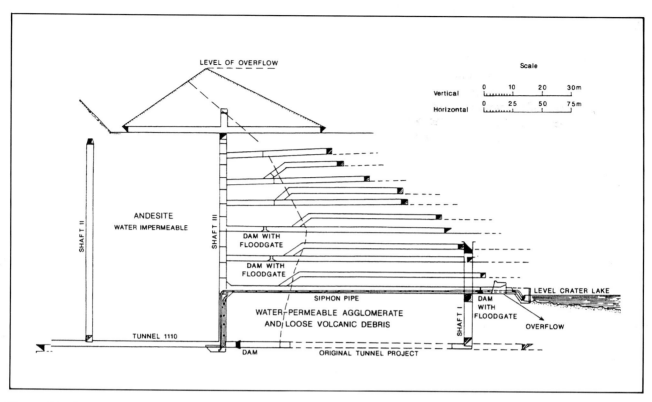

Figure 11 Cross section of the tunnel works on the southwest slope of the Kelud volcano, Eastern Java (simplified after IJzendoorn, 1952).

however, being swept away by the mudflows accompanying the ill-famed eruption of 1919 that spread over an area of more tha 130 km², wholly or partly destroying more than a hundred villages and killing 5110 people.

Engineering works to drain the crater lake by way of a tunnel, to diminish the mudflow hazard, started in September 1919 (van Bemmelen, 1949). The works took longer than expected because of unforeseen difficulties. The aim was to connect the crater bottom with the Badak ravine by a tunnel; the boring operation was able to start at both ends, as the crater lake had not yet re-established itself after the eruption. Tunnelling westward from the crater had to be given up, however, when a lava plug rose near the boring site in December 1920. The eastward tunnelling from the Badak ravine in the lavas of the crater rim progressed slowly due, among other reasons, to the high temperature (46°C) and humidity at the working front. The volcanic debris of the inner crater slope was reached in November 1922 and from then onward the tunnel had to be lined with concrete. By that time, however, the crater was half-full with water again (22 million m³) and the pressure of the water column resulted in an inrush of water, mud, and debris into the tunnel in April 1923, killing five workers in the process. A new plan then was devised aiming at a step-wise lowering of the lake by way of seven parallel tunnels at different levels and by applying a syphon

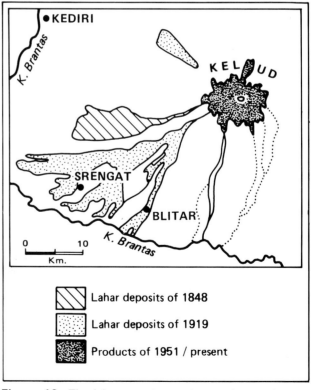

Figure 12 The lahars on the southwest slopes of Kelud volcano, Eastern Java, and the tephra deposits in the top area developed since 1951.

system for draining off the water (Figure 11). This project ultimately was successfully implemented and the volume of the lake was reduced to less than 2 million m³.

The eruption of 1951 proved the effectiveness of the tunnel system: no sizeable mudflows developed and the tephra products of the eruption were deposited mainly on the upper slopes of the volcano (Figure 12), whereas formely they were washed down by the blown-out lake water (IJzendoorn, 1952). The tunnels thereafter had to be cleaned of mud and debris, but they functioned adequately during the eruption of 1966 and again recently during the eruption of February 1990. The death toll is now incomparably less than prior to the construction of the tunnel works (32 victims in 1990). Rain lahars may gain somewhat in importance because of the ash deposits on the slopes; but they are much less violent than those caused earlier by the lake water. The case of the Kelud mudflows demonstrates clearly that concerted action can strongly reduce this type of volcanic hazard (Bercy and others, 1983).

Conclusions

The three examples given clearly show the diversity of lahars as a volcanic hazard. They also show that their disastrous aspects can be reduced by concerted effort. The geomorphological terrain configuration and the eruption characteristics of the volcanoes concerned are two basic aspects of lahar disaster mitigation, but should be coupled with a survey of the land utilization and population density distribution of potentially endangered areas. Mitigation programmes should comprise the following four elements:

(a) lahar hazard zoning (as part of general volcanic hazard zoning);
(b) monitoring of lahar flows, coupled with an early-warning system and including rainfall monitoring for coping with rain lahars;
(c) building protective structures where appropriate; and
(d) awareness-raising among the population.

The programme should ultimately result in a framework for land-use planning aiming at low population densities in endangered zones and in a scenario for emergency/rescue operations to be implemented in case of an eruption. The issue is of particular importance where active volcanoes occur in developing countries, because farmers will always be tempted to occupy fertile land even if it is dangerously situated.

References

Bemmelen, R.W. van, 1949. *The Geology of Indonesia*. Vol 1A, 220 p, The Hague, Government Printing Office.

Bercy, C., and others, 1983. Underwater noise survey in the crater lake of Kelut volcano (Indonesia). In: H. Tazieff and J.C. Sabroux (eds), *Forecasting Volcanic Events*, Elsevier Publishers, Amsterdam, p 529–543.

Directorate of Volcanology, 1977. *Hazard zoning map of Mount Merapi.*

Maruyama, Y., 1980. Applied geomorphological land classification on debris control planning in the area of Mt Merapi, Indonesia. *Proceedings of the 10th International Cartographic Association Conference*, Tokyo, 19 p.

Naranjo, J.J., and others, 1986. The November 13, 1985 eruption of the Ruiz volcano, Colombia: tephra fall, jökulhlaups and lahars.*Science*, vol 232, p 964–967.

Rivera, A., 1986. Events leading to Colombia's Ruiz volcano eruption and its results described. *The Mining Engineer*, vol 2, p 99–100.

Sigurdsson, H., and S. Carey, 1986. Volcanic disasters in Latin America and the 13th November 1985 eruptions of Nevado del Ruiz volcano in Colombia. *Disasters*, vol 10 (3), p 205–240.

Tazieff, H., 1983. Monitoring and interpretation of activity on Mt Merapi, Indonesia, 1977–1980: an example of practical "civil defense" volcanology. In: H. Tazieff and J.C. Sabroux (eds), *Forecasting Volcanic Events*, Elsevier Publishers, Amsterdam, p 485–492.

Ulate, C.A., and M.F. Corrales, 1966. Mudflows related to the Irazo volcano eruptions. *Journal of the Hydraulics Division, ASCE*, vol 6, p 117–423.

Verstappen, H.Th., 1982. *Applied Geomorphology*. Elsevier Publishers, Amsterdam, p 417–423.

Verstappen, H.Th., 1988. Geomorphological surveys and natural hazard zoning, with special reference to volcanic hazards in Central Java. *Zeitschrift fur Geomorphologie, Neue Folge*, Supplement Band 68, p 81–101.

VSTC, 1985. *Debris Flow Forecasting and Warning at Mt Merapi.* Volcanic Sabo Technical Centre report, p 1–6.

VSTC, 1986. *Volcanic Debris Control in the Area of Mt Merapi.* Volcanic Sabo Technical Centre report, p 1–54.

IJzendoorn, M.J. van, 1952. The eruption of Gunung Kelud on August 31, 1951, proved the utility of the Kelud tunnelworks. *Indonesian Journal of Scientific Research*, vol 1, p 8–9.

Zen, M.T., 1983. Mitigating volcanic disasters in Indonesia. In: H. Tazieff and J.C. Sabroux (eds), *Forecasting Volcanic Events*, Elsevier Publishers, Amsterdam, p 219–236.

6 The 1985 Nevado del Ruiz eruption: scientific, social, and governmental response and interaction before the event

M.L. Hall

Abstract   The 1985 eruption of Nevado del Ruiz Volcano, Colombia, which resulted in 23,000 deaths, has been studied by South American scientists, sociologists, and engineers, who evaluated the responsibilities and actions of the groups that played important roles before the eruption. The study is based upon interviews of government, scientific, media, and civil defense participants and the detailed perusal of all reports, publications, and newspapers prior to the eruption. This article deals with the actions related to 1) volcano hazard evaluation, 2) volcano monitoring activities, 3) media involvement, and 4) governmental interaction and response.

Introduction

During the past 15 years the field of volcanology has made great advances, especially in our understanding of eruptive processes as well as the precursory phenomena that frequently precede and accompany eruptions. Consequently, the prediction of a volcanic eruption has become increasingly more feasible and will likely be demanded by the public in the near future.

Thus, the disaster caused by the eruption of Nevado del Ruiz volcano in Colombia on November 13, 1985, which resulted in 23,000 deaths when the town of Armero was overrun by debris flows, represents a special tragedy, not only for Colombia but also for volcanology. This event was the second worst volcanic disaster of the century, the first being the eruption of Mont Pelée, Martinique, in 1902, which resulted in nearly 30,000 deaths.

Today, volcanologists understand more clearly what types of studies are required to mitigate the impact of volcanic eruptions upon the community. Generally, two types of programs should be carried out, more or less simultaneously.

The first program consists of an evaluation of the volcanic hazards most likely to occur during a future eruption. Potential volcanic hazards are forecast on the assumption that future eruptions will generally follow the pattern of past eruptions. The evaluation is based upon a knowledge of the volcano's history, the nature and frequency of its eruptions, its explosivity, and the drainage channels used by flowage deposits (Crandell and Mullineaux, 1975).

The second involves a program of permanent monitoring of dangerous active volcanoes. Such a program should include seismic vigilance, geodesic measurements of the volcano's shape, periodic temperature and geochemical analyses, as well as repeated visual observations and photographic documentation. The operation of a net of at least four high-gain seismographs, installed around the volcanic structure and operated in real time via telemetry, offers the best monitoring capability both in the long and short term, as this permits the detection of the first seismic events associated with the reactivation of the volcano as well as those that often announce the beginning of an eruption (Tilling, 1989).

During the 12 months of precursory activity of Nevado del Ruiz, parts of these programs were carried out by the Colombians and visiting volcanologists. However, the best intentions and efforts failed, a terrible tragedy resulted, and we are left to ask what went wrong.

Given these circumstances, the Centro Regional de Sismologia para America del Sur (CERESIS) formed a study group comprising three volcanologists, two sociologists, and three engineers*, all from Latin American countries including Colombia, to determine and evaluate the scientific, social, and governmental actions that led to the fateful eruption of 13 November 1985. The study is based upon all written information available to the group, including newspapers, magazines, and technical reports written during the pre-eruption period, as well as upon interviews with many participants and upon personal knowledge of the events. Part of this article is drawn from that study titled "Volcanic Risk: Evaluation and Mitigation in Latin America", which was sponsored by CERESIS and with support from the Canadian International Development Research Centre (IDRC) contract No 3-P-86-0232, which is gratefully acknowledged. A detailed chronology of events with supporting references is given in Hall (1990).

The study led to many notable observations and far-reaching conclusions (CERESIS, 1989), of which only a few will be dealt with here. This article will focus on the actions related to:

(a) Volcano hazard evaluation
(b) Volcano monitoring
(c) Media involvement, and
(d) Governmental interaction and response.

Volcano hazard evaluation

Introduction

Colombia contains the most northerly volcanoes of the great Andean chain of stratovolcanoes that extends 6000 km from southern Chile. Nevado del Ruiz volcano, one of the northernmost of this chain, is situated on the crest of the Cordillera Central, which separates the Magdalena River basin in Tolima Department to the east from the Cauca River basin (Caldas Department) to the west (Figure 1). The volcano lies 30 km southeast of Manizales and 150 km northwest of Bogota, Colombia's capital. Most of the intermediate and upper slopes of the cone are cold, uninhabited grasslands; however, a significant population and rich agricultural lands occupy the distant lower slopes of this Cordillera.

Before the 1985 eruption, Nevado del Ruiz volcano was well known to Colombians as a majestic, glacier-clad mountain. Unlike most other volcanoes in Latin America, Ruiz had been the focus of many detailed studies, including two PhD dissertations (Herd, 1982; Jaramillo, 1980) and a comprehensive geothermal report consisting of nine tomes put together by Italian scientists in 1983. In 1984–5, new investigations of the volcano were initiated by the University of Grenoble (France) and the Instituto Nacional de Investigationes Geologico et Mineras (INGEOMINAS, the Colombian Geological Survey). Unfortunately, none of these studies foresaw future volcanic activity of Ruiz, and consequently no attempt was made to assess the potential hazards.

The history books mention the destruction of the Armero region by mudflows in 1595 and 1845, and as recently as 1935 and 1950 torrential rains generated debris flows that passed through Armero, causing considerable damage. Clearly scientists and historians should have been more aware of the implications of the precarious situation.

*Members of the study group included: Alberto Giesecke, Patricia Anzola, Benjamin Fernandez, Oscar Gonzalez-Ferran, Minard Hall, Bruno Podesta, Alejandro Rodriguez, and Alberto Sarria.

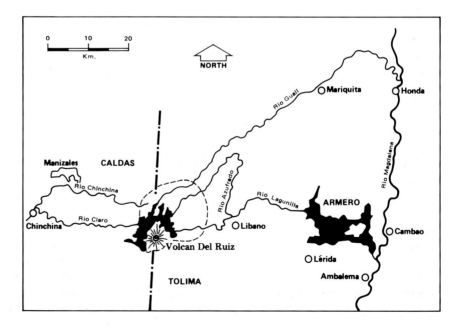

Figure 1 Sketch map showing Nevado del Ruiz volcano's position on the crest of the north-south oriented Cordillera Central between Tolima Department to the west and Caldas Department to the east. The Rios Guali, Azufrado, and Lagunilla drain eastward to the Rio Magdalena. Westward the Rios Chinchina and Claro pass near but safely by Manizales, capital of Caldas Department. Armero is located at the eastern foot of the mountains, where the Rio Lagunilla leaves its canyon and spreads out onto an alluvial fan.

Chronological development of the eruption

Beginning in December 1984, increased fumarolic activity and numerous felt earthquakes prompted visits to the crater by local mountaineers and geologists. Immediately a concerned citizens group was formed in Manizales to investigate the activity, evaluate the danger, and seek support from the local and national governments. It recommended that Caldas and neighboring departments undertake a survey of the volcanic and seismic risk of Ruiz. A geological team from INGEOMINAS was sent in late February 1985 to assess the problem; it concluded that the anomalous activity was 'normal' in the life of a volcano.

A staff member from the United Nations Disaster Relief Office (UNDRO) visited the volcano by chance in mid-March, assessed the situation as potentially dangerous and concluded "to protect the population near the volcano from a possibly big eruption, INGEOMINAS is obligated to effect a monitoring program and a hazards evaluation, while civil defense is obligated to prepare the logistical support, alarm systems, and if necessary the evacuation of persons in the high-risk areas" (Tomblin, 1985). Fortuitously a seminar on volcanic risk, planned the previous year, was held in Manizales at that time, and this concluded that the national Government was responsible for assessing the volcanic hazards of Ruiz.

In early May a follow-up visit by an UNDRO scientist confirmed that the volcano remained in an abnormal state, and that none of the recommended studies had been initiated. This lack of interest was due in part to the fact that the danger was not immediately recognized. Because the high region around the volcano is essentially uninhabited, little forethought was given to possible danger far down the rivers that drain the cone. Another misconception was that Ruiz could not be an active volcano, since it was completely mantled by ice and snow.

In early August, INGEOMINAS had prepared a proposal for UNESCO requesting funds to support a program of volcanic risk evaluation and mapping; it was either not sent or not acted upon.

The situation remained unchanged until 11 September when a small phreatic eruption spewed ash westward over Manizales and generated a moderate-sized lahar that traveled 27 km down the Rio Azufrado valley. These events prompted the national government to reconsider the potential danger and consequences. So, on 17 September, the Minister of Mines ordered that a volcano hazards map be prepared immediately, promising the funds to accomplish the task. Geologists from INGEOMINAS, Central Hidroelectrico de Caldas (CHEC, the regional electric company), and the University of Caldas worked together to make the map in the very short period of three weeks, turning over the map to the government and civil defense authorities at a press conference on 7 October. Although preliminary, the map clearly and accurately showed those areas that could potentially be over-run by lahars (Figure 2). Furthermore, INGEOMINAS warned that Armero ran a 100% probability of flooding, if lahars were generated by an eruption of Ruiz.

Unfortunately the usefulness of the map was not appreciated. Only 10 copies of the map were prepared, thereby limiting their distribution to a few Ministries and civil defense. Furthermore, the government being skeptical of an impending eruption and worried more about the consequences of the map itself, ordered that the map be re-checked and re-submitted by 12 November, effectively "tabling" the map for another month. The revised map, without important changes, was not ready until a few days after the fateful eruption.

Conclusions

With respect to the recognition of potential danger and its evaluation, the following conclusions seem warranted:

1. The preliminary hazard map showed the likely types of volcanic phenomena that might occur around Ruiz and in the areas of high risk, especially along the drainages where mudflow activity was expected. Unfortunately its short lead time (from 7 October to 13 November) and restricted distribution limited its usefulness as a mitigation tool. If, for example, civil defense had received the map earlier, they could have been better prepared.

2. Valid recommendations concerning a hazard evaluation of Ruiz were made as early as February and March 1985. If they had been heeded and a hazards map prepared, it could have had the positive effect of promoting general interest, greater public awareness, and a more thorough civil defence preparedness.

3. As a consequence of the map's publicity on 7 October, it met with hostile opposition and discrediting attacks in the press by economic interests in Manizales and Caldas, who were concerned by the threat of devaluation of property values. These factors apparently had no effect upon the veracity of the hazards evaluation, but did add to the public's confusion and skepticism that reigned in Manizales prior to the eruption.

Volcano monitoring

Chronological development

Unlike most volcanoes where the reactivation interval typically lasts a few weeks or months (Simkin and Siebert, 1984), Ruiz experienced almost a full year of ominous signs, including felt earthquakes on the cone and notable fumarolic activity in the crater. Beginning in

Figure 2 Basically identical to the volcanic hazard map made public on 7 October, this, the first published map, was released after the tragedy. The dark heavy lines leading away from the volcano at the center show accurately the predicted routes of mudflow activity and that Armero, Mariquita, and Chinchina were located in high risk zones.

December 1984, this unusual activity prompted visits by local geologists and mountaineers. Visual and photographic documentation of the abnormal state of the crater began with the visits made in January and February 1985, which was given wide publicity when it appeared in the local Manizales newspaper. In their report to the authorities, local scientists strongly recommended that permanent surveillance of Ruiz be initiated as soon as possible. The INGEOMINAS team also echoed this sentiment in their report of late March.

In their visits of March and May, UNDRO experts emphasized the need for seismic monitoring as well as other surveillance techniques in order to evaluate the state of the volcano properly.

However, no serious attempt at monitoring was initiated until late July when, after numerous difficulties, INGEOMINAS and the Instituto Geofisico de los Andes (Bogota) installed four portable seismographs at sites located about 12 km from the crater and began to record the seismic activity. For lack of telemetry, the instruments had to be visited once a day to change records, a process that required a full day and delayed seismogram inspection until the following day.

At the request of UNDRO and the Swiss Disaster Corps, a seismologist from the Swiss Seismological Service went to Colombia on 11 August and deployed three additional seismographs in collaboration with CHEC. Unfortunately the two groups, INGEOMINAS and CHEC, tended to work independently and not combine their data. Because there was no seismologist in Manizales, the INGEOMINAS seismograms were sent to a contracted seismologist in Bogota, which resulted in a delay of several weeks before they were properly

interpreted. In fact, the first written seismic report did not appear until shortly before the 13 November eruption, although by 7 October it was generally known that the seismic activity was centered under the north flank of the cone at 5–10 km depth.

On 6 September unusual seismic activity began, preceding the 11 September phreatic eruption. Recognized only in hindsight, it consisted of intervals of harmonic tremor an hour long followed by longer periods of quiet. If it had been studied by experienced seismologists, it might have warned of a change in the state of the volcano.

Although a geochemist from CHEC had been collecting and analysing water samples from the crater since February, the arrival of an experienced New Zealand geochemist on 22 September began an attempt to establish a systematic program of geochemical monitoring. However, the samples had to travel to New Zealand for analysis, again delaying their interpretation. The results were not available until after the 13 November eruption.

During October the seismic activity continued, but at a lower level than September, which may have tended to calm the population and delay public preparedness. An attempt was made in early October to unite the efforts of INGEOMINAS and CHEC in the reading and interpretation of the seismograms, so that they could be processed within 24 hours in Manizales before being sent to Bogota. Furthermore, by combining their data, much greater precision and reliability in the seismic results could have been achieved. Unfortunately this brief collaboration had effectively ended by mid-October, and during the month prior to the eruption little seismographic information was obtained or processed.

Figure 3 Following the 13 November 1985 eruption, the US Geological Survey rapidily installed many seismic/deformation stations around the volcano, whose data were telemetered continuously to Manizales. If there had been an international cache of these remote stations on permanent call, their timely deployment in early 1985 might have gone a long way toward mitigating the disaster.

Beginning on 4 October, a specialist in volcano deformation spent three weeks on the volcano helping national surveyors to establish dry tilt nets and install electronic tiltmeters, with which to detect swelling of the cone due to magma intrusion. No abnormal variation was detected. Because the electronic tiltmeters were brought by the specialist from a neighboring country, and thus were only on loan for the duration of his visit, this equipment could not be left in place where it might have helped to predict the main eruption. This administrative problem points out the need for regional caches of loanable equipment and experts (Figure 3). Indeed, following the 13 November eruption, the US Geological Survey rapidly installed many seismic/deformation stations around the volcano, these data being telemetered continuously to Manizales. If there had been an international cache of these remote stations on permanent call, their timely deployment in early 1985 might have gone a long way toward mitigating the disaster.

Shortly before the 13 November eruption, INGEOMINAS published a report summarizing their seismic monitoring activities, which noted that bands of harmonic tremor were often detected several days before phreatic explosions, and that these tremors might be useful as a forecasting tool. This possibility was not appreciated until much later.

Finally, at the regular meeting of the Caldas committee of volcanic studies on the night prior to the fateful eruption, there was obvious concern about the situation since, though abnormal seismicity continued, both the seismic and deformation monitoring programs had died out, and the owners of the seismographs used by INGEOMINAS had requested that they be returned for use in other projects.

Conclusions

In terms of the monitoring activities on Nevado del Ruiz prior to the eruption of 13 November 1985, the following conclusions seem relevant:

1. During the 11-month period prior to the eruption the sulfur and steam emission and the seismicity increased significantly; these were recognized early as possible precursory activities. No change in the temperature of several hot springs around the cone was detected during this period. In hindsight there was no recognizable change in the deformation of the cone.
2. The monitoring activities were initiated too late to have established true baseline conditions and were too limited in scope and continuity to have been effective. As an example, the seismic monitoring would have been more valuable if it had been telemetered. Clearly, it is important to establish some permanent monitoring on all high-risk volcanoes, long before any reactivation.
3. Inexperience in seismic and geochemical data processing and interpretation, as well as unnecessary delay in getting the results back to the scientists at Manizales, impeded an effective monitoring program. Again, countries with high-risk volcanoes should have a core of experienced scientists upon whom they can depend in times of crisis.

4. All of these problems jointly precluded a realistic attempt to understand the activity and predict an eruptive event.
5. Scientifically it is still impossible to distinguish between a sequence of volcanic precursory signals that lead to an eruption and those that do not. There are many more volcanic crises than there are eruptions. And therein lies the great difficulty for volcanologists—to be able to determine unequivocally that an eruption is pending. Volcanoes that have been well-monitored for a long time offer the best chance of timely, accurate eruption forecasting.

Media involvement

Introduction
The media (newspapers, magazines, radio, and television) played very important roles during the crisis. In Colombia newspapers are based in populated centers throughout the country. Typically big-city newspapers stress national and international events, while the regional papers lean toward local and regional news. Magazines are generally published monthly, are based only in the big cities, and feature stories of great interest; because of their nature they cannot handle fast-breaking news. Television, because of its limited coverage and reception outside the urban areas, was not an important factor in the crisis. The radio service is highly decentralized in Colombia and widely received both in rural and urban areas, and thus it undoubtedly played an important role in the emergency. Unfortunately because of the fleeting nature of radio news, it was impossible to review what had been said and what precisely had been its effect.

Interestingly, the media's effectiveness in the crisis depended upon inherent differences in the culture, traditions, and awareness of the two regions directly affected by the emergency. The two departments, Caldas to the west and Tolima to the east, sit on opposite sides of the volcano. Tolima, located closer to the republic's capital, Bogota, depended upon that city for much of its commerce, culture, and news. Caldas, having always been more isolated from Bogota and other cities by the mountains, had traditionally been more independent, dynamic, and shown greater initiative.

Thus, it was no surprise that in Caldas there was an effective local newspaper, La Patria, which gave excellent coverage of the developing Ruiz crisis beginning in January 1985. Also Caldas, having a greater degree of electrification and an average higher educational level than Tolima, was in a more favorable position to obtain and understand both printed and electronic news. Thus, the populations of Caldas and Manizales were more aware and better prepared for the emergency.

Tolima, on the other hand, had not developed a strong local news service over the years, but depended upon Bogota's two daily newspapers which, being more concerned with the politics and commerce of the capital city, relegated news of the volcano's situation, if any, to their back sections, thereby playing down their importance. Their limited coverage of the crisis began in mid-September, following the 11 September eruption.

These two daily newspapers, El Tiempo and El Espectador, published 11 and 20 articles of varying size, respectively, in the pre-event period—a small contribution considering the 40–80 pages published daily by each one. Conversely, La Patria published more than 60 articles, most of them of greater depth and detail and often placed on or near the front page. Only two articles were published by magazines in this period; both had the effect of playing down the idea that a dangerous situation might be developing and curtailing any questioning by the population about what the government was doing to reduce the danger.

Conclusions
1. The media played a significant role in the pre-eruption period, but its coverage was only partial, superficial, and at times distorted. Problems common to reporting elsewhere also surfaced at Ruiz, including the difficult conveyance of seemingly incomprehensible data and interpretations from the scientists to the journalists, divergent or opposing opinions published by competing papers, and inevitable wrong conclusions written by newsmen with little knowledge of the subject. A single spokesman representing the collective opinion of both the scientists and civil defense would have greatly improved the quality of news reporting and its impact through daily briefings.
2. Possibly the torpid behavior of the national press (and others) reflects in part the role it chose of being prudent and not wanting to be an alarmist, when it was unclear if an eruption was imminent or even probable. Conversely, the local paper in Manizales, with its greater coverage, was accused repeatedly of being sensationalist and alarmist. Clearly, there exists a very narrow line indeed between these two extremes.
3. Because almost all of the news concerning the volcano originated in Manizales, it was generally thought that the danger existed only there. Subsequently, when the scientists announced that Manizales was not in danger, this was widely interpreted to mean that the volcano itself was no longer dangerous.
4. The relatively well informed Caldas community was better able to act independently as well as to question and subsequently provoke the national Government into action. Precisely the opposite situation existed in Tolima.

Governmental interaction and response

Introduction

In its simplest form, the executive power of the Colombian Government consists of three levels: the municipal, the departmental, and the national. In 1985 the municipal governments were administered by mayors appointed by the President of the Republic; their success as mayors depended in part upon their own initiative as well as their relationship with the Governor of the Department. The latter was also nominated by the President and, depending upon his influence with the President (for obtaining funds, for example) as well as his stature as a local leader in the Department, he might give considerable direction and force to the Department. The President of Colombia, elected by the people, presided over the running of the nation. He in turn had his immediate staff of advisers and close associates who counseled him in matters of national concern. Under his direct control were the Ministers of the 13 ministries, each with its numerous dependencies, such as INGEOMINAS and the Colombian Civil Defense (Ministry of Defense). Finally, it should be mentioned that there is a bicameral Congress, its members being elected by the populace of each Department.

Chronological development

In Caldas Department and its capital, Manizales, the Governor and Mayor played important roles from the beginning. As early as late January 1985, a concerned citizens group was formed by business leaders and scientists, who subsequently involved both the Mayor and Governor. By 4 March this Civic Committee with its scientific commission had become a legal entity and had recommended that a program of surveillance and studies of the volcano be established immediately by CHEC and the local universities. In early February the Caldas civil defense leader told the civil defense chief in Bogota about the situation, who in turn advised INGEO-MINAS. They, representing the national Government, made an inspection of the volcano but concluded that the anomalous activity did not warrant concern.

During the third week of March, the National University in Manizales and CHEC jointly sponsored a seminar concerning volcanic risk on Ruiz, which had been planned and organized long before the crisis had developed. Its main conclusions were:

(a) The consequences of volcanic risk were the responsibility of the national Government;
(b) the abnormal activity of Ruiz totally justified any and all efforts that might help to mitigate a possible catastrophe of volcanic origin;
(c) both a hazards evaluation and a monitoring program should be initiated quickly; and
(d) a public awareness program was needed immediately.

By 12 April the Caldas civil defense leader had prepared a tentative emergency plan for the Department. The Mayor and Governor were advised of the implications of Ruiz's unrest by a visiting UNDRO expert in early May.

UNDRO, UNESCO, and the World Organization of Volcano Observatories (WOVO), being aware of the growing crisis at Ruiz and its possible consequences, maintained contact with different Colombian entities, urging them to make a formal request to UNESCO for technical aid. Finally on 26 June the Colombian Ambassador to UNESCO in Paris wrote to the Colombian Minister of Foreign Relations in Bogota, stating that UNESCO was ready to send experts, equipment, and aid; he stressed the urgency of making a formal request to the agency. In one of the more notable blunders, the Ambassador's letter got lost in Colombian bureaucracy, resurfacing two months later in a distant office of the Ministry of Public Education.

In early July the Civic Committee presented their report of the state of the volcano and their recommendations concerning a program of monitoring and hazard evaluation. The report was well received by the Governor of Caldas and the Mayor of Manizales, who jointly contributed 1.5 million pesos (US$2500) to begin the program, and made a formal request to the Swiss Disaster Corporation for technical aid. CHEC was given the task of contacting the Italian Government about assistance.

The actions of the three governmental levels were comparatively slow during the months of July, August, and September, until the phreatic eruption of 11 September. During the long period from January to September 1985, in which Caldas and Manizales were very attentive to the problem, Tolima on the other side of the volcano was poorly informed of the volcano's unusual activity, skeptical of any possible danger, and basically remained unconcerned and uninvolved in hazard mitigation efforts.

The 11 September eruption deposited minor ash in Manizales and sent a moderate-sized lahar 27 km down the Rio Azufrado in Tolima, which clearly showed that the volcano had to be taken seriously. This event finally put the volcano crisis on the front page of Colombia's principal newspapers, thus giving it a wider coverage nationally as well as within official circles in Bogota.

Prior to this eruption the national government had no policy or programs aimed at mitigating volcano crises. There was widespread skepticism in the Government and in the population in general about the possibility of an eruption or catastrophe. This skepticism was exceeded only by the public's concern as to whether the Government was doing anything about the problem.

On 12 September the Caldas Governor and the Manizales Mayor met with civic leaders, the police chief, and civil defense authorities to discuss the emergency plan and, if necessary, the evacuation of people along the Rio Chinchina. Furthermore, on the advice of the Civic Committee, they reported to the national press that the Rio Claro in Caldas and the Rios Azufrado, Lagunilla, and Guali in Tolima, all ran high risks of mudflow inundation; these were precisely those rivers that were later devastated by lahars (Figures 1 and 4). Again demonstrating leadership, the Caldas Governor requested an immediate meeting of the governors of the four Departments (including Tolima) that could potentially be affected by an eruption of Ruiz. That meeting, held on 16 September and attended only by representatives of the three other governors, concluded that each Department would manage its own emergency plans autonomously, thereby precluding a combined emergency effort sought by the Caldas Government.

On 17 September the national Government held its first meeting to confront the volcano crisis on Ruiz. It was attended by the directors of the principal technical and scientific state institutions, including INGEOMINAS and the national Civil Defense Office, and presided over by the Minister of Mines. Its purpose was to adopt a prevention and emergency plan, and it established six subcommissions of wide scope to formalize the plan. As a consequence, INGEOMINAS, CHEC, and the University of Caldas were instructed to initiate hazards evaluation and mapping; meanwhile, civil defense declared a state of alert for the Ruiz region.

On 18 September, while the Caldas civil defense office began its volcano awareness program at the elementary and high school levels, the Tolima Red Cross warned through a national newspaper that "because of its location Armero with its 38,000 inhabitants could disappear, if there is a flood along the Lagunilla and other rivers that head on the volcano".

Unimpressed by the national Government's response, the Caldas Representative to the National Congress provoked a debate on its floor about the dangers of an eruption of Ruiz, the high probability of devastating mudflows, and the lack of governmental action to prevent a catastrophe. He verbally attacked several Ministers for not clearly seeing the problem and for taking little action, which he characterized as improvised and irresponsible.

Although the Tolima Government essentially maintained a passive role through the crisis, it did convene a meeting of mayors and other local authorities on 1–2 October to advise them of the actions to be taken in case of an eruption. The importance of having radio communication between the volcano and the threatened towns was stressed, but no radios were available at the time. The emergency committee in Manizales distributed a hand-out indicating what one should do in case of an eruption: these recommendations were subsequently published by a national newspaper in Bogota, thereby giving them a limited distribution in Tolima as well.

In a widely publicized press conference held on 7 October, the Minister of Mines presented the Ruiz hazards map that had been prepared since 17 September. It correctly showed the zones of greatest risk, especially the Armero and Mariquita areas. INGEOMINAS' official report stated that "there is a 100% probability of

Figure 4 The broad summit of Nevado del Ruiz, with its 17 km² of snow and ice, is here covered by volcanic debris and ash which was erupted on the night of 13 November 1985. In the foreground are seen the headwaters of the Rio Azufrado down which the lahars began their two-hour trip to Armero.

lahar formation if, because of an eruption, Ruiz's snow cap should melt, which would greatly endanger the Tolima towns of Armero, Mariquita, Honda, Ambalema, and the lower parts of the Rio Chinchina (Caldas). However, towns such as Armero would have approximately 2 hours in which to evacuate".

A small photograph of the hazard map was published on the front page of a leading Bogota newspaper on 9 October. The map was returned to INGEOMINAS for rechecking and its formal presentation set for 12 November. As a result the Mayor of Armero stated that many inhabitants were uncertain whether to stay or leave Armero. Furthermore he stated that Armero's emergency committee did not have the equipment nor financial support to confront a possible catastrophe. "The inhabitants have lost faith in the truthfulness of the news and essentially have put their fate into the hands of God", he concluded.

On 29 October Tolima's emergency committee revised their emergency plans for Herveo and Libano as well as the Red Cross's communication system. It should be pointed out that at the time of the eruption no radio link existed between the headwaters of the Azufrado and Lagunilla rivers and the threatened towns, which could have provided an early warning and permitted an evacuation, since the mudflows required about two hours to travel the 50 km route.

Most of the eruption day, 13 November 1985, was cloudy and rainy. Explosions occurred at 1506 hours, and during the subsequent hours ash fell over Armero and other towns to the northeast. This event prompted Tolima's Red Cross to order the evacuation of Armero in the early evening, but it remains uncertain if this order was ever transmitted or, if so, why the alarm was not sounded in Armero. The principal eruptions occurred at 2108 and 2130 hours, during a major lightning storm in Armero and a televised soccer game. They went unnoticed except by those scientists who were near the summit of the volcano. The last eruption is thought to have been responsible for the pyroclastic flows which in turn generated the devastating lahars which, traveling down the west slopes, arrived in Chinchina at 2240 hours, and traveling down the east side arrived in Armero at 2335 hours. Upon leaving the narrow canyon of the Rio Lagunilla the mudflow broke out over the alluvial fan upon which Armero was located (Figure 5). The town was overridden by at least three successive fast-moving flows, which essentially carried away or buried 22,000 inhabitants and 80% of the town's structures. Additionally 1000 fatalities occurred along the Rio Chinchina.

Conclusions

1. Prior to the tragedy there was no national policy or program directed toward volcano crises in Colombia. Nor were there institutions or experts dedicated to volcanic studies or monitoring. Similarly, civil defense had never had to contend with a progressively developing emergency such as a volcanic crisis. Consequently there was no systematic or sustained civil defense preparedness, except partially in Caldas. As an example, no program of permanent vigilance with radio communication had been ordered or set up in the headwaters of the Rio Lagunilla above Armero, which could have easily warned of descending lahars.

Figure 5 The devastated remains of Armero testify to its having been overridden by three consecutive lahars that poured out of the Rio Lagunilla canyon to the right of the photo. The poor siting of the town near the mouth of the canyon had subjected it many times in the past to floods and small lahars.

2. Repeated mitigation recommendations, made by nationals as well as invited experts, as early as 10 months prior to the eruption, were not addressed by the national Government, until it was too late for civil defense and others to respond adequately.

3. The scientific role, both national and international, was aimed more toward verifying, through limited monitoring, whether an eruption was likely, and if so, when it might occur. Its contribution would have been greater if, simultaneously, the scientists had established a continuing dialogue with civil defense, presenting explicit and detailed descriptions of the hazards map and possible eruption scenarios, in addition to collaborating with civil defense on simulations and evacuations. The two factors that curtailed the development of these activities were the late preparation of the hazards map and the slowness with which the national civil defense understood and acted upon the problem.

4. The overall responsibility of the Ruiz crisis management and resulting tragedy lies with the national Government. The factors that hindered the Government's role include its skepticism about an eruption, inadequate leadership, a non-responsive bureaucracy, and other distracting emergencies.

5. The local initiatives originating in Manizales proved to be the most important ones for mobilizing interest, volcanic studies, public awareness, civil defense preparedness, and outside assistance. This corroborates well the motto that 'civil defense begins at home'.

6. After the publicity given to the hazards map and the stated danger, neither the national Government nor the affected population can argue that they were ignorant of the situation. However, both remained skeptical about the possibility of an eruption and more so about any catastrophe . . . until it occurred.

Epilogue

A volcano emergency brings together different groups and institutions to form a complex working group which attempts to carry out diverse and often new activities, aimed at mitigating the uncertain socio-economic impact of an actual or threatened eruption. The situation is usually one where previous volcanological or crisis experience, as well as governmental support, are lacking. This working group arrangement, usually brought together in a hurry, is difficult and nearly impossible; but it can be greatly improved upon.

Volcanologists and volcanological programs often hold the key for succussful mitigation of volcanic disasters. Volcano hazard evaluations and maps, prepared long before an emergency, force civil defense to recognize the risk and prepare emergency plans; they awaken public awareness about nearby volcanoes; and encourage land-use planning of high-risk zones. Permanent telemetered monitoring of all high-risk volcanoes insures against surprise eruptions, crystallizes and trains a group of national volcanologists, establishes the normal background activity of a volcano, and improves the chances of timely and accurate eruption forecasting. Through these activities, local volcanologists gain the experience needed better to advise and help civil defense in its prevention and preparedness program, as well as how to deal with the press and other media.

References

CERESIS, 1989. Riesgo Volcanico: Evaluacion y Mitigation en America Latina (Volcanic Risk: Evaluation and Mitigation in Latin America). Unpublished study, Canadian IDRC contract No 3-P-86-0232.

Crandell, D.R., and D.R. Mullineaux, 1975. Technique and rationale of volcanic hazards appraisals in the Cascade Range, Northwestern United States. *Environmental Geology*, vol 1, p 23–32.

Hall, M.L., 1990. Chronology of the principal scientific and governmental actions leading up to the November 13, 1985 eruption of Nevado del Ruiz, Colombia. In: Ruiz Volume, *Journal of Volcanology and Geothermal Research*, vol 42 (1/2), p 101–116.

Herd, D., 1982. *Glacial and Volcanic Geology of the Ruiz-Tolima Volcanic Complex—Cordillera Central, Colombia.* Special Publication, INGEOMINAS, Bogota.

Jaramillo, J.M., 1980. *Petrology and Geochemistry of the Nevado del Ruiz volcano, Northern Andes, Colombia.* Doctoral dissertation, University of Houston, 167 p.

Simkin, T., and L. Siebert, 1984. Explosive eruptions in space and time. In: *Explosive Volcanism*, National Academy of Sciences Press, p 110–121.

Tomblin, J., 1985. Unpublished UNDRO Mission Report.

Tilling, R., (ed.), 1989. *Volcanic Hazards: Short Course in Geology*: Vol 1. American Geophysical Union, 123 p.

7 Lahars of Cotopaxi Volcano, Ecuador: hazard and risk evaluation

Patricia A. Mothes

Abstract   Cotopaxi Volcano in north central Ecuador is a steep-sided stratocone covered by abundant snow and ice, which has produced about 30 eruptions since the early 1500s. Lahars generated by eruptions have swept through the surrounding populated valleys on numerous occasions, but the volcano has largely been quiescent since 1877. Despite this, the volcano is under continual monitoring at the present time, and both scientific and civil defense activities are aimed at preparing the population for eventual eruptions by employing hazard and risk maps of the volcano, eruption scenarios, simulated evacuations and educational seminars.

Introduction

Ecuador's two main mountain ranges, the Cordillera Occidental to the west and the Cordillera Real to the east, are both noted for their spectacular snowcapped volcanoes. The two cordilleras are separated by the densely populated Interandean Valley and extend northwards from central Ecuador into Colombia (Figure 1).

In Ecuador the Cordillera Occidental is 270 km long and comprised of volcanic centers which are principally breached calderas of dacitic composition spaced approximately 30 km apart (Hall and Wood, 1985). Pichincha Volcano, upon whose eastern flanks is located Ecuador's capital, Quito, is the only volcano of this cordillera to have produced verifiable historical eruptions. At the southern extent of this range, Chimborazo Volcano with its voluminous snow and icefields is the highest (6300 m) and largest volcano in the country.

On the opposite side of the graben-like Interandean Valley, 40–50 km to the east, is the parallel Cordillera Real that also extends northwards into Colombia. For more than 370 km it is adorned by scattered centers of andesitic stratocones. The more notable snowcapped volcanoes include, from south to north, Sangay, Altar, Tungurahua, Cotopaxi, Antisana, and Cayembe, of which all but Altar and Cayembe have had historical activity. The upwelling of moist air masses coming from the Amazon basin against these volcanoes assures the presence of permanent snow/ice fields, especially on the eastern side of each cone.

About three million people live within the Interandean Valley between the two volcanic chains; they will be directly or indirectly affected by future eruptions.

With the exception of remote Sangay Volcano, which has daily eruptions, no other Ecuadorian volcano has produced more eruptions in historical times than Cotopaxi. Because of the large volume of snow and ice that cover its flanks, many of these past eruptions have generated voluminous debris flows. Given its unique position between two densely populated valleys, future eruptions will almost certainly generate lahars which could have devastating impacts upon the large population and infrastructure now located in those valleys. As such, the threatened areas around Cotopaxi offer the most vivid example in Ecuador of human settlement and infrastructure residing in the paths of major mudflows.

Cotopaxi Volcano

With a diameter of 22 km, almost 3000 m of relief and an elevation of 5897 m, Cotopaxi is one of the great stratovolcanoes that dominates the Sierran landscape (front cover). Its symmetrical cone is topped by a crater 800 m in diameter and 160 m deep. Its snow and ice fields are estimated to cover about 20 km^2, and have a volume of 0.5–1.0 km^3. Three important drainage systems head

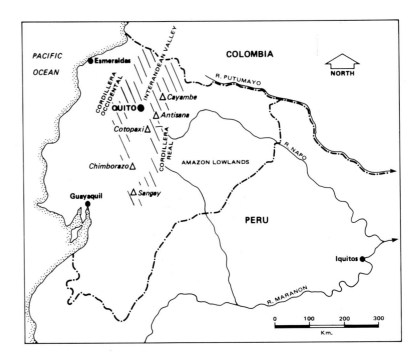

Figure 1 Map of Ecuador showing the Cordilleras Occidental and Real, the Interandean Valley, and several volcanoes.

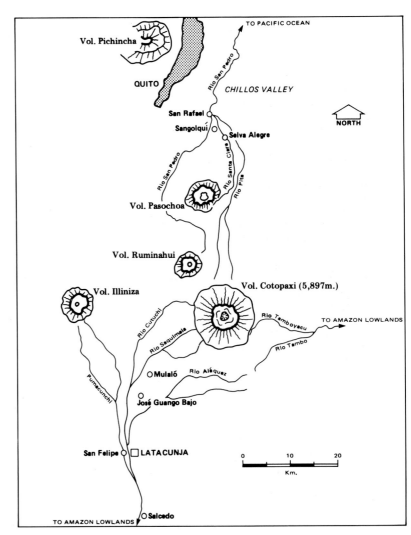

Figure 2 Regional map showing the three main drainage systems that head on the volcano—the Rio Pita to the north, the Rio Cutuchi to the west and southwest, and the Rios Tambo-Tamboyacu to the east-southeast.

54

upon the volcano: they are the Rio Pita to the north, the Rio Cutuchi system to the west and southwest, and the Rio Tambo-Tamboyacu system to the southeast (Figure 2). Although these streams cut deeply into the outer flanks of the volcano, the cone exhibits a young morphology with mere traces of incision by contemporary glaciers.

Attesting to its very active nature, Cotopaxi's flanks are covered by lava flows, pyroclastic flows, ash falls and lahar deposits. The volcano has had at least seven distinct periods of activity during the last 40,000 years (Hall, 1979, 1987). Of the more recent periods, Period 6, possibly starting 4000 years BP, includes the formation of dacitic pyroclastic flows, avalanches from a sector collapse of the cone, and many voluminous lahars, in addition to 26 other eruptions (Hall, 1987 and unpublished field data; Smyth and Clapperton, 1986). Period 7 embraces the recent historical eruptions whose deposits are typified by variable decimeter-thick accumu-

lations of pumice and ash-fall deposits, small pyroclastic flows, small lava flows averaging 60 million m³, and lahars (Hall, 1987). Recent lava flows from Cotopaxi are olivine andesite with two pyroxenes, and andesitic basalt with SiO_2 values of 54–58%.

Cotopaxi's historical eruptions

Since the Spanish arrived in Ecuador in 1532, Cotopaxi has had about 30 eruptions (Figure 3). The first of these in 1533 or 1534 was reported by the Spanish conquerors Ceiza de Leon and Benalcazar. The eruption supposedly caused great fright to the Imperial Incas, then defending the Quito area, weakening their resistance; hence the Spanish easily gained control (Wolf, 1904). Following a dormant period of 208 years for which no verifiable eruptions have been noted, the volcano reawakened in 1742, producing three eruptions in June, July, and December of

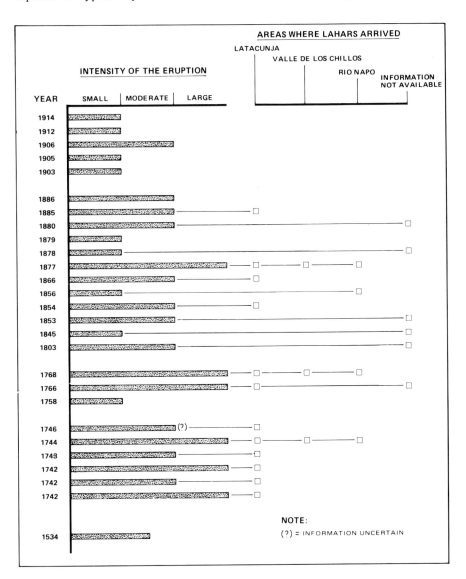

Figure 3 Cotopaxi has had about 30 eruptions since the beginning of the historical era in 1532. These can be divided into roughly three periods: the mid 1700s, the mid and late 1800s and early 1900s. Lahars have traveled down all three drainages, but especially the Rio Cutuchi, inundating the Latacunga Valley.

55

Figure 4 Lahar deposits underlying Latacunga are principally fine-grained and homogeneous, derived from hyperconcentrated flows. The lahar deposits of the 1877 eruption buried a three-story textile factory that once prospered on the west bank of the Rio Cutuchi in Latacunga.

that year. All three were notable for the destruction caused by lahars to the haciendas, livestock, and people in the Latacunga Valley on the volcano's southwest side (Paredez, 1982). Significant lahar-producing eruptions also occurred in 1766 and 1768, inundating the main drainages with lahars and reportedly dropping a thickness of 30 cm of pumice and ash on haciendas located to the west of the volcano (Wolf, 1904; Egred, 1989).

Between 1768 and 1877, an estimated seven eruptions were reported which produced small to moderate lahars that traveled down the Rio Cutuchi drainage, but apparently caused minimal damage (Paredez, 1982; Egred, 1989).

The most destructive historical eruption occurred on 26 June, 1877, preceded by small ash emissions on 24 and 25 June. While it is difficult to compare the magnitude of this eruption or the volume of its lahars to those of the previous century, the 1877 eruption is the most thoroughly recorded of all Cotopaxi events. Important observations were compiled by Wolf (1878), Sodiro (1877), and Whymper (1892), although not one of them actually witnessed the eruption. They arrived shortly thereafter and collected data and accounts from survivors and through fieldwork.

In this, the last major eruption of Cotopaxi to date, lahars traveled down the three principal rivers and destroyed textile factories in the Chillos and Latacunga Valleys (Figure 2), damaged haciendas, towns and bridges, and swept away an indigenous settlement along the Rio Napo, east of the volcano in the Amazon lowlands. Wolf (1878, p 17) wrote ". . . on this terrible day (26 June 1877), hundreds of people and thousands of animals lost their

lives and Cotopaxi changed the countryside into deserts of sand and stones and destroyed in one hour the work of many generations". Padre Sodiro (1877, p 13) presented one of the more lucid descriptions of the advancing lahars when he stated "the road between Mulalo and Latacunga was full of travelers, and given the great velocity of the lahar, many people, some on galloping horses, were overcome by the advancing wave, and thus disappeared". The lahar that flowed down the volcano's north side reportedly arrived 18 hours later at the Pacific port of Esmeraldas, 270 km from its source, where "the Rio Esmeraldas rose several feet and the current brought cadavers and pieces of houses and furniture from the Chillos Valley" (Wolf, 1878, p 25).

Characteristics of Cotopaxi's lahars

Lahars generated on Cotopaxi since the 1500s have formed by the instantaneous melting of the snow and ice cap by pyroclastic flows and lavas. The 1877 eruption produced many pyroclastic flows that, according to eyewitnesses in the town of Mulalo, 25 km southwest of the volcano, "came over the crater rim like the foam on the top of a pot of rice that is boiling over" (Wolf, 1878, p 20). These are now recognized as scoria-rich pyroclastic flows that swept down all quadrants of the cone, melting the snow, and generating the lahars that descended all of the drainages.

A vast depositional plain has formed at the foot of the volcano's north and east sides, while the west and south flanks are steep and little deposition occurs until the

Figure 5 This group of buildings, mostly residences, are concentrated along the east bank of the Rio Cutuchi in Latacunga. All are built upon historical lahar flow deposits.

rivers exit the steep-walled canyons and cross the plain near Mulalo. Lahar deposits close to the volcano are generally coarse and bouldery, with 20–40% of matrix made up of fine to coarse sand and only a low percentage of silt and clay. Further downstream, on the depositional plain, more than 25 km from the crater, most of the large boulders have dropped out of the flows and thus the size range tends to be smaller and more homogeneous. At distances greater than 25 km, especially in the valleys where river water was mixed into the advancing lahar, the larger clasts were progressively left behind due to the diminishing yield strength of the flow. Thus the deposits show a sequential transition from clast-abundant debris-flows to hyperconcentrated stream-flows (HCF) (Figure 4). These latter flows were first described at Mount St Helens in Washington State, USA (Janda and others, 1981) where they formed at distances of 27–43 km downstream. Similarily, the HCFs at Cotopaxi tend to form massive homogeneous deposits and contain an abundance of sand and gravel with little silt or clay-size particles. On the depositional plains around Cotopaxi, lahar deposits dominate the stratigraphy and few incipient soils have formed in the sequence, suggesting almost continual lahar activity over the past centuries.

Two streams, the Rio Santa Clara and Rio Pumacunchi, neither of which originate on Cotopaxi, are also main avenues for lahars. The northward-flowing Rio Santa Clara which rises on the extinct Pasochoa volcano and flows on a course that parallels the Rio Pita, has carried Cotopaxi lahars on numerous occasions. At times, lahars flowing down the Rio Pita have super-

elevated on a sharp right turn located 25 km downriver from the volcano, spilled over the 35-m-high bank, and entered the Santa Clara drainage (Mothes, 1988). In the Latacunga Valley a similar situation occurs with the Rio Pumacunchi, which flows parallel to the Rio Cutuchi but several kilometers to the westward of it. On numerous occasions lahars traveling down and across the Latacunga depositional plain have spilled over into the lower Pumacunchi channel. The lahars of the 1877 eruption inundated both drainages described above, surprising many people, particularly in the towns of San Felipe, a suburb of Latacunga through which the Rio Pumacunchi flows (Sodiro, 1877), and in Selva Alegre located on the banks of the Rio Santa Clara.

The present-day population centers of Latacunga and Salcedo on the Rio Cutuchi, and Quito's suburbs of San Rafael and lower Sangolqui along the Rio Pita are underlain primarily by hyperconcentrated flow deposits (Figure 5). However, it is very common to see large boulders in the yards of recently built homes along Rios Pita and Cutuchi (Figure 6), a reminder of the great force that lahars have.

Besides the historic lahar deposits, which generally are restricted to stream channels and their adjacent flood plains, a more extensive deposit is that of the prehistoric "Chillos Valley Lahar". This attained a maximum height of 110 m above the Chillos valley floor and mantled an area of 226 km². Its deposit is generally two meters thick and has a pumiceous composition in which silt and clay-size particles account for 15–20% of the matrix, which allowed it to maintain a sufficient yield strength to transport meter-size boulders more than 250 km from

Figure 6 An example of the new housing tracts that are being constructed upon bouldery lahar deposits along the margins of the Rio Pita, near San Rafael.

the source. The Chillos Valley lahar is believed to have been generated by a sector collapse of Cotopaxi's cone and a major pyroclastic flow, events which occurred more than 2400 years ago (Hall, 1987; Hall and Mothes, unpublished field data). Due to the special circumstances that produced this exceptionally large lahar, its reoccurrence is not considered likely.

Estimates of the velocity of the 1877 lahar have been made, although they are hampered by the poor preservation of the deposits and the lack of good time control. Nonetheless an approximation has been derived by comparing lahar velocity data from other volcanoes as well as by studying the historical accounts (INECEL, 1989). Given that the gradients of the steep gulleys on the middle and upper flanks of the cone average 28%— although in the first 5 km from the crater the quebradas have an even steeper gradient—it is reasonable to believe that the lahar speed could have attained 60–100 km/hr. Similar velocities were measured on the Pine Creek drainage on Mount St Helens volcano, where slopes have a similar steep gradient and the canyons are narrow and chute-like (Pierson, 1982). On the volcano's lower flanks, the stream gradients become less extreme, and here presumably the velocity diminished, although there are several sites 20 km or more downstream from the cone where the flow had enough velocity to pass over topographic barriers 35 m high, and may have still maintained a velocity of about 60 km/hr (R. Janda, personal communication, 1988). The surrounding depositional plains, particularly on the west and south side, are wide and flat with an average gradient of only 1.2%. According to our estimates, supported by

historical accounts, lahar flows with volumes of 45,000–70,000 m^3/s could reach Latacunga, 43 km from the crater, in approximately 40–60 minutes (INECEL, 1989). A similar arrival time is estimated for San Rafael on the Rio Pita.

Hazard and risk assessment

Given the high vulnerability of the inhabitants of the Chillos and Latacunga valleys, and prompted by notable restlessness of Cotopaxi in 1975–1976 (increased number of volcanic quakes, increased vapor emission), the first preliminary hazard map was prepared by Hall (1976) and presented to the National Civil Defense office. A more detailed version of this map was later published by the US Geological Survey (Miller and others, 1978). Both maps clearly delimited the lahar hazard zones associated with the three drainages and provided a document on which civil defense could base their emergency planning. Fortunately the volcano calmed down and remains that way to the present day.

In 1988 an updated hazard map was published in Spanish at a scale of 1:50,000 (Hall and von Hillebrandt, 1988). This map, published in two sheets (North and South) (Figure 7) has been widely distributed to national and regional civil defense offices and governmental authorities and is used as a basis for a variety of mitigation exercises. Each map has an accompanying text written for the non-scientist which describes the typical eruptive phenomena that would be expected in a future Cotopaxi eruption. The map itself delimits the worst-case and typical-case scenarios for lava flows, pyroclastic flows, lahars, ashfalls, and rock avalanches, based upon the mapped deposits of past eruptions.

While the intent of a hazard map was to show the potential danger of different volcanic phenomena, it does not adequately demonstrate the vulnerability or volcanic risk of an area. Thus, ideally, a volcanic risk-map should also be prepared. It should show in detail the population distribution and infrastructure that

occupy the high- and low-risk zones. In the case of Cotopaxi, the hazards maps show that future lava and pyroclastic flows are expected to remain within the uninhabited areas around the cone, and that the probability of sector collapses and lateral blasts is very low.

In contrast, future lahar flows are considered to offer the greatest danger to the most people, since many thousands of people live along the threatened

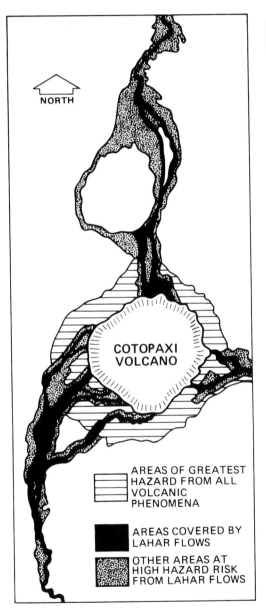

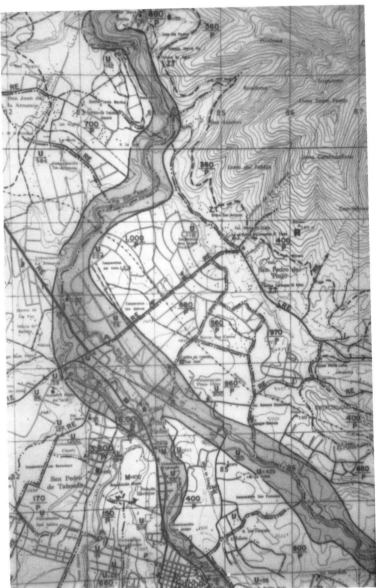

Figure 7 Volcanic hazards map of Cotopaxi. The volcanic cone is in the center of the map. The area of greatest risk from all volcanic phenomena is shown in gray surrounding the volcano. The Rios Pita, Santa Clara, and the Rio Cutuchi system with four tributaries, all with high lahar hazard, are darkened.

Figure 8 Portion of the lahar risk map for the populated areas of San Rafael and Sangolqui in the Chillos Valley. The junction of the Rios Santa Clara and Pita is shown. The darkened zones are those of highest hazard from lahars. The population of an area is indicated by a number/P. Additional data include the number of lots in subdivisions, school attendance, major infrastructure, roads that could serve as evacuation routes (RE), and refuge areas (R).

drainages. Consequently a study was undertaken to show in greater detail the social/economic impact of future lahars. The results of the inventory were published as a series of ten maps (at a scale of 1:25,000) that cover the Rio Pita drainage north of Cotopaxi and the Rio Cutuchi drainage south of the volcano (Mothes, 1988). These maps utilize the same lahar hazard boundaries that were established in the hazards map. The risk maps have complemented the hazards maps by showing more ground detail at the larger scale (Figure 8).

In both the high- and low-risk zones of the map, the following information is provided:

(a) The approximate number of people concentrated in towns and built-up areas, the dispersed rural population, the number of houses, the number of subdivision lots, the schools with their attendance, health centers, markets and commercial centers, and tourist attractions.

(b) Vital infrastructure including highways, railroad lines, airports, bridges, reservoirs and sources of potable water, hydroelectric plants and electricity transmission lines, and the Trans-Ecuadorian oil pipeline and smaller pipelines.

(c) The economic base which includes industries, agricultural land use, and military bases.

(d) On the perimeter of the hazard zones are indicated the nearest elevated areas which could provide immediate security to the population from advancing lahars, as well as the provisional evacuation routes which would most effectively allow people to travel away from the hazard zones rapidly.

Several generalities can be drawn about the human settlement in the lahar hazard zones. Both valleys, but especially the Chillos Valley located only 20 minutes by road from Quito, are experiencing high annual growth rates (around 6%) and the systematic development of housing tracts upon the lahar plains alongside the Rios Pita and Cutuchi. For example, in the Chillos Valley there are some 1500 new housing lots located upon the 1877 lahar deposits, while in the Latacunga area approximately 3500 lots are being developed upon those same deposits. New industries, a new four-story regional hospital, schools, and other infrastructure have been heedlessly built within the high-risk zones during the past ten years.

The population living within the area of greatest lahar hazard in both valleys is estimated at about 30,000 people. In the Latacunga valley some 20,500 people live in the zones of highest risk, the majority of them clustered around Latacunga or Salcedo; however, there is also a substantial dispersed rural population of 6000 individuals living upon the lahar's depositional plain.

The population (9000 people) at highest risk in the Chillos Valley is primarily concentrated in and near the towns of San Rafael, Selva Alegre and the lower sectors of Sangolqui that adjoin the Rio Santa Clara.

Along those rivers that drain the east side of the volcano and enter the Amazon basin, the population living in the high-risk zone is estimated at 2200 people. This region is more than 150 km downstream and nearly 5000 m lower than Cotopaxi's summit.

The zone of low lahar hazard is that delimited by the Chillos Valley Lahar. The probability of recurrence of a lahar of this magnitude is very remote and thus it has not received the same concern given to the high-risk zones. The estimated population within this low hazard zone is 133,000 people, of which 51,000 live in the Latacunga Valley and 82,000 in the Chillos Valley.

Civil Defense actions

Since the initiation of the program, jointly organized by the United Nations Disaster Relief Organization (UNDRO) and the US Agency for International Development (USAID) for the mitigation of volcanic disasters in Ecuador in 1987, in which the Instituto Geofísico of the Escuela Politécnica Nacional and the National Civil Defense office have been the main actors, a number of public education and preparedness activities have been promoted to increase the level of public awareness and understanding about the lahars in the threatened region. These activities resulted in the transferral of scientific data from the Instituto Geofísico to Civil Defense and their application toward relevant mitigation exercises.

Starting in mid-1987, the inter-institutional advisory group of Civil Defense, comprising professionals in many fields, visited all six provinces which could be affected by Cotopaxi lahars. In each provincial capital, a seminar of several days' duration was presented in which the lahar problem was discussed and analyzed for that particular region, and a field trip was carried out for local public officials and interested citizens to illustrate what lahars are and which areas would likely be affected. Finally the seminar concluded with the preparation of provincial- and county-level emergency plans by local civil defense authorities for dealing with an eventual lahar crisis.

Besides these regional seminars, which have greatly increased awareness about lahars for the populace living near the volcano, as well as those living hundreds of kilometers downstream, an acted-out eruption scenario and two simulated evacuations have also been sponsored by the National Civil Defense office and their provincial counterparts.

An eruption scenario of a future hypothetical Cotopaxi eruption (Mothes and Hall, 1988) was prepared, based

Figure 9 Simulated evacuations, such as this one in Latacunga in 1989, in which 5000 people participated, have helped to raise the awareness level of the population that would be most affected by future lahar flows.

upon the volcano's history and characteristics and adapted to the local conditions and culture. It was subsequently acted out by the governors, provincial civil defense officers and other authorities from the six threatened provinces. During the 10-hour exercise, each provincial committee was physically isolated, and could only maintain communication with the National Civil Defense officials and scientists monitoring the volcano by means of one telephone and one transceiver. Each committee was thus advised sequentially of the hypothetical eruption and the formation and advance of the lahars, while National Civil Defense radioed instructions and recommendations to each province where the local committee then had to set in motion the corresponding actions of their emergency plan.

During the following year, simulated evacuations were carried out in Latacunga and Sangolqui (Figure 9). Each evacuation was preceded by a short training session for the local leaders and teachers. At the sound of a siren, approximately 5000 students and government workers were evacuated from their schools and offices and walked to the designated refuge areas, usually within 1 km or less of their school or office. Afterwards a general critique was held by the leaders and evaluators of the event. Both evacuations were considered an educational success, which served to inform students and the public about the lahar hazards and what preventative actions should be taken.

These seminars and evacuation exercises are being reinforced with the publication of evacuation posters for the Chillos and Latacunga Valleys. These posters, part of the Instituto Geofísico's UNDRO program, represent a collaborative effort with Civil Defense; they would be posted once Cotopaxi has shown clear signs of reactivation.

These educational campaigns have reached many residents but certainly not all. A recent poll taken by D'Ercole (1989) showed that those least aware of the lahar problem were the wealthier and more educated people living upon the lahar plain in the Chillos Valley. Those in San Rafael are primarily new residents from Quito, who have no rural roots or collective memory of the lahars that swept through this community 113 years ago. Thus, when asked about the greatest threat to their lives from the volcano, located 50 km up-valley, but in clear view, the majority of San Rafael residents replied that the major threats were lava flows, ash-fall and lahars (mudflows); and only 30% knew when the last major eruption had occurred. In comparison, residents of Jose Guango Bajo, a rural community southwest of the volcano that is infrequently affected by lahars, replied that ash-fall was the greatest hazard, followed by lahars—an accurate appraisal. This community retains a better recollection of the 1877 event, an aspect directly related to the collective memory of older family members and the knowledge passed down through the generations about the history of their area and the volcano's activity.

Future educational campaigns must focus on a more comprehensive and lengthier program within the schools, at local community halls and in churches in order to disseminate broadly the pertinent hazard information. Socio-economic differences of the people living in rural areas and in the more urbanized areas indicate a need to diversify the communication approach.

Wealthier, better educated families receive the majority of their information from newspapers and television. In contrast, rural people, primarily agriculturalists, obtain their information concerning Cotopaxi mainly from the radio or from conversations. Furthermore, simulated exercises serve mainly to inform the residents of the affected town, in this case Latacunga and Sangolqui; others may become aware of the event only if, by chance, they tune into the media that very day or the next. For this reason, a permanent and broad educational campaign is necessary, so that people remain informed, if only marginally, of the hazards which could endanger their area. Starting in 1990, Partners of the Americas have joined in a strong collaborative effort with Civil Defense and the Ministry of Education to ameliorate these educational deficiencies.

Ongoing monitoring

Monitoring of Cotopaxi began in 1975 when scientists of the Escuela Politécnica Nacional set out portable seismographs and established dry-tilt deformation triangles in response to increased vapor emission and seismic activity. Much has advanced since then. Presently a net of telemetered seismographs continually transmits its seismic data in real time to the Instituto's computer center. An extensive electronic distance measuring (EDM) net has been established and is periodically reoccupied to detect any inflation of the cone. These advances were possible through the UNDRO/USAID program as well as the US Geological Survey's Volcano Crisis Assistance Program. Future initiatives (given availability of funding) will test lahar flow-detectors on Cotopaxi as well as install alarm systems in the high-hazard populated zones, in collaboration with Civil Defense.

Summary

Cotopaxi's eruptive history since the 1500s consists of one long period of repose (1534–1742) and three periods of activity. During these periods of reactivation, eruptions occurred approximately every four years, although the range varies from zero to 30 years. No significant activity has been observed since the last eruption in 1914; thus the probability of a future eruption soon is steadily increasing.

The situation in Ecuador is quite different than the circumstances surrounding the 1985 Nevado del Ruiz lahar tragedy (Hall, this volume). This is not to say that the people and property in the Chillos and Latacunga valleys would be unaffected by future lahars: they certainly will be. However, the following circumstances and mitigation activities should assure that the impact is reduced.

(a) The vast majority of people living in the lahar hazard zones have a greater awareness of the problem since they have a view of the volcano and are reminded frequently of the volcano's presence. There are, however, people living hundreds of kilometers downstream in the Amazon basin or on the Pacific lowlands who have a less developed concept of the volcano and its hazards, given its distance away and their poor knowledge of it.

(b) At all levels, government officials are aware of the general volcanic hazards, and most are probably aware of the lahar dangers.

(c) Very high quality hazard and risk maps for Cotopaxi are available and in circulation, and are presently being used at the national, provincial and county level to refine civil defense emergency plans.

(d) Civil Defense personnel in the six provinces have visited the high-risk lahar zones, have seen the deposits, and are aware of the consequences of future lahar flows to populated areas and to the important infrastructure.

(e) Cotopaxi volcano, being only 60 km south-southeast of Quito, is clearly visible on an average day. Should Cotopaxi reactivate, given the proximity of the volcano, the Quito-based media would probably give priority coverage to the volcanic phenomena, the monitoring efforts, and civil defense actions, and not cast this information in the back pages as was often the case of Nevado del Ruiz in the national Colombian newspapers. Additionally, the local newspaper in Latacunga publishes observations about the volcano on a regular basis. If the volcano eventually reactivates, this newspaper will undoubtedly serve as an important source of information for people living in the Rio Cutuchi drainage.

(f) The news media can be both friend and foe. More often than not, news reporters in Ecuador have distorted and sensationalized natural disaster events. Post-analysis of these situations shows that both scientists and Civil Defense officials must be strict disciplinarians of the press, demanding that the reporting is correct, timely, and not distorted.

(g) The next eruption, like many of the previous events, might be preceded by weeks or months of minor ash emissions and the generation of small lahars which are good warning devices and cause only minimal damage. Given the sometimes long periods of minor precursory activity, such as the 11 months in the case of Nevado del Ruiz in Colombia, it may be difficult to hold the interest and attention of the population for such a period so as to assure an adequate state of continued awareness and preparedness for a major eruption.

Conclusions

Given its long eruptive history in prehistoric times and during the last 450 years, Cotopaxi volcano will surely erupt again. However, unlike other eruptions, there should be a significant warning period, thanks to the modern monitoring techniques implemented on the volcano. An ever-growing population of 30,000 people presently live upon the lahar deposits of the 1877 eruption; these same areas of deposition will almost certainly be overrun again by future lahars. Thus overall public awareness through educational programs is essential in order that the population remain prepared for an eventual eruption. The use of lahar detectors in the drainages in conjunction with alarm systems in the towns would also be greatly advantageous, but have not been implemented for lack of funds. Great strides have been made by the Ecuadorian Civil Defense in collaboration with the Instituto Geofisico in raising the awareness level of the population by sponsoring eruption scenarios, simulated evacuations, other educational forums, and the publication of hazard and risk maps and informative posters. However, it is recognized that the message has not reached everyone in the high-risk zones, and that a much broader approach to volcano hazard education must be implemented so that eruption-caused destruction to life and property can be minimized.

References

D'Ercole, R., 1989. La castastrofe del Nevado del Ruiz; una ensenanza para el Ecuador? El caso del Cotopaxi (The Nevado del Ruiz catastrophe; a lesson for Ecuador? The case of Cotopaxi). *Estudios de Geografia*, vol 2, p 5–32, Quito.

Egred, J., 1989. Recopilacion historica sobre erupciones volcanicas: volcanes—Tungurahua, Cotopaxi and Pichincha (Historical compilation of volcanic eruptions: Tungurahua, Cotopaxi and Pichincha). Unpublished manuscript, Instituto Geofisico, Escuela Politecnica Nacional, Quito, 30 p.

Hall, M.L., 1976. Mapa preliminar de los peligros volcanicos del Volcan Cotopaxi (Preliminary map of volcanic dangers of Cotopaxi Volcano). Unpublished map, Escuela Politécnica Nacional, Quito.

Hall, M.L., 1979. Riesgos potenciales de las erupciones futuras del Volcan Cotopaxi (Potential risks of future eruptions of Cotopaxi Volcano). Unpublished manuscript, Empresa de Agua Potable-Quito.

Hall, M.L., 1987. Peligros potenciales de las erupciones futuras del Volcan Cotopaxi (Potential hazards of future eruptions of Cotopaxi Volcano). *Politécnica-Monografia de Geologia*, vol 5, p 41–80, Quito.

Hall, M.L., and C. von Hillebrandt, 1988. Mapa de los peligros volcanicos potenciales asociados con el Volcan Cotopaxi, Zonas Norte y Sur; Escala 1:50,000 (Map of potential volcanic hazards associated with Cotopaxi Volcano, North and South Zones; scale 1:50,000). Instituto Geofisico, Escuela Politécnica Nacional, Quito.

Hall, M.L., and C.A. Wood, 1985. Volcano-tectonic segmentation of the northern Andes. *Geology*, vol 13, p 203–207.

INECEL, 1989. *Informe Final de Volcanologia. Estudios de Factibilidad (Final Report on Volcanology.* Feasibility Studies). Prov. Hidroeléctrico San Francisco, Quito.

Janda, R.J., K.M. Scott, K.M. Nolan, and H.A. Martinson, 1981. Lahar movement, effects, and deposits. In: P.W. Lipman and D.R. Mullineaux (eds), *The 1980 Eruptions of Mount St Helens, Washington*, US Geological Survey Professional Paper 1250, p 461–478.

Miller, C.D., D.R. Mullineaux and M.L. Hall, 1978. Reconnaissance map of potential volcanic hazards from Cotopaxi Volcano, Ecuador. *US Geological Survey Miscellaneous Investigation Series* Map I-1702.

Mothes, P.A., 1988. *Riesgos Laharicos del Volcan Cotopaxi, Ecuador (Lahar Risks of Cotopaxi Volcano, Ecuador)*. Report and 10 maps, scale 1:25,000. Instituto Geofisico, Escuela Politécnica Nacional, Quito.

Mothes, P.A., and M.L. Hall, 1988. *Escenario de una Posible Erupcion del Volcan Cotopaxi, Ecuador (Scenario of a Possible Eruption of Cotopaxi Volcano, Ecuador)*. Instituto Geofisico, Escuela Politécnica Nacional, Quito, 13 p.

Paredez, E., 1982. *Cotopaxi: Documentos de Oro (Cotopaxi: Documents of Gold)*. Editiones Cotopaxi, Latacunga, 190 pp.

Pierson, T.C., 1982. Flow behavior of two major lahars triggered by the May 18, 1980 eruptions of Mount St Helens, Washington. *Proceedings of the Symposium on Erosion Control in Volcanic Areas*, Vancouver, Washington, p 99–129.

Sodiro, L., 1877. *Relacion Sobre la Erupcion del Cotopaxi Acaecida del Dia 26 de Junio, 1877 (Account of the Eruption of Cotopaxi of 26 June, 1877)*. Imprenta Nacional Quito, 40 p.

Smyth, M.A., and C. Clapperton, 1986. Late Quaternary volcanic debris avalanche at Cotopaxi, Ecuador. *Revista CIAF*, vol 11, no 1, p 24–38.

Whymper, E., 1892. *Travels Amongst the Great Andes of the Equator*. Peregrine Smith Books, Salt Lake City, Utah, 456 p.

Wolf, T., 1878. *Memoria Sobre El Cotopaxi y su Ultima Erupcion Acaecida el 26 de Junio de 1877 (Memoir of Cotopaxi and its Final Eruption of 26 June, 1877)*. Imprenta del Comercio, Guayaquil.

Wolf, T., 1904. *Cronica de los Fenomenos Volcanicos y Terremotos en el Ecuador (Chronicle of Volcanic and Earthquake Phenomena in Ecuador)*. Imprenta de la Universidad Central, Quito, 167 pp.

8 Seismic monitoring of Lake Nyos, Cameroon, following the gas release disaster of August 1986

Alice B. Walker, D.W. Redmayne and C.W.A. Browitt

Abstract On 21 August 1986 a cloud of dense gas was emitted from Lake Nyos in Cameroon causing death, by asphixiation, of over 1700 people in the nearby villages. An international relief effort was initiated and scientific teams from a number of countries were sent to Lake Nyos in order to determine the cause of the gas release. It was agreed that the carbon dioxide, which was the predominant gas in the cloud, had a deep-seated magmatic origin and that the lake was charged with it to near-saturation levels prior to the gas outburst. Uncertainty remained, however, about the mechanism of its eruption from the lake; in particular, whether the lake overturned in response to a small external trigger or whether a phreatic volcanic eruption occurred.

In an attempt to resolve this problem, and following the report of a second 'eruptive event' on 30 December 1986, the British Geological Survey installed a network of hydrophones and geophones in and around the lake during February 1987. During the following six months, regional earthquakes and small local seismic events, interpreted as rockfalls, were detected but no significant crustal earthquakes were found beneath or close to the lake. No clear evidence of magma movement or other signs of volcanic activity were found, suggesting that direct volcanic activity was not involved. Noise outbursts detected, mainly at night, may have been due to gas emissions from rock fissures.

The lack of tectonic seismicity at the lake during the monitoring period argues against an earthquake trigger for the gas release. In the wider volcanic province, however, many earthquakes were detected and such a trigger cannot be completely ruled out.

The future safety of people who live around the lake depends either on controlled degassing or lowering of the lake level with monitoring of CO_2 levels in the lake on a continuous basis. The data recording techniques used in this study could readily be applied to such a broadly-based monitoring system.

Introduction

The Lake Nyos disaster

On 21 August 1986, at about 21:30 hours, a cloud of carbon dioxide gas was emitted from Lake Nyos (Plate 1), a crater lake in Cameroon, central Africa. In the immediate area of the lake many people were reported to have experienced a warm sensation, smelled an odour of rotten eggs or gunpowder, and rapidly lost consciousness. The resultant aerosol of water and carbon dioxide flowed down river valleys, causing the death by asphixiation of at least 1700 people and killing about 3000 cattle as it passed through the villages of Nyos Market, Cha and Subum (Clark and others, 1987). The distribution of dead animals indicated that a lethal concentration of the gas reached a height of 120 m above the lake surface.

Following the outburst, the lake level was reported to be about 1 m lower than its spillway which, as it occurred in the rainy season, represented a sudden loss of volume. The lake turned dull red in the centre, owing to the formation of a surface layer containing iron hydroxide and plant debris. The significance of the reddening is not clear: it has been claimed that it could be an indication of volcanic activity—this discoloration is said to be a common phenomenon amongst crater lakes in Cameroon.

65

A technical team from the British Geological Survey (BGS) took photographs of Lake Njupi, which is about 1 km northeast of Lake Nyos, on 21 January 1987 (Plate 2) and 14 February 1987 (Plate 3). The former shows the lake red in colour while in the latter it is more normal. Red discoloration was also observed in both Lake Nyos and Lake Njupi by a French scientist (Chevrier, 1987), on the day following a second explosive event at Lake Nyos on 30 December 1986.

In the weeks following the disaster, a number of scientific teams visited the area in an attempt to discover the cause of the outburst and evaluate the risk of future occurrences. The British team, sponsored by the British Government's Overseas Development Administration (ODA), consisted of S.J. Freeth, R.L.F. Kay and P.J. Baxter, from University College of Swansea, BGS and Cambridge University respectively; in their recommendations, they noted a need for seismic monitoring at the lake. This paper describes the results of this monitoring.

A similar outburst event had occurred two years earlier in Cameroon. On 15 August, 1984, a lethal gas outburst issued from Lake Monoun, 95 km south of Lake Nyos, killing 37 people. Sigurdsson (1987) and Sigurdsson and others (1987) concluded that the most likely explanation for this earlier gas release was that an overturning of the lake had occurred, releasing dissolved carbon dioxide from the bottom waters. They concluded that it was triggered by a landslide which may itself have been initiated by a small earthquake.

Tectonic and geological setting

Lake Nyos (Lockwood, 1987) is a *maar* lake, a crater lake formed by shallow explosive eruptions, which may have been formed as recently as a few hundred years ago by a phreatic eruption as a result of heating of groundwater by an underlying magmatic source (Kling and others, 1987). It is situated in the Cameroon Volcanic Line, 1400 km long, of which Mount Cameroon, on the coast 270 km to the southwest, is a presently active centre. Mount Cameroon has erupted four times this century, the last in 1982. Fitton (1980) and Fitton and Dunlop (1985) reported morphological evidence of volcanic activity within the last million years throughout the length of the volcanic line, and they concluded that the whole region is still potentially active.

Lake Nyos (Figure 1) is almost 2 km long, 1 km wide and 208 m deep. The crater has steep walls and a flat, smooth floor; no changes in the bathymetry have been

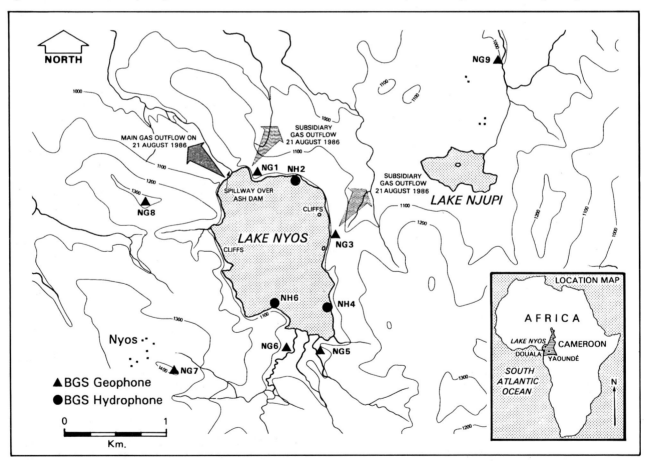

Figure 1 Location of seismic monitoring stations around Lake Nyos.

Plate 1 Lake Nyos viewed from the southeast, early September 1986 (photograph by S. J. Freeth).

Plate 2 Lake Njupi, which lies northeast of Lake Nyos, showing red discoloration on 21 January 1987.

Plate 3 Lake Njupi on 14 February 1987: the red discoloration has cleared.

67

Plate 4 A geophone recording site at Nyos. All equipment is contained in the fibreglass box.

Plate 5 A radio repeater site for the Nyos network.

Plate 6 Unstable cliffs on the west side of Lake Nyos.

detected in surveys following the gas outburst compared to an earlier one (Hassert, 1912). The water level is controlled by a natural dam of pyroclastic deposits, 40 m high, at its northern end: this in itself is a potential hazard, owing to the general lack of integrity of the material and the flooding potential should it fail.

Cause of the disaster

Several hypotheses have been presented as to the cause of the emission of gas at Lake Nyos. An overturning of the lake waters certainly occurred, but the trigger for this remains unknown. It is also clear that, as in the case of Lake Monoun, the bottom waters were saturated with carbon dioxide derived from a mantle source. The United States scientific team (Clark and others, 1987) favoured a cause arising from a slow build-up of carbon dioxide, which dissolved in the lake waters through groundwater; possibly the event was triggered by a small earthquake or landslide or even by heavy rain or wind. The British team (Freeth, Kay and Baxter), whilst essentially accepting a similar mechanism, suggested that a pulse of cold CO_2 could have been the trigger which caused the lake to overturn and release the already dissolved gas (Freeth and others, 1987; Freeth and Kay, 1987). The French team, however, concluded that the gas outburst had been a phreatic eruption of a volcanic nature and that further volcanic activity could be expected (Tazieff, 1987). The occurrence of the second event on 30 December 1986, reported by Chevrier (1987), added possible support to this hypothesis.

The need for seismic monitoring

In the light of the remaining uncertainty, the British Geological Survey was asked by ODA to monitor the lake area for seismic activity. The intention was to detect any activity associated with gas or magma movements, indicating the risk of a volcanic eruption, and to reveal any tectonic activity in the region of the lake, which would support the hypothesis that an earthquake may have triggered the gas release.

It was also anticipated that the detection of other localised activity might provide information about the lake's gas emission. Of further interest was the extent of seismic activity in the North West Province of Cameroon, in which Lake Nyos and many other maar lakes are situated. The seismicity of the Mount Cameroon area, 270 km to the southwest, has been studied in detail in a joint project by the Institut de Recherches Géologiques et Minières (IRGM) of the Cameroon Republic and the University of Leeds, England (Ambeh and others, 1989). The seismic activity of the volcanic highlands around Lake Nyos is less well known and its study would give further information on potential areas of future volcanic activity throughout the North West highlands.

Data acquisition

Equipment installed at Lake Nyos

In February 1987, BGS installed a seismic monitoring network of two hydrophones and seven geophones in and around Lake Nyos (Figure 1). Geographical co-ordinates of these stations are given in Walker and others (1988). The geophone type used was the SM6 (Model B) velocity transducer, manufactured by Sensor, and the hydrophones were the MP-24 pressure sensitive transducer, manufactured by Geospace. Output from these sensors was amplified, frequency modulated and transmitted by UHF radio to a ridge crest 5 km south of the lake. These instruments, a solar panel, batteries and antenna were all contained in a small fibreglass box to facilitate rapid deployment by helicopter (Plate 4).

On the ridge, 625 m above the lake level, a radio-repeater station was built (Plate 5) to relay the signals from the lakeside to the recording site at Bamenda, 50 km to the south. A Willmore MK III seismometer was also installed at the repeater site, to help in the identification of events. It was adopted later as part of the IRGM/Leeds University regional seismic monitoring network.

In addition to seismic monitoring, an oxygen-deficiency sensor was installed at station NG5. It transmitted a signal each minute, which appeared as a superimposed negative pulse on the seismic signal to confirm that the atmospheric oxygen level was above 18%. If the level of oxygen at the lakeside were to fall below 18% (normal level 21%) then an additional negative pulse would be superimposed on the first. As no continuous monitoring of simple parameters such as oxygen level, temperature, salinity and pH from around and within the lake had been carried out previously, this demonstrated the feasibility of transmitting such data.

The network remained unaltered until the beginning of May 1987, when another hydrophone (NH6) was installed replacing NG6. In collaboration with IRGM and Leeds University a further four seismometers were installed during May, with the aim of providing a broader regional coverage of the area. The data presented in this paper are based on the original Nyos network and cover the period up to the end of July 1987.

Recording equipment at Bamenda

At Bamenda, a Geostore magnetic tape recorder continuously registered ten seismic channels, an internal clock, a radio time-channel and two flutter channels. Absolute time was obtained from the Liberian transmitter of the Omega navigation system. The Geostore recorder with a dynamic range of better than 40 db, was operated at 15/160 inches per second (2.4 mm/sec), giving a bandwidth to 32 Hz. In addition, an event-triggered digital seismic detection system was also

Table 1 P-wave velocity structure and Vp/Vs ratios (Vp/Vs = 1.73 for both regional and lake events)

Regional events		Lake events	
Velocity of layers (km/s)	Depth to top of layer (km)	Velocity of layers (km/s)	Depth to top of layer (km)
5.4	0.0	1.7	0.0
5.9	0.5	3.0	0.20
6.2	8.0	5.9	2.52
6.4	15.0	7.0	18.87
8.0	32.0	8.0	34.15

installed (Houliston and others, 1985). This provided a wider dynamic range, 72 db, and recorded frequencies up to 40 Hz onto digital tape. However, in the rainy season the digital system was continually triggered during thunderstorms, and recording was suspended at those times. Three channels were also displayed on a Streckeisen paper-loop recorder to give an immediate record of incoming data.

Analysis procedures

The analogue tapes were replayed on a Racal Store 14 tape-deck, and were monitored at 80 times real-time, bringing the signal into the audio range. From a list produced during listening to these signals, paper play-outs were made on an ink-jet recorder to enable the phase arrival times to be picked. These data were then processed by the computer program HYPOCENTRE (Lienert and others, 1986) to determine epicentres and depths. In the field, the location program was run on an IBM PC computer.

Two crustal velocity models were used for location purposes, one for regional events and a second for lake events. The crustal velocity structure adopted for regional events is that given by Stuart and others (1985) for the Adamawa Plateau on the Cameroon Volcanic Line, 400 km east of Lake Nyos. For events at Lake Nyos, a shallow structure was employed, based on the local geology. This was tested and adjusted using trial rock-drops as artificial seismic sources, until the most accurate locations were obtained. Details of these structures are given in Table 1.

Duration magnitudes (MD), determined from the duration of the oscillation, were calculated because uncertainties in the epicentral distances of regional events, and the proximity of lake events to the sensors, made distance-based magnitude scales inappropriate. Tsumura's (1967) formula was used, based on the total duration of oscillation on the seismogram. The formula takes the form:

$$MD = 2.85 \log(F - P) + 0.0014\Delta - 2.53$$

where $F - P$ is the total duration in seconds and Δ is the epicentral distance in kilometres (the distance term is effectively zero for distances up to 100 km and thus has little effect upon events located within 100 km of the instruments). Ambeh and others (1989) used a duration formula based on Lee and others (1972) for the central California area. Applying their formula to our data, duration magnitudes were increased by an average of 0.3 MD.

Field observations and experiments

To determine the local velocity structure, a series of rock-drops was arranged at different locations around the lake. The size of rock varied from about 25 kg to large boulders weighing approximately one tonne. Figure 2 shows the signature of a typical rock-drop which was recorded by the network, the phase arrival approximately two seconds after the first being an acoustic arrival. The position of the rock-drop on 3 April 1987 at 14:35 GMT (Figure 3) was determined several times, varying the velocity structure and the ratio of P-wave and S-wave velocities (Vp/Vs ratio); the structure and Vp/Vs ratio which gave the most accurate location, within 100 m of its actual position, was then used to locate subsequent rock-drops and lake events. This velocity structure and Vp/Vs ratio are shown in Table 1.

Events detected by the Nyos network

During the monitoring period events were detected from many sources. The concentrated nature of the network gave detection of very low signals in the vicinity of the lake itself, minimum magnitudes of locatable events being as low as −2.5 MD during quiet conditions, although much higher during thunderstorms. Local and regional earthquakes were also detected, the detection threshold rising with distance from the network. Teleseisms (distant earthquakes) were recorded, but no detailed analysis was carried out on these. For the following discussion, events which were recorded by the network have been divided into different categories.

Regional earthquakes

During the six-month period of monitoring, over 30 regional earthquakes within 200 km of Lake Nyos were located by the network, with magnitudes up to 3.5 MD, their positions being shown on Figure 4. The accuracy of the locations reduces with distance from the monitoring network. A seismogram of an earthquake approximately 140 km southeast of Lake Nyos is shown in Figure 5.

The general scatter of the seismicity throughout the region shows that the area is seismically active. This

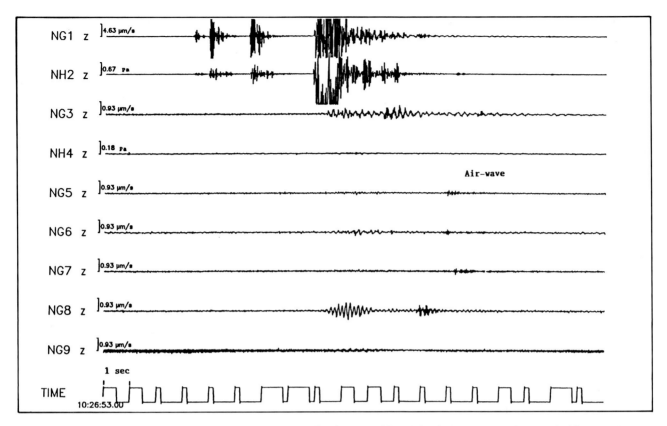

Figure 2　Seismogram of a rock-drop near NG1 and NH2, 16 April 1987 10:27 GMT. Length of record is 19 seconds.

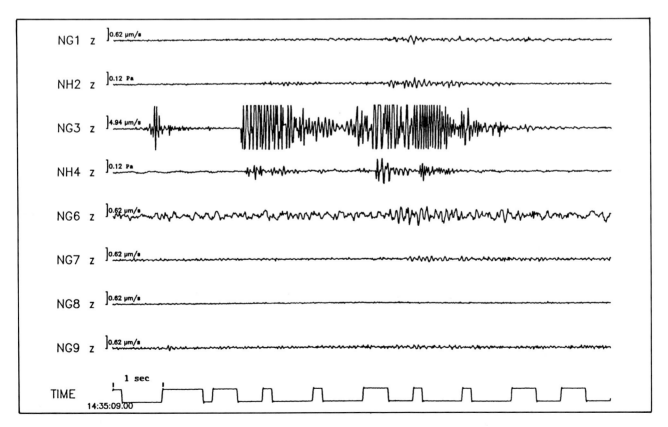

Figure 3　Seismogram of a rock-drop near NG3, 3 April 1987 14:35 GMT. Length of record is 10 seconds.

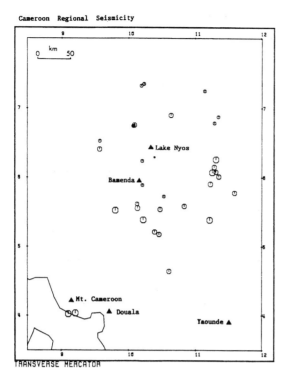

Figure 4 Regional seismic events located by the Nyos network in the period February to July 1987 (symbol size is proportional to magnitude).

is probably associated with the recent volcanism of Cameroon.

Lake events

A number of minor events were located near and in Lake Nyos but they were very small (-2.5 MD to 0.1 MD) and often only detected on one or two stations. Most could be attributed to large and small rocks falling from the unstable cliffs which surround the lake (Plate 6).

The majority of these small events originated at the northern end of the lake and were recorded most clearly by NG1 and NH2 (Figure 1). The largest, with magnitudes approaching 0.1 MD, were also recorded by other stations. Figure 6 shows an example of an event which can be seen clearly on geophone NG1 and hydrophone NH2, stations which are located at the north end of the lake near the spillway and close to the area which Freeth and Kay (1987) suggested as the source of the gas release. It is thought that these events may be natural rockfalls from the steep cliffs nearby, or debris moving near the spillway.

The signal recorded from the artificially induced rock-drop (Figure 2) has some similarities to the natural lake events. They both have precursory signals, which in the case of the induced rock-drop corresponds to the boulder bumping as it rolled towards the cliff edge, and

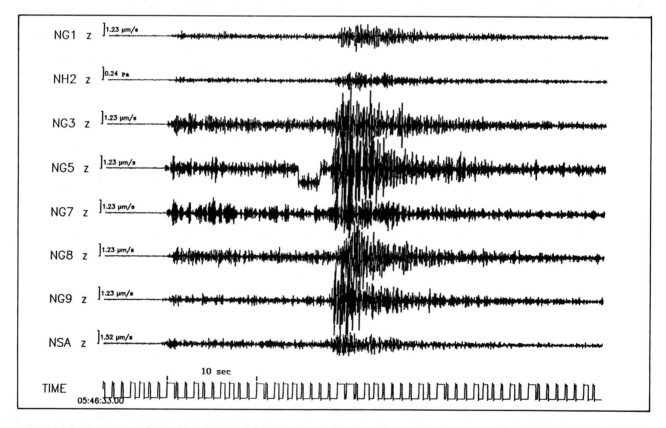

Figure 5 Seismogram of a regional earthquake 140 km southeast of Lake Nyos, 29 April 1987 05:46 GMT, 3.4 MD. Length of record is 54 seconds.

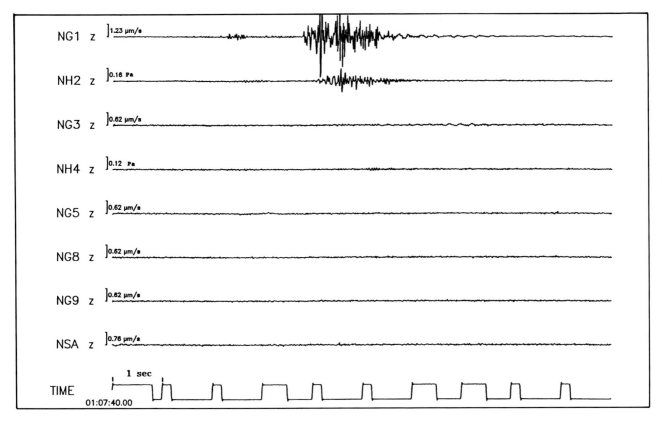

Figure 6 Seismogram of a lake event, 26 February 1987 01:07 GMT. Length of record is 10 seconds.

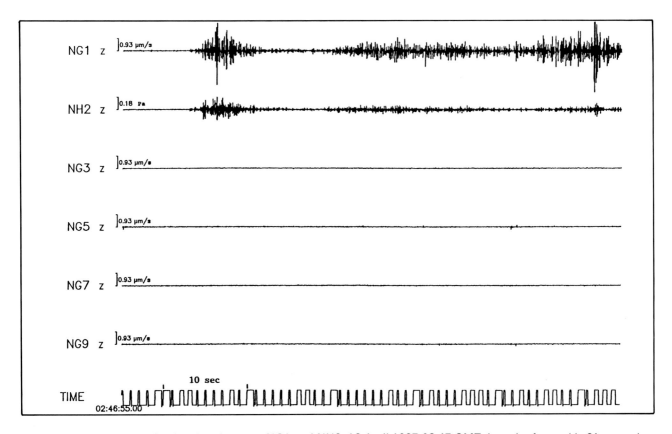

Figure 7 Seismogram of a signal outburst on NG1 and NH2, 16 April 1987 02:47 GMT. Length of record is 61 seconds.

Table 2 Unexplained signal outbursts at the north end of Lake Nyos

Date dd mm yy	(GMT) Time hh mm ss	Duration minutes	Comments
15 04 87	05 53 51	004	Slightly increased noise on NG1 and NH2 for several minutes prior to main start time, when an abrupt increase in noise occurred lasting 4 minutes at varying levels before dying away (NG3 down).
15 04 87	22 02 22	004	Slightly increased noise on NG1 and NH2. Small NG1/NH2 event at 22:06:17 GMT. Possibly heavy rain also (NG3 down).
15 04 87	22 37 47	010	Slightly increased noise on NG1 and NH2. Possibly heavy rain also (NG3 down).
16 04 87	02 47 02	003	Abrupt start on NG1 and NH2. Very strong signal, stronger on NG1. Varied in level throughout duration. No disturbance of the lake noticed at 09:30 GMT the same day and oxygen levels normal (NG3 down).
16 04 87	04 53 35	007	Abrupt start on NG1 and NH2. Strong signal, stronger on NG1. Varied in level throughout duration then died away. No disturbance of the lake noticed at 09:30 GMT the same day and oxygen levels normal (NG3 down).
16 04 87	22 04 10	003	Slight noise increase on NG1 and NH2. Possibly heavy rain noise also.
17 04 87	05 07 36	022	Increased noise at varying levels, initially mainly on NG1. Sharp increase on NG1 and NH2 at 06:24:18 GMT before dying away.
18 04 87	06 07 45	005	Abrupt start on NG1, NH2 and NG3. Strong signal on NG1 and NH2, strongest on NH2, NG3 weak. NG1 and NH2 at 23 Hz but NG3 at about 6 Hz.
19 04 87	00 28 40	013	Gradual build-up to varying levels of noise. NG1 and NH2, NG1 stronger. Thirty-second signal burst on NG5 at 00:32:08 GMT. Rain noise also.
19 04 87	01 43 40	009	Increased noise, mainly on NG1. Seven-second noise burst on NG1 and NH2 at 01:50:57 GMT, just prior to the end of outburst.
19 04 87	05 03 49	022	Increased noise at varying levels, mainly on NG1, where strong, but also on NH2.
19 04 87	06 07 14	016	Increased noise at varying levels on NG1 and NH2. Slightly stronger on NG1.

both display surface waves. The instability of rocks on some of the shoreline cliffs is evident; Plate 6 shows evidence of some recent landslides on the western side of the lake.

A number of other events were detected in the lake area, usually on one station only and of low signal amplitude. Many of these are likely to be man-made or animal noise, there being several Fulani villages with cattle herds in the area and several of the sites (NG7, NG8 and NG9) are on the higher ground around the lake where the cattle frequently graze. A number of low-frequency monotonic (7 Hz) events were seen on NG6 alone, and later on NH6, throughout the recording period. It was noted that at times these merged into periods of continuous 'noise'. Nothing was seen on the NH4 hydrophone at that time or on nearby geophone NG5. This was obviously a very localised source, one possible explanation being stream noise during periods of increased water flow.

Signal outbursts

In the middle of April 1987 a number of unexplained signal outbursts were detected on the sensors at the north end of the lake, NG1 and NH2 (Figure 7). They were

frequently at night and typically continued for several minutes (Table 2). They had a peak frequency of 23 Hz and varied from an abrupt start, as shown in Figure 7, to a gradual build-up of signal lasting approximately ten minutes. It is thought that they could represent gas movements within and from the underlying rocks, as there are carbonated springs in the region of Nyos. Other explanations are possible, however, such as water movements near the spillway, landslide noise or water leaking through the spillway, but little evidence of these could be found on visits to the lake.

Following the first occurrence of unexplained signal outbursts recorded on NG1 and NH2, a site visit was made the next day (16 April 1987). On arrival at the lakeside the lake appeared calm and was blue in colour. Oxygen levels were tested using a hand-held oxygen monitor and found to be normal and there was no evidence to suggest that there had been any recent landslides at the northern end of the lake.

Complex lake events

In addition to the events which were located using the Lake Nyos network, three significant events occurred which had complex signatures and arrivals too emergent for location purposes. They have been interpreted as multiple rockfall events in which a number of discrete bursts of activity represent larger rock units. These events all occurred during May when heavy rains would have increased the instability of the cliffs. The events were:

12 May 1987 16:23 GMT This event (Figure 8), with frequencies from 15 to 28 Hz, appeared from the arrival pattern to have had its source on the west side of the lake. The total duration of the event measured on NH6 was about 13 seconds and it comprised at least five discrete bursts of activity. The most likely cause of this event was a multiple rockfall on the west side of the lake in the vicinity of the unstable cliffs (Plate 6). We cannot, however, rule out a deeper source, the arrival pattern being consistent with focal depths from 0 to 3 km. It must be pointed out that the exceptionally high amplitudes on NH6 are due to the sensor being undamped and do not imply excessive signal at that site.

30 May 1987 05:10 GMT This event (Figure 9) was also emergent in character and could not be precisely located. The dominant frequency was 12 Hz and its signal duration was approximately nine seconds. The earliest arrival time and highest amplitudes were recorded on NG3, indicating a source near to this,

probably a multiple rockfall on the steep cliffs nearby.

The attenuation of the signal from this event is similar to that from a calibrating rock-drop at NG3 (Figure 3), leading to the conclusion that its source was also shallow and close to NG3. An origin close to the lake bed, however, is not ruled out although the source could not have been much below that.

31 May 1987 02:13 GMT This event (Figure 10) had its source near NG1 and NH2. The signal was also detected on stations NG3, NG8 and on the hydrophone NH6. A location could not be determined because of the emergent nature of the onsets. The duration of the event was about ten seconds and lower-frequency surface waves, similar to those generated by artificial rock-drops, are evident on NG1 and NH2 in the latter part of the seismogram. The signal on NG1 and NH2 had a peak frequency of 23 Hz, which is very similar to the signal outbursts detected during April 1987. This 23 Hz signal may have been due to sensor site effects—i.e. they may have been installed on less solid material than other sensors. Two main amplitude maxima, with a separation of two seconds, can be seen in Figure 10, suggesting two discrete events. A marked signal attenuation across the network and the presence of surface waves implies a shallow source, so this may represent a multiple rockfall at the north end of the lake.

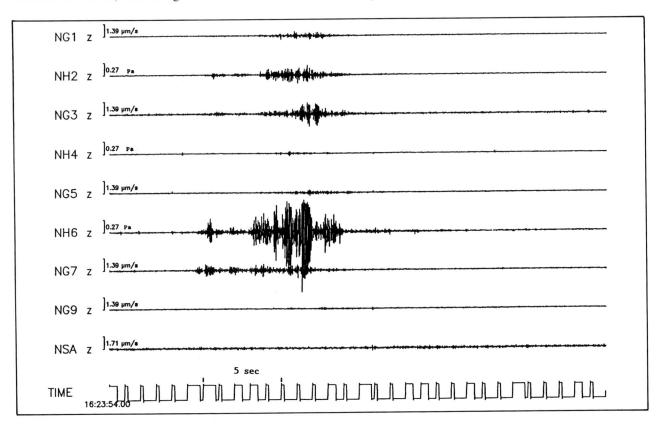

Figure 8 Seismogram of event at the west side of the lake, 12 May 1987 16:23 GMT. Length of record is 32 seconds.

Non-seismic events

At times, violent thunderstorms were experienced during which vibration from heavy rain lowered the detection capabilities of the seismic network. Figure 11 shows the effect lightning had, breaking the transmission signal and showing up as a spike. As thunder produces air-waves, which couple into the ground, the signals are more distinctive on the geophones than on the submerged hydrophones. The slow travel time of 0.3 km/s across the array made identification of thunder easy.

Discussion

Volcanic activity at Lake Nyos and in Cameroon

The highlands of North West Cameroon are an area of recent volcanism, as is clearly indicated by extensive eruptive rock types and the existence of little-eroded cinder cones. Mount Cameroon on the Gulf of Guinea, 270 km southwest from Lake Nyos, is presently active with its most recent eruption in 1982. Lake Nyos itself, a maar lake, probably came into existence in the very recent past as the result of a violent phreatic eruption. Kling and others (1987) suggested an age of "a few hundred years" for Lake Nyos, based on geomorphological evidence.

Seismicity associated with Mount Cameroon has been studied in detail by Ambeh and others (1989). They showed that many earthquakes occurred in the area of the volcano, and indeed a number of the larger Mount Cameroon events were detected by the Nyos network. The identification of regional seismicity in the North West highlands of Cameroon in this study is indicative of continuing stress in this region which also may similarly be associated with magma movements at depth.

No significant tectonic earthquakes were detected in the vicinity of Lake Nyos during the six-month monitoring period from February to July 1987. We would expect to have seen tectonic activity at a moderate to high level if a magma body was ascending close to the surface; the absence of such activity, in our view, argues against such active volcanism and a phreatic cause to the outgassing.

Harmonic tremor was not observed on the Nyos network nor were related B-type earthquakes which are often associated with magma movements (McNutt, 1986). Typical frequencies of B-type events are from 0.5 to 5 Hz, which extends below the theoretical response range of the geophones. In practice, however, the geophones did register teleseisms with frequencies in the region of 1 Hz, although at a much reduced level. The Willmore Mk III seismometer at the repeater site was fully effective in this range.

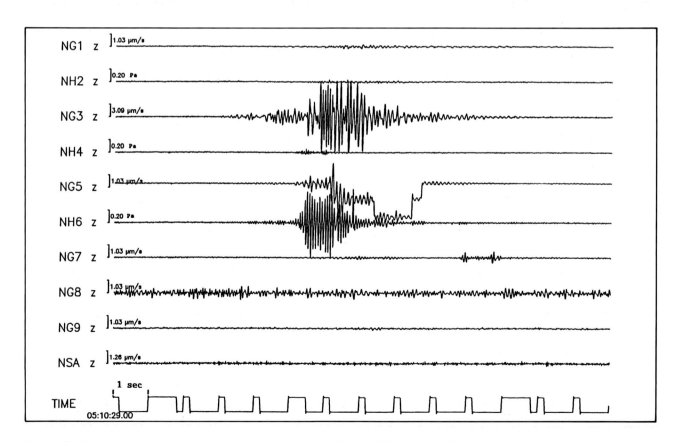

Figure 9 Seismogram of event on the east side of the lake, 30 May 1987 05:10 GMT. Length of record is 14 seconds.

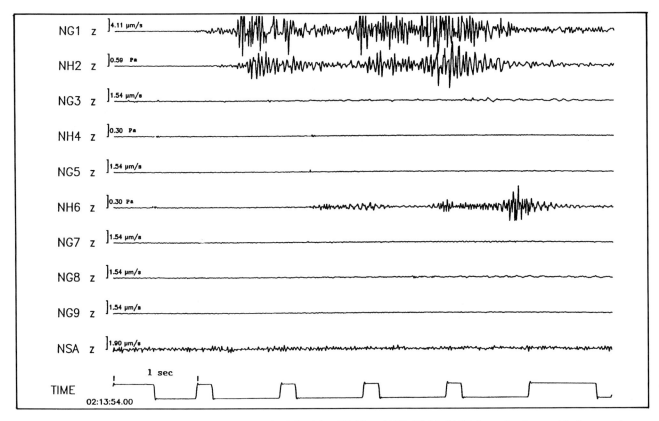

Figure 10 Seismogram of event at the north end of the lake, 31 May 1987 02:13 GMT. Length of record is 6 seconds.

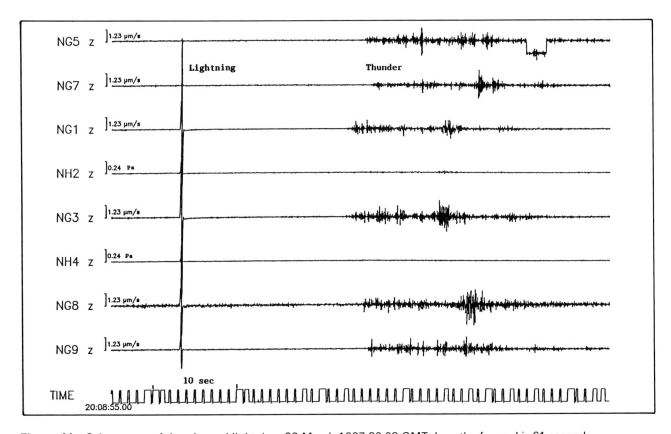

Figure 11 Seismogram of thunder and lightning, 22 March 1987 20:08 GMT. Length of record is 61 seconds.

The stability of the cliffs and spillway

The Nyos seismic network has detected many events which have been interpreted as being rockfalls from cliffs and the occasional more extensive landslide. This is in itself an indication of the young age of Lake Nyos with its steep surrounding cliffs. It is possible that such an occurrence acted as a trigger for the major outgassing event of 1986, given the finely balanced nature of the lake at that time.

A number of the lake seismic events appear to be from the vicinity of the spillway and may represent movements within it or rocks falling from its face. Given the unstable nature of the spillway which has already been identified (Clark and others, 1987), it is significant that movement may be present here, although the resolution of the network was not sufficient to be certain of this.

The future safety of Lake Nyos

Any assesment of the future safety of Lake Nyos depends critically upon which theory of CO_2 recharge is adopted. If the only mechanism of recharge is through slow groundwater percolation, then the recharge rate can be determined through occasional monitoring of the lake; long before saturation is reached in the deep waters, measures could be taken to discharge the CO_2 in a controlled way (Clark and others, 1987).

Should incipient volcanism be the cause of the lake's outgassing, the lake must be regarded as a potential hazard at all times, with a risk of increased volcanism and even the potential for violent phreatic eruptions such as the one which created the lake in the first place. Seismic evidence cannot entirely rule out this possibility, but the lack of crustal seismicity at the lake argues against it.

Between these two opposing views is the possibility that the lake is being recharged by an irregular series of pulses of gas from a sub-surface reservoir in addition to percolation. This would increase the recharge rate and render it less predictable. It is also possible that such a pulse could have acted as a trigger to the outgassing event of 21 August 1986, as suggested by Freeth and Kay (1987). Such gas pulses might explain the signal outbursts detected at the northern end of the lake during April 1987, although at that time the lake did not appear to be affected.

Future work at Lake Nyos

It is clear that Lake Nyos still presents a real but uncertain threat to the surrounding area, but a number of steps could be taken to monitor the situation and mitigate the risk.

A number of suggestions have been made for making the lake safe. Clark and others (1987) and Lockwood and others (1989) suggested a controlled degassing of the lake by pumping up the deep water and allowing the gas to exsolve naturally on reaching the surface. They also advised that the spillway should be lowered or the lake drained below the spillway level in order to reduce the risk of serious downstream flooding should the spillway collapse. If engineering solutions such as these are not adopted then it is recommended that the lake and its environs be monitored continuously for all relevant parameters. In this study, the continuous monitoring of 10 seismic sensors, with the high data rates involved, demonstrates that many other parameters which need to be sampled less frequently can easily be handled. Samples at one-minute intervals are infrequent compared with the seismic sampling rate. Furthermore, the multiplexing of an oxygen deficiency signal onto a seismic trace has shown how high and low data rates can be effectively combined.

In the long term, seismic activity at the lake might be economically monitored using only one sensor (or, ideally, at three stations) as part of the regional seismic network. To identify movement in the spillway rocks, however, a more intensive exercise would need to be conducted.

Conclusions

During the six month's monitoring period of the Nyos network, regional tectonic earthquakes and small local events, interpreted as rockfalls, were detected; but no significant crustal earthquakes were found beneath or close to the lake. There was no clear evidence of magma movement or other signs of volcanic activity, suggesting that direct volcanism was not involved, and the lack of tectonic seismicity at the lake during the monitoring period argues against an earthquake trigger for the gas release. In the wider volcanic province, however, many earthquakes have been detected and such a trigger cannot be completely ruled out.

Localised noise outbursts with a peak frequency of 23 Hz were detected on a number of occasions by geophones and a hydrophone at the northern end of the lake. These may represent gas emissions from rock fissures and, if so, would lend support to a gas-pulse trigger for the outburst, which in turn would give rise to greater uncertainty in the CO_2 recharge rate.

The future safety of people who live around the lake and downstream from it depends either on controlled degassing or lowering of the lake level with accompanying careful monitoring of levels of CO_2 and other physical and chemical parameters of the lake on a continuous basis. The data transmission and recording techniques used in this study, including the multiplexing of an oxygen deficiency sensor into the seismic data stream, could readily be applied to such a broadly-based monitoring system.

Acknowledgements

Our thanks are due to D. Yeoman and the late S. Bartolski of Helimission, Bamenda, for helicopter transport to Lake Nyos and to Mr E. Gamnje, the regional geologist from the Ministry of Mines and Power, Government of Cameroon, for helpful assistance and advice. The British Embassy in Yaoundé provided support and advice on logistic matters throughout the mission.

This work was supported by the UK Overseas Development Administration and the Natural Environment Research Council and is published with the approval of the Director of the British Geological Survey (NERC).

References

Ambeh, W.A., J.D. Fairhead, D.J. Francis, J.M. Nange and S. Djallo, 1989. Seismicity of the Mount Cameroon Region. *Journal of African Earth Sciences*, vol 9 (1), p 1–7.

Clark, M.A., H.R. Compton, J.D. Divine, W.C. Evans, A.M. Humphrey, G.W. Kling, E.J. Koenigsberg, J.P. Lockwood, M.L. Tuttle and G.N. Wagner, 1987. *The 21 August 1986 Lake Nyos Gas Disaster, Cameroon: Final Report of the United States Scientific Team to the Office of US Foreign Disaster Assistance.*

Chevrier, R.M., 1987. Phenomenologie: Lac Nyos: explosions phreatiques 30 decembre 1986. Unesco, *International Scientific Conference of the Lake Nyos Disaster*, Yaoundé, 16–20 March 1987.

Fitton, J.G., 1980. The Benue trough and Cameroon line—a migrating rift system in West Africa. *Earth and Planetary Science Letters*, vol 51, p 132–138.

Fitton, J.G., and H.M. Dunlop, 1985. The Cameroon line, West Africa, and its bearing on the origin of oceanic and continental alkali basalt. *Earth and Planetary Science Letters*, vol 72, p 23–38.

Freeth, S.J., and R.L.F. Kay, 1987. The Lake Nyos gas disaster. *Nature*, vol 325, p 104–105.

Freeth, S.J., R.L.F. Kay, and P.J. Baxter, 1987. *Reports by the British scientific mission sent to investigate the Lake Nyos disaster.* Commissioned report by the Disaster Unit of the Overseas Development Administration, Foreign and Commonwealth Office, London, UK.

Hassert, K., 1912. Seenstudien in Nord-Kamerun. *Zeitschrift der Gesellschaft für Erdkunde zu Berlin*, vol 7–41, p 135–144 and 203–216.

Houliston, D.J., J. Laughlin and G. Waugh, 1985. Event-triggered seismic detection systems developed by the British Geological Survey. *Earthquake Engineering in Britain*, Thomas Telford Ltd, London.

Kling, G.W., M.A. Clark, H.R. Compton, J.D. Devine, W.C. Evans, A.M. Humphrey, E.J. Koenigsberg, J.P. Lockwood, M.L. Tuttle and G.N. Wagner, 1987. The 1986 Lake Nyos gas disaster in Cameroon, West Africa. *Science*, vol 236, p 169–174.

Lee, W.H.K., R.E. Bennet, and K.L. Meagher, 1972. A method of estimating magnitude of local earthquakes from signal duration. *US Geological Survey Open File Report 28.*

Lienert, B.R., E. Berg and L.N. Fraser, 1986. HYPOCENTER: an earthquake location method using centered scaled and adaptively damped least squares. *Bulletin of the Seismological Society of America*, vol 76 (3), p 771–783.

Lockwood, J.P., 1987. Geological setting and origin of Lake Nyos, Northwest Cameroon. Unesco, *International Scientific Conference of the Lake Nyos Disaster, Conclusions and Recommendations*, Yaoundé, 16–20 March 1987.

Lockwood, J.P., J.E. Costa, M.L. Tuttle, J. Nni and S.G. Tebor, 1989. The potential for catastrophic dam failure at Lake Nyos maar, Cameroon. *Bulletin of Volcanology (1988)*, vol 50, p 340–349.

McNutt, S.R., 1986. Observations and analysis of B-type earthquakes, explosions and volcanic tremor at Pavlof volcano, Alaska. *Bulletin of the Seismological Society of America*, vol 76 (1), p 153–175.

Sigurdsson, H., 1987. Reporter general, Unesco, *International Scientific Conference of the Lake Nyos Disaster, Conclusions and Recommendations*, Yaoundé, 16–20 March 1987.

Sigurdsson, H., J.D. Divine, F.M. Tchoua, T.S. Presser, M.K.W. Pringle and W.C. Evans, 1987. Origin of the lethal gas burst from Lake Monoun, Cameroon. *Journal of Volcanic and Geothermal Research*, vol 31, p 1–16.

Stuart, G.W., J.D. Fairhead, L. Dorbath and C. Dorbath, 1985. A seismic refraction study of the crustal structure associated with the Adamawa Plateau and Garoua Rift, Cameroon, West Africa. *Geophysical Journal of the Royal Astronomical Society*, vol 81, p 1–12.

Tazieff, H., 1987. Nyos disaster: limnic overturn or phreatic eruption? Unesco, *International Scientific Conference of the Lake Nyos Disaster, Conclusions and Recommendations*, Yaoundé, 16–20 March 1987.

Tsumura, K., 1967. Determination of earthquake magnitude from total duration of oscillation. *Bulletin of the Earth Resources Institute, Tokyo*, vol 45, p 7–18.

Walker, A.B., D.W. Redmayne and C.W.A. Browitt, 1988. Seismic monitoring of Lake Nyos, Cameroon, following the gas release disaster of August 1986: final report to the Overseas Development Administration by the British Geological Survey. *Global Seismology Report No WL/88/14*, Natural Environment Research Council open-file report, British Geological Survey, June 1988, 24 p.

Part Two

Earthquake Hazards

9 Long-term seismic hazard in the Eastern Mediterranean region

N.N. Ambraseys

Abstract Study of seismicity in the Eastern Mediterranean region shows that analysis of long-term observations provides a more sound basis for seismic hazard assessment and prediction than statistics based on recent events. Slip rates for plate movement along major boundary faults are only partly accounted for by seismic motions of modern events, and areas presently thought to be asesimic are more probably merely quiescent. Study of historical and archaeological data are valuable, in many cases confirming the seismicity derived from 20th-century data but also indicating other areas of potential seismicity. The magnitude of early earthquakes can be assessed from the intensity distribution and reported fault displacement, leading to enhanced data for parameters for the design of earthquake-resistant structures.

Introduction

The study of macroseismic data for large earthquakes that occurred in the region of the Eastern Anatolia, the Dead Sea Fault System and Northwest Arabia during the last millenium shows that a considerable number of large events occurred in groups, in relatively short periods, with repeat times of 200 to 350 years. Some of these events, of surface-wave magnitude M_s greater than 7.0, most probably broke segments of fault systems and were responsible for great loss of life.

The lack of such large or even smaller events during the present century should not, however, be interpreted to indicate a reduction in the potential earthquake hazard in this region. The analysis of long-term observations demonstrates that statistics based on short-term data alone do not provide a reliable assessment of seismic hazard, and that clustering of major events renders simple statistical models inadequate. This analysis also shows that, for certain regions, the total seismicity over a period of a few decades accounts for little of the slip-rate that can be calculated from plate motions over a much longer period of time. It is obvious that a detailed long-term perspective will give a more accurate picture of seismicity than will a short-term one.

In particular, a long-term view shows that areas where

there is presently little or no earthquake activity may be *quiescent* rather than *aseismic*. The intention of this paper is to draw attention to the importance of long-term seismicity in assessing earthquake hazard, and show how magnitudes and ground motions of historical earthquakes can be assessed.

Case histories

Macroseismic information comes from many and disparate sources. For the earlier events, one of these sources is inscriptions explicitly mentioning earthquake destruction or extensive repairs after an earthquake, and others referring to remission of tribute following an earthquake. Figure 1 shows the distribution of sites that have provided such information in the Eastern Mediterranean region during the period 13th century BC to 6th century AD. This distribution of sites, with few exceptions, coincides with the seismicity distribution of the 20th century. It confirms that modern earthquake occurrence is largely consistent with early seismicity patterns but also, what is more important, that some of these early events have happened where their occurrence could not be predicted from short-term, 20th-century seismological data (Ambraseys, 1988a).

For later events, historical sources may be used. The northwest border of Arabia and northern Hijaz (Figure 2), like the rest of the Arabian plate, has been considered one of the least seismic parts of the world. During this century not a single significant earthquake has been located teleseismically and until recently there has been no long-term evidence for a damaging shock in this area. However, results from a recent study of the seismicity of the region seems to call for a revision of the accepted view (Ambraseys and Adams, 1988; Ambraseys and Melville, 1988). More specifically, we have found evidence for at least three large earthquakes during the last 11 centuries. In addition there is evidence of volcanic activity in the area in the mid-7th century AD. All three earthquakes were undoubtedly large events of $M_s > 7.0$, indicated by the significant damage and high number of casualties they caused in such a sparsely populated region, the large area over which they were felt, and the long-period effects reported hundreds of kilometres away. The survival of this information over such a long period of time is itself an indication of the large size of these events, which are atypical of the short-term seismicity of the region (Ambraseys and Melville, 1989).

More detail becomes available for later events. For instance, the analysis of macroseismic data based on primary sources for large historical earthquakes for the period 1100–1988 ($M_s \geqslant 6.5$), along a segment about 350 km long of the northern part of the Dead Sea fault system, shows that ten such events occurred in three relatively short periods with repeat times of 200 to 350 years (Figure 3) (Ambraseys, 1989; Ambraseys and Barazangi, 1989). The larger of these events broke the north segment of the Dead Sea fault system, possibly including the western-most segment of the East Anatolian fault system near the border between Syria and Turkey. The $M_s = 7+$ earthquake in the Bekaa valley northwest of Damascus on 25 November 1759 and its foreshock (Figure 4) almost certainly produced surface faulting, probably along the Yammouneh fault in the valley (Figure 5), a region almost totally quiescent during this century.

One arrives at the same conclusion by analysing historical data over a larger area. For instance, it is generally recognised that the Border Zone 1000 km long that defines the boundary between the Arabian and Turkish plates has been remarkably inactive, with only three $M_s = 6.7$ magnitude earthquakes during this century. The seismic hazard map of this zone, based

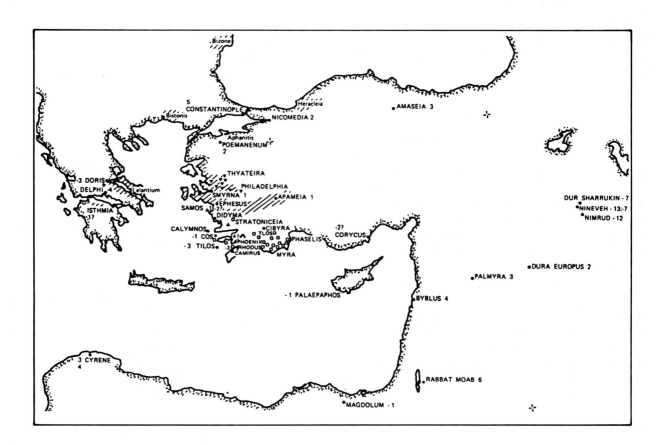

Figure 1 Distribution of archaeological sites that are mentioned in inscriptions as having suffered earthquakes before the 6th century AD. Numbers refer to centuries, with BC centuries shown as negative.

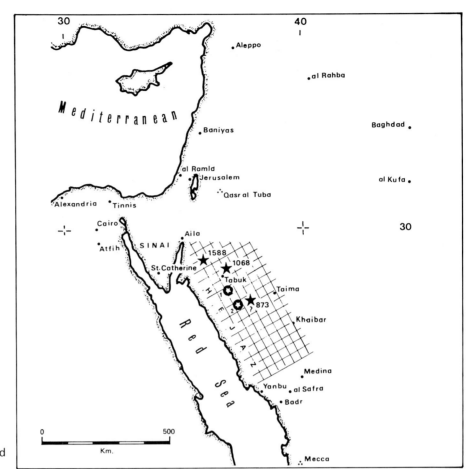

Figure 2 Location map of the Hijaz study area (hatched). Stars show locations of adopted epicentres of the earthquakes of September 873, 18 March 1068 and 4 January 1588. Open stars show the location of Hala'l Ishqa (1) and Hala'l Badr (2) associated with the lava flows of 641 AD.

on 20th-century data (Burton and others, 1984) shows negligible activity. Also, the predicted velocity between Arabia and Turkey is two orders of magnitude greater than that obtained from a summation of seismic moments of events in the 20th century. What is not so generally recognised, however, is that the apparent quiescence of this zone is only temporary and that it does not represent the real tectonic activity that accommodates almost all the Arabia-Turkey motion, and part of that between Arabia and Eurasia along this boundary. The fact that all of our 20th-century records are for a quiescent period can be demonstrated by extending observations backwards in time by a few centuries. Figure 6 shows the locations of the earthquakes identified so far in the Border Zone with $M_s \geqslant 6.6$ for the period 1500–1988 (Ambraseys, 1989). This establishes beyond doubt that, in this large area, all of our 20th-century assessment of seismicity is based on data from a quiescent period.

Even for more recent times, macroseismic data are invaluable. Recent seismicity maps show the West

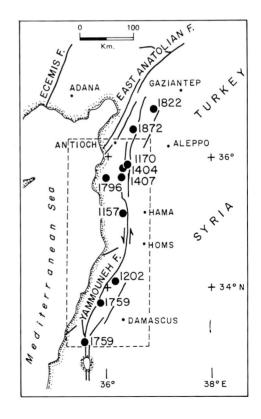

Figure 3 Location map of large historical earthquakes along the Yammouneh fault and its extension in the Bekaa. See also Figures 4 and 5.

85

African region to be almost aseismic. In fact there is significant activity in the coastal areas of Guinea, along the Ivory Coast, and particularly in Ghana and in Cameroon (Figure 7, Table 1). Historically there has been activity at some inland locations such as near Timbuktu (Figure 8). East of the Gulf of Guinea, activity extends inland through Gabon, Cameroon and the Central African Republic towards more active areas in Zaïre (Ambraseys and Adams, 1986).

The uncritical use of macroseismic or instrumental data quite often leads to errors and, in assessing seismicity, great care must be taken to evaluate critically the information available. Quite often earthquakes are incorrectly attributed to a particular region by gross error. In our study of West African seismicity (Ambraseys and Adams, 1986), for instance, out of a total of nearly 200 reported events from published sources, nearly 70 turned out to be either completely spurious, because of printing or seismological errors, or could not be substantiated because of the very poor data on which they were based. Mislocation of early events is another source of error while modern misintrepretations are quite frequent. An example of such seismological errors comes from the automatic analysis procedure at the Large Aperture Seismic Array (LASA) in Montana. These events have been located by beam-forming from LASA only, at the far range of P-wave distance. Inspection shows that many of these reported events follow large earthquakes in the South Pacific, and the arrivals are likely to be steeply arriving core phases, coming via an antipodal path from the opposite azimuth from the Pacific. The false interpretation of these arrivals as P-phases, gives rise to a broad band of earthquakes running through Africa, Sudan, the Red Sea and Saudi Arabia to the Persian Gulf. These spurious events could easily be associated with the Cameroon volcanic line separating the cratons of West and Central Africa, and with other tectonic elements (Figure 9).

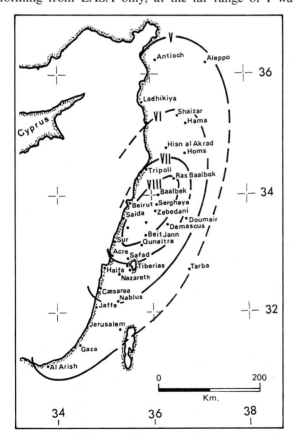

Figure 4 Intensity distribution of surface effects of the main Bekaa valley earthquake of 25 November 1759.

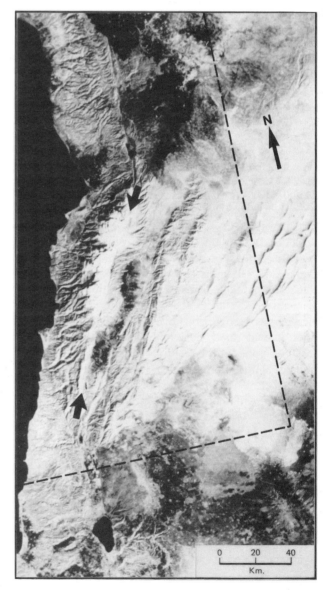

Figure 5 Landsat image of the area outlined by dashed lines in Figure 3. The arrows indicate the Yammouneh fault.

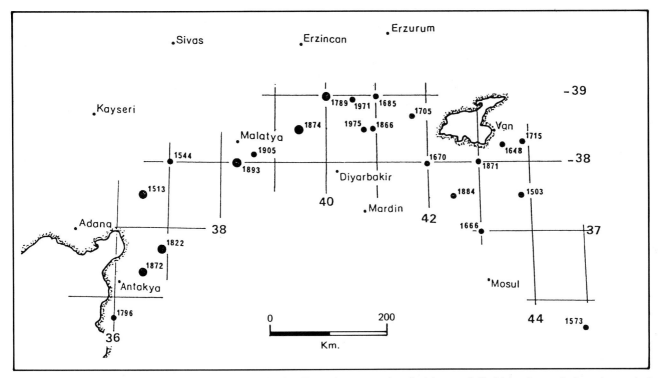

Figure 6 Distribution of earthquakes in the Border Zone between the Arabian and Turkish plates during the period 1500–1988, marked with year of occurrence. Large dots correspond to magnitude 7.0, smaller ones to 6.0.

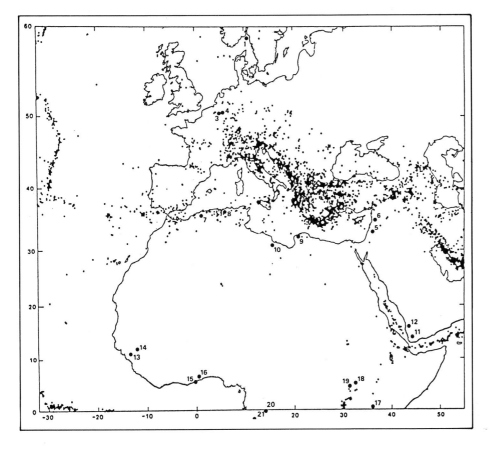

Figure 7 Earthquakes in Europe, Middle East and North Africa from mid-1963 to mid-1983, reported by the US Coast and Geodetic Service (NEIS). Numbers indicate pairs of significant earthquakes in seismically quiet areas, referred to in Table 1.

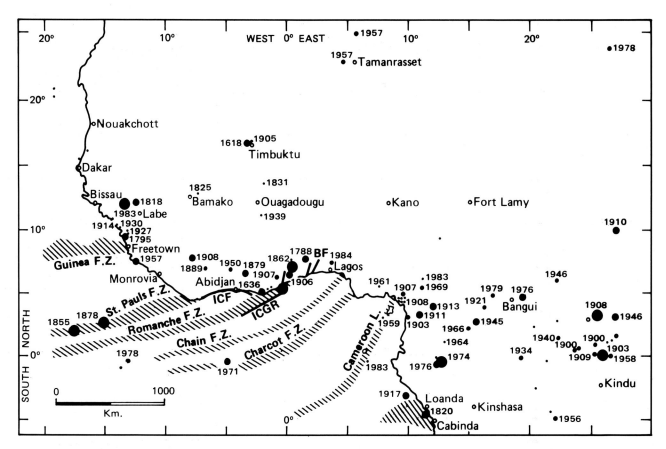

Figure 8 Earthquake distribution in West Africa showing magnitude and date; a = $M_s \geqslant 6.0$; b = $6.0 > M_s \geqslant 5.0$; c = $5.0 > M_s \geqslant 4.0$; d = $M_s < 4.0$.

Table 1 Earthquakes occurring in low seismicity areas

	Date	Time	Epicentre		I_o	M_s	Country
1	1904 Oct 23	1027	59.0°N	10.5°E	(VII)	5.8	Norway
2	1819 Aug 31	1330	66.7°N	15.5°E	VIII	6.2	Norway
3	1983 Nov 8	0050	50.6°N	5.5°E	VIII	4.3	Belgium
4	1692 Sep 18	1420	50.6°N	5.6°E	VIII	4.7	Belgium
5	1927 Jul 11	1304	31.8°N	35.5°E	VIII	6.0	Israel
6	1759 Nov 25	1623	33.6°N	35.8°E	IX	7.2	Lebanon
7	1980 Oct 10	1225	36.1°N	1.4°E	IX	7.3	Algeria
8	1856 Aug 22	1140	36.7°N	6.1°E	IX	7.3	Algeria
9	1963 Feb 21	1715	32.7°N	21.0°E	VIII	5.4	Libya
10	1935 Apr 19	1523	31.4°N	15.4°E	IX	7.0	Libya
11	1982 Dec 13	0913	14.7°N	44.2°E	IX	5.9	Yemen
12	1941 Jan 11	0831	16.6°N	43.3°E	IX	5.8	Yemen
13	1983 Dec 22	0411	11.9°N	13.5°E	IX	6.3	Guinea
14	1818 Jan –	–	12.1°N	12.4°W	(VII)	5.9	Guinea
15	1939 Jun 22	1919	5.3°N	0.2°W	IX	6.5	Ghana
16	1862 Jul 10	0815	7.0°N	0.4°E	IX	6.5	Ghana
17	1928 Jan 6	1932	0.2°N	36.2°E	VIII	6.8	Kenya
18	1915 May 21	0419	4.4°N	31.9°E	VIII	6.3	Sudan
19	1850 Jan –	–	4.2°N	31.6°E	IX	6.0	Sudan
20	1974 Sep 23	1928	0.3°S	12.8°E	–	6.1	Gabon
21	1820 – –	–	4.5°N	11.6°E	IX	6.2	Gabon

Assessment of magnitude and ground motions of historical earthquakes

The size or magnitude of early earthquakes may be assessed from the extent of the area over which the shock was felt with a given intensity I_i, and approximately from the length of fault rupture and displacement which can then be calibrated against macroseismic information about similar, 20th-century, earthquakes for which instrumental data are available.

The size of a historical earthquake, in terms of surface wave magnitude M_s, may be calculated in terms of the mean radius, R, in kilometres of the isoseismal of intensity, I_i, in the Medvedev-Sponheuer-Karnik (MSK) scale (see panel, p. 90–91). Regional empirical formulae have been derived of the type

$$M_s = A + B(I_i) + C(R_i) + D\log(R_i) \qquad (1)$$

for shallow earthquakes (h ≤ 25 km) by correlating teleseismic M_s values with observed macroseismic estimates of R_i and I_i for 20th-century earthquakes. Thus, for the Bekaa valley earthquake of 25 November 1759, from Figure 4, and for attenuation conditions similar to those in Turkey, this gives $R_8 = 51$ km, $R_7 = 82$ km,

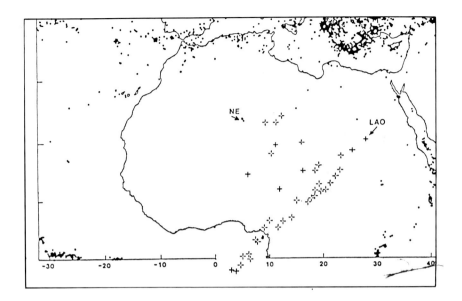

Figure 9 Spurious earthquake locations in Africa, given by LASA array (LAO—crosses; see text). Magnitudes assigned to these events by LASA were between 4 and 6, mostly 5+. The linear extension of these events in a northeasterly direction through the Gulf of Guinea could easily be associated with the zone separating the cratons of West and Central Africa. Sites of nuclear explosions in the Sahara are indicated by NE.

$R_6 = 136$ km and $R_5 = 230$ km. If we assume an error in I_i of ± 0.5, the above equation gives $M_s = 7.2 \pm 0.3$.

The constants A, B, C and D in equation (1) for different regions are:

	A	B	C	D	r_2	n
Balkans	−0.902	0.578	1.10×10^{-3}	2.11	0.78	354
Turkey	−0.529	0.583	1.96×10^{-3}	1.83	0.94	494
NE Europe	−1.100	0.620	1.30×10^{-3}	1.62	0.92	300

with $r_2 =$ coefficient of determination, $n =$ sample size.

An estimate of magnitudes from the dimensions of fault rupture for Eastern Mediterranean and Middle Eastern earthquakes may be made from the major axis solution

$$M_s = 4.63 + 1.43\log(L) \qquad (2)$$

where L is the observed length of fault-break in kilometres and M_s is the surface wave magnitude of the associated earthquake. A better estimate of magnitude from the dimensions of rupture and dislocation for events in the same region may be made from

$$M_s = 1.1 + 0.4\log(L^{1.58}R_2) \qquad (3)$$

where L and R are the length of the fault-break and relative displacement respectively, in centimetres (Ambraseys, 1988b). Thus, for the Spitak, Armenia, earthquake of 7 December 1988, with a rupture length of 30 km and average horizontal and vertical displacements of 70 and 110 cm respectively, we obtain $M_s = 6.8$.

The assessment of the peak horizontal ground accelerations, a, and velocities, v, at a site located at a distance,

d, from the causative fault may be estimated using a revised version of the attenuation law by Joyner and Boore (1981) in terms of surface wave magnitude M_s (Ambraseys, 1990). The values of a and v, in cm/sec^2 and cm/sec respectively, are given by

$$\log(a) = -0.789 + 0.213(M_s) - \log(r) - 0.00255(r) + 0.25P, \text{ and} \qquad (4)$$

$$\log(v) = -0.376 + 0.433(M_s) - \log(r) - 0.00255(r) + 0.22P + 0.17S \qquad (5)$$

in which for accelerations $r = (d_2 + 52)^{\frac{1}{2}}$, and for velocities $r = (d_2 + 16)^{\frac{1}{2}}$.

The nearest distance of the site from the fault, d, is in kilometres, and P is zero for 50 percentile values and one for 84 percentile values. S takes on the value of zero at rock sites and one at soil sites.

Results

The study of the long-term seismicity of the Eastern Mediterranean region suggests that on large strike-slip faults much of the slip is associated with rather infrequent clusters of large-magnitude shocks separated by relatively long periods of quiescence. This attitude is less well defined in normal zones where there is no evidence, so far, that earthquakes have exceeded magnitudes of about $M_s = 7.0$.

There is also evidence that, in some of the tectonic units in the region, much of the deformation is taken up aseismically with only a small part of it being associated with regularly occurring medium-magnitude events.

MSK-81 Seismic Intensity Scale
(Medvedev-Sponheuer-Karnik)

Classification of the scale

1. Types of structures (buildings not anti-seismic)

A: Buildings of fieldstone, rural structures, adobe houses, clay houses;

B: Ordinary brick buildings, large block construction, half-timbered structures, structures of hewn blocks of stone;

C: Pre-cast concrete skeleton construction, pre-cast large-panel construction, well-built wooden structures.

2. Definition of quantity

Single, few: $= 10\%$

Many: $= 20–50\%$

Most: $= 60\%$

3. Classification of damage to buildings

Grade 1: Slight damage: Fine cracks in plaster; fall of small pieces of plaster.

Grade 2: Moderate damage: Small cracks in walls; fall of fairly large pieces of plaster; pantiles slip off; cracks in chimneys; parts of chimneys fall down.

Grade 3: Heavy damage: Large and deep cracks in walls; fall of chimneys.

Grade 4: Destruction: Gaps in walls; parts of buildings may collapse; separate parts of the buildings lose their cohesion; inner walls and filled-in walls of the frame collapse.

Grade 5: Total damage: Total collapse of buildings.

4. Arrangement of the scale

(a) Persons and surroundings

(b) Structures

(c) Nature.

Intensity degrees

I   *Not noticeable*

(a) The intensity of the vibration is below the limit of sensibility; the tremor is detected and recorded by seismographs only

(b) —

(c) —

II   *Scarcely noticeable* (very slight)

(a) The vibration is felt only by individual people at rest in houses, especially on upper floors of buildings.

(b) —

(c) —

III   *Weak*

(a) The earthquake is felt indoors by a few people, outdoors only in favourable circumstances. The vibration is weak. Attentive observers notice a slight swinging of hanging objects, somewhat more heavily on upper floors.

(b) —

(c) —

IV   *Largely observed*

(a) The earthquake is felt indoors by many people, outdoors by a few. Here and there people awake, but are not frightened. The vibration is moderate. Windows, doors and dishes rattle; floors and walls creak; furniture begins to shake; hanging objects swing slightly. Liquids in open vessels are slightly disturbed. In stationary motor cars, the shock is noticeable.

(b) —

(c) —

V   *Strong*

(a) The earthquake is felt indoors by most people, outdoors by many. Many sleeping people awake; a few run outdoors; animals become uneasy. Buildings tremble throughout; hanging objects swing considerably; pictures swing out of place; occasionally pendulum clocks stop; unstable objects may be overturned or shifted; open doors and windows are thrust open and slam back again; liquids spill in small amounts from well-filled open containers. The vibration is strong, resembling sometimes the fall of a heavy object in the building.

(b) Damage of Grade 1 in a few buildings of Type A is possible.

(c) Sometimes there is a change in the flow of springs.

VI   *Slight damage*

(a) The earthquake is felt by most indoors and outdoors. Many people in buildings are frightened and run outdoors; a few lose their balance; domestic animals run out of their stalls. In a few instances, dishes and glassware may break; books fall down; heavy furniture may possibly move and small steeple bells may ring.

(b) Damage of Grade 1 is sustained in single buildings of Type B and in many of Type A; damage in a few buildings of Type A is of Grade 2.

(c) In a few cases, cracks up to a width of 1 cm are possible in wet ground; in mountains, occasional landslips may occur. Changes in the flow of springs and of levels of well water are observed.

VII   *Damage to buildings*

(a) Most people are frightened and run out of doors; many find it difficult to stand. The vibration is noticed by persons driving motor cars. Large bells ring.

(b) In many buildings of Type C, damage of Grade 1 is caused; in many buildings of Type B damage is of Grade 2; many buildings of Type A suffer damage of Grade 3, a few of Grade 4. In single instances, landslips of roadways occur on steep slopes; locally cracks in roads and stone walls.

(c) Waves are formed on water, and water is made turbid by mud stirred up. Water levels in wells change, and the flow of springs changes; in a few cases, dry springs have their flow restored, and existing springs stop flowing. In isolated instances, parts of sandy or gravelly banks slip off.

VIII   *Destruction of buildings*

(a) General fright; a few people show panic, also persons driving motor cars are disturbed. Here and there branches of trees break off; even heavy furniture moves and partly overturns; hanging lamps are damaged in part.

(b) Many buildings of Type C suffer damage of Grade 2, and a few of Grade 3; many buildings of Type B suffer damage of Grade 3, and a few of Grade 4; many buildings of Type A suffer damage of Grade 4, and a few of Grade 5. Memorials and monuments move and twist; tombstones overturn; stone walls collapse.

(c) Small landslips occur in hollows and on banked roads on steep slopes; cracks form in the ground up to widths of several centimetres. New reservoirs come into existence; sometimes dry wells refill and existing wells become dry; in many cases, the flow and level of water in wells changes.

IX   *General damage to buildings*

(a) General panic. Considerable damage to furniture; animals run to and fro in confusion and cry.

(b) Many buildings of Type C suffer damage of Grade 3, and a few of Grade 4; many buildings of Type B show damage of Grade 4, and a few of Grade 5; many buildings of Type A suffer damage of Grade 5. Monuments and columns fall. Reservoirs may show heavy damage. In individual cases, railway lines are bent, and roadways damaged.

(c) On flat land, overflow of water, sand and mud is often observed. Ground cracks of widths up to 10 cm form, in slopes and river banks more than 10 cm; furthermore, a large number of slight cracks occur in the ground. Falls of rock occur, and many landslides and earth flows. Large waves occur on water.

X   *General destruction of buildings*

(b) Many buildings of Type C suffer damage of Grade 4, and a few of Grade 5; many buildings of Type B show damage of Grade 5. Most of Type A collapse; dams, dykes and bridges may show severe to critical damage; railway lines are bent slightly; road pavement and asphalt show waves.

(c) In ground, cracks form up to widths of several decimetres, sometimes up to one metre; broad fissures occur parallel to watercourses; loose ground slides from steep slopes; considerable landslides are possible from river banks and steep coasts. In coastal areas, there is displacement of sand and mud; water from canals, lakes, rivers etc. is thrown on land; new lakes are formed.

XI   *Catastrophe*

(b) Destruction of most, and collapse of many buildings of Type C; even well-built bridges and dams may be destroyed, and railway lines largely bent, thrusted or buckled; highways become unusable; underground pipes destroyed.

(c) Ground is fractured considerably by broad cracks and fissures, as well as by movement in horizontal and vertical directions; numerous landslides and falls of rock. The intensity of the earthquake requires special investigation.

XII   *Landscape changes*

(b) Practically all structures above and below ground are heavily damaged or destroyed.

(c) The ground surface is radically changed. Considerable ground cracks with extensive vertical and horizontal movement are observed. Falls of rock and slumping of river banks occur over wide areas; lakes are dammed; waterfalls appear, and rivers are deflected. The intensity of the earthquake requires special investigation.

The available evidence also suggests that normal faulting earthquakes of about $M_s = 7.0$ are generally multiple events, with individual sub-events apparently not larger than about $M_s = 6.5$. This also appears to be the case with thrust faulting where multiplicity is replaced by clustering of large magnitude "aftershocks" and "foreshocks".

The earthquake of 7 December 1988 in Armenia was a multiple event of magnitude $M_s = 6.7$, not exceptionally large by global standards (Bommer and Ambraseys, 1989). Its seismic moment was about 1.1E26 d-c, and it was associated with a fault-break about 30 km long, showing oblique thrust with an average throw of 110 cm to the south-southwest. Spitak, a rapidly growing town with a population of about 20,000, situated near the fault-break, was totally destroyed; and Leninakan, the second largest city in Armenia with a population close to 300,000, located 23 km from the other end of the break, was ruined. The total loss of life was about 30,000 and the immediate financial loss is estimated at over 15,000 million roubles (approximately US$15,000 million), a record for a 20th-century earthquake of this magnitude. The average population density of Armenia is 117 per km^2; however, one third of Armenia's population lives in the capital Yerevan, about 80 km south of the epicentral area.

The comparison of the short-term (50 year) and long-term (500 years) seismicity dictates that, until we understand from regional and local tectonics why the larger events occurred where they did, and how other parts of a region are genuinely different, the subdivision of a region into discrete seismo-tectonic provinces on short-term observations is not a realistic proposition.

Furthermore, the indiscriminate use of a Poisson model (random generations) in earthquake statistics may lead to unrealistic results. It is simply not logical to ignore the possibility that all of our instrumental or short-term observations may be for a quiescent period in the seismic activity of a region. This is one alternative that must be borne in mind when making an assessment of seismicity, and this is the chief reason why statistics based on short-term data alone do not provide a reliable assessment of the true earthquake hazard.

Finally, the question of multiplicity of earthquakes or "dense clustering" is an important consideration. Large aftershocks or multiple events become a critical design parameter for the construction of earthquake resistant structures.

References

Ambraseys, N.N., 1988a. Engineering seismology. *Journal of Earthquake Engineering and Structural Dynamics*, vol 17, p 1–105.

Ambraseys, N.N., 1988b. Magnitude-fault length relationships for earthquakes in the Middle East. In: W.H.K. Lee and K. Shimazaki (eds), *Historical Seismograms and Earthquakes of the World*, Academic Press, San Diego, p 309–310.

Ambraseys, N.N., 1989. Temporary seismic quiescence: SE Turkey. *Geophysical Journal*, vol 96, p 311–331.

Ambraseys, N.N., 1990. Uniform magnitude re-evaluation of European strong-motion records. *Journal of Earthquake Engineering and Structural Dynamics*, vol 19, p 1–20.

Ambraseys, N.N., and R.D. Adams, 1986. Seismicity of West Africa. *Annales Géophysiques*, vol 4B, p 679–702.

Ambraseys, N.N., and R.D. Adams, 1988. The seismicity of Saudi Arabia and adjacent areas, Part B. *Engineering Seismology and Earthquake Engineering (ESEE) Publication no 88/11*, Imperial College, London.

Ambraseys, N.N., and M. Barazangi, 1989. The 1759 earthquake in the Bekaa Valley: implications for earthquake hazard assessment in the eastern Mediterranean region. *Journal of Geophysical Research*, vol 94, p 4007–4013.

Ambraseys, N.N., and C.P. Melville, 1988. The seismicity of Saudi Arabia and adjacent areas, part A. *Engineering Seismology and Earthquake Engineering (ESEE) Publication no 88/11*, Imperial College, London.

Ambraseys, N.N., and C.P. Melville, 1989. Evidence for intraplate earthquakes in northwest Arabia. *Bulletin of the Seismological Society of America*, vol 79.

Bommer, J.J., and N.N. Ambraseys, 1989. The Spitak, Armenia USSR earthquake of December 1988. *Journal of Earthquake Engineering and Structural Dynamics*, vol 18, p 921–925.

Burton, P.W., R. McGonigle, K. Makropoulos, and S.B. Ucer, 1984. Seismic risk in Turkey, the Aegean and the eastern Mediterranean. *Geophysical Journal of the Royal Astronomical Society*, vol 78, p 475–506.

Joyner, W.B., and D.M. Boore, 1981. Peak horizontal acceleration and velocity from strong-motion records. *Bulletin of the Seismological Society of America*, vol 71, p 2011–2038.

10 The ROA *Earthquake Hazard Atlas* project: recent work from the Middle East

M.R. Degg

Abstract In 1989 the Reinsurance Offices Association (ROA) of London initiated an *Earthquake Hazard Atlas* project. The principal aim of this is to facilitate access to information concerning exposure to earthquake hazard, presenting standardised data in a format that enables them to be accessed and incorporated into hazard and risk assessments with relative ease, and hopefully ensures their usefulness to a wide range of people.

This paper provides an overview of the project and illustrates its various components through reference to recently published parts for the Middle East. These include a regional hazard assessment (based upon a review of tectonic setting and historical and 20th-century earthquake data), a national analysis for Israel (based upon a new hazard zonation for the state) and a new database of historical earthquake activity in the region. New atlas parts are currently being compiled, and it is intended that with time the atlas will build up to provide comprehensive, standardised coverage for all seismic regions of concern. As such, it may help to address one of the key issues that has hindered earthquake hazard mitigation to date, that of data dissemination: the fact that so often the data needed to ensure sensible hazard management practice exist but are not readily accessible to those who most need them.

Introduction

There has been a marked rise in the number of natural catastrophes recorded in recent years. The Munich Reinsurance Company (Munich Re.) has shown that between the 1960s and 1980s, the number of major disasters reported increased by a factor of 5.0. By extrapolating figures to today's prices they have calculated that the economic losses associated with these events increased by a factor of 3.1 and the insured losses by 4.8 (Munich Re., 1990).

Such trends have caused genuine concern about the lack of reliable information (even at the global and regional scales) regarding exposure to natural hazards. This concern is particularly justifiable for countries of the Third World, many of which are subject to severe natural hazards. The rapid rates of population growth, urbanisation and industrialisation that have characterised many developing countries in recent decades have often served to increase the likelihood of large human and economic hazard losses. Recent catastrophes (e.g.

the 1985 Chilean earthquake, in which 145 people died and 500,000 were made homeless, and the Mexican earthquake of the same year—see Degg, this volume) have reinforced the need to obtain reliable hazard data for these countries and to make them as widely available as possible.

In that context, a problem encountered at all levels of hazard assessment is the lack of reliable hazard data that can be readily accessed on a day-to-day working basis by those who, potentially, are in the best situations to ensure sound hazard management practice, e.g. planners, policy makers, direct insurers and reinsurers, engineers, environmental agencies and educationalists. This, at first, seems incongruous in view of the large amounts of hazard research taking place in academic institutions and industry. However, the reality of the situation seems to be that little of this work filters through to the people who most need it, except by way of specially commissioned reports. There is, therefore, a genuine need to make fundamental hazard data widely and cheaply available, and thereby to induce a greater awareness of exposure to natural hazards.

A number of research initiatives have addressed this problem in recent years, and amongst those particularly concerned with this type of work have been many of the major international insurance and reinsurance companies. Very often it is they who are left to bear the brunt of the economic losses caused by natural disasters; it is, therefore, understandable that they should seek to foster increased awareness within their own industry and related professions concerning exposure to natural hazards. In 1978 the Swiss Reinsurance Company (Swiss Re.) published an *Atlas on Seismicity and Volcanism* and the Munich Re. (1978, 1988) produced a *World Map of Natural Hazards*. Both publications have been extremely successful and are widely used as means of obtaining hazard data on the global/regional scale. The global nature of the studies (e.g. the World Map has a scale of 1:30,000,000) means that the amount and refinement of the hazard data shown is somewhat limited. However, the advantages to be gained from presenting uniform hazard data for all regions in a format that enables them to be readily accessed and interpreted cannot be denied.

More recently, the Reinsurance Offices Association (ROA) of London has initiated an *Earthquake Hazard Atlas Project*. This is a continuing effort which seeks to explore means of improving accessibility to essential data on the earthquake hazard, whilst at the same time avoiding many of the problems of scale that have characterised and limited previous global studies. This paper examines the various facets of the project through recourse to the first published parts, which concern the Middle East (Degg and Doornkamp, 1989, 1990).

The Middle Eastern region has a long and relatively well documented history of earthquake activity that has taken a heavy toll in lives and damage. Recent disasters in the region, such as the 1982 Dhamar earthquake in the Yemen which killed approximately 2500 people and left tens of thousands homeless (Arya and others, 1985; Plafker and others, 1987), have served to emphasise that this threat continues and to reinforce the need for reliable evaluations of the earthquake hazard.

The *Earthquake Hazard Atlas* project

The *Earthquake Hazard Atlas* has four main parts. Part A contains introductory text and essential background information for use in conjunction with Parts B, C and D. These, in turn, relate to the three main components of the atlas project and contain:

Part B: Regional evaluations
Part C: National analyses
Part D: Historical catalogues.

The structure of the atlas and the way in which the parts are coded is such that new sections can be added to it in a logical manner as they are produced (or requested by the user). For example, the regional evaluation of the Middle East is coded B1, and other regional evaluations B2, B3, etc. The country analyses of the Middle East begin with the code C1 and are numbered C1.1 (Israel), C1.2 (Egypt) etc. The historical catalogues for the Middle East begin with the code D1 and are numbered in an identical manner to the national analyses, e.g. D1.1 (Israel). The division of the atlas into sections with regional and national analyses is to enable hazard evaluations to be presented at a range of scales within an overall global framework. Parts B, C and D of the atlas are examined in greater detail below.

Part B of the Atlas: regional evaluations
To assess earthquake hazard in any one part of the world it is important to understand the regional setting (geological, tectonic and seismological) of the area. The purpose of the regional evaluations, therefore, is to summarise the available information concerning these aspects in order to set the context within which larger-scale hazard assessments may be considered. The evaluations involve reviews of published literature and draw upon data from many different sources.

The starting point for the regional assessments is a review of tectonic setting. Such an analysis for the Middle East is summarised in Figure 1, which draws together information from a range of sources concerning the different tectonic provinces of the region (e.g. McKenzie, 1970; McKenzie and others, 1970; Nowroozi, 1971; Le Pichon and Francheteau, 1978; Nur and Ben-Avraham, 1978; Papazachos and Comninakis, 1978; Sengor, 1979; Adams and Barazangi, 1984; Lovelock, 1984; Mart and Rabinowitz, 1986; Rotstein and Ben-Avraham, 1986; Kashai and Croker, 1987; Lyberis, 1988; El-Isa and Al Shanti, 1989) and attempts to provide a coherent summary. Inevitably the diagram over-simplifies the situation, but nevertheless provides a meaningful basis for a regional analysis of the earthquake hazard.

Information concerning the nature and distribution of 20th-century seismic activity must then be related to this tectonic analysis. Data of this type are relatively easily obtained from institutions such as the United States Geological Survey (USGS), the International Seismological Centre (ISC) in Newbury, Berkshire, UK, the British Geological Survey (BGS) and the various seismological stations that operate around the world (e.g. the Helwân Observatory, Cairo). Twentieth-century seismic data for the Middle East are summarised in Figure 2, which shows the epicentral locations of earthquakes of Richter magnitude (M_s) > 5.0 recorded in the region between 1900–1983.

Such data, when related to the tectonic setting of the region (Figure 1), provide an initial impression of the distribution of earthquake hazard. However, this impression is often of limited reliability because of the relatively short time span that the 20th-century data cover. Ambraseys (1971, 1975, 1989, and in this volume) and Ambraseys and Barazangi (1989), amongst others, have shown that for regions such as the Middle East where low-to-moderate rates of seismic activity are experienced, historical (i.e. pre-20th century) data need to be included in hazard evaluations in order to enhance their meaningfulness. Access to these data, however, tends to be much more difficult and costly (particularly in terms of time) than 20th-century data. The atlas project has, therefore, devoted considerable attention to the collation of historical earthquake data and to their inclusion in the hazard analyses. The work, which is described in greater detail below (see Part C: Historical Earthquake Catalogues), has involved the compilation of computerised databases of pre-20th century earthquake activity.

A preliminary regional analysis (spatial) of historical data for the Middle East is summarised in Figure 3. This shows the distribution of historical earthquake effects for the period 1 AD to 1899 AD. The effects include the locations of cities damaged by earthquakes and regions experiencing earthquake damage (for which specific affected localities are not always given, e.g. "damage throughout upper Galilee"). The occurrence of earthquake-related hazards, such as faulting, ground failure, seismic sea waves (tsunamis, and seiches in the Dead Sea and Lake Tiberias) and volcanic activity, is also shown. Finally, the epicentral locations of some large earthquakes are plotted; for earthquakes prior to the 19th century, epicentral estimates are very approximate. Epicentral locations of 19th-century earthquakes are more precisely defined but considerable error is still to be expected (the first effective seismographs were not constructed until the latter part of the 19th century).

The figure compares reasonably well with the map of historical seismicity produced by Ambraseys (1975), delineating the epicentral areas of destructive or

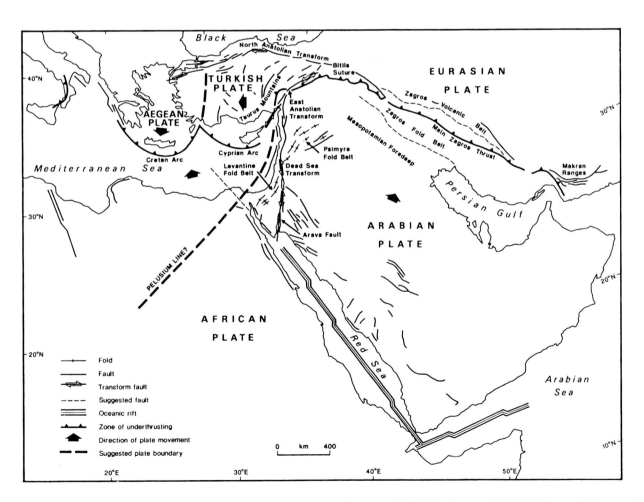

Figure 1 The tectonics of the Middle East, based upon maps by McKenzie (1970), Sengor (1979), Adams and Barazangi (1984), Lovelock (1984) and Kashai and Croker (1987).

damaging events experienced in the Middle East during the first 17 centuries AD. There are, however, important differences between the two maps, due largely to the difference in emphasis between them. In particular, Ambraseys' map focusses attention upon the correlation between historical earthquake activity and tectonic setting, whereas Figure 3 serves to emphasise that the earthquake threat is not restricted to the immediate vicinities of the major tectonic belts but can extend for considerable distances away from them (in terms of both the direct effects of earthquake ground shaking and secondary effects such as inundation by seismic sea waves). In that context, two areas that history has shown are not particularly seismically active in themselves, but which have experienced earthquake damage on numerous occasions, are the Nile delta of Egypt and the Mesopotamian plains of Iraq. For centuries, these areas have been foci of civilisation and continue to be amongst the most densely populated parts of the Middle East. The historical records show that much of the earthquake damage experienced in these areas has been caused not by local events, but by high-magnitude distant events.

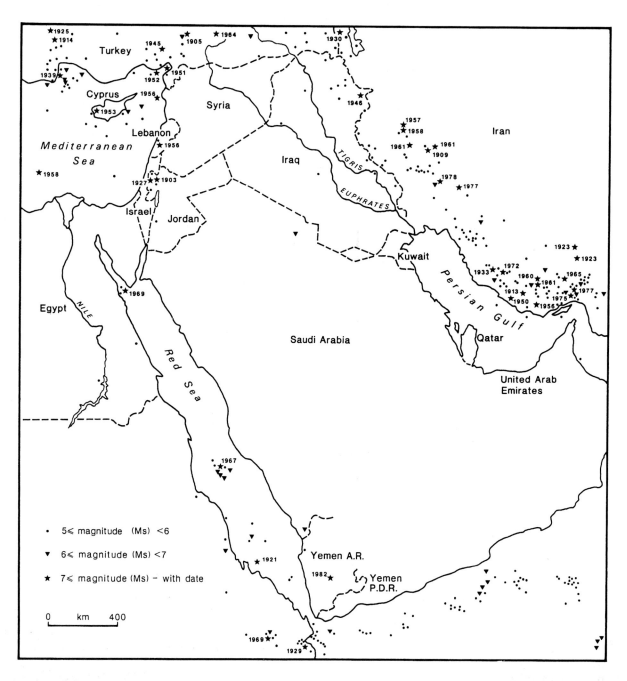

Figure 2 Earthquake activity in the Middle East (1900–1983) (source of data: Riad and others, 1985).

The delta region of Egypt has proved susceptible to large-magnitude earthquakes along the Hellenic arc in the eastern Mediterranean, and also to similar events in the northern Red Sea region, whilst earthquakes along the Zagros collision zone of the Iran/Iraq border have inflicted damage upon cities of the Mesopotamian plains (see Figures 1 and 2). Some of this latter damage has been associated with liquefaction (see Figure 3).

It is interesting to consider that, during the 20th century,

trends towards high-rise construction in the delta and plains regions have probably served to increase vulnerability to earthquakes of this type. This is because high-rise buildings are characterised by lower natural frequencies of vibration than low-rise ones and therefore tend to be much more sensitive to the low-frequency ground motions that predominate at distance from an earthquake epicentre (high-frequency earthquake motions are usually selectively attenuated with increasing distance from an epicentre,

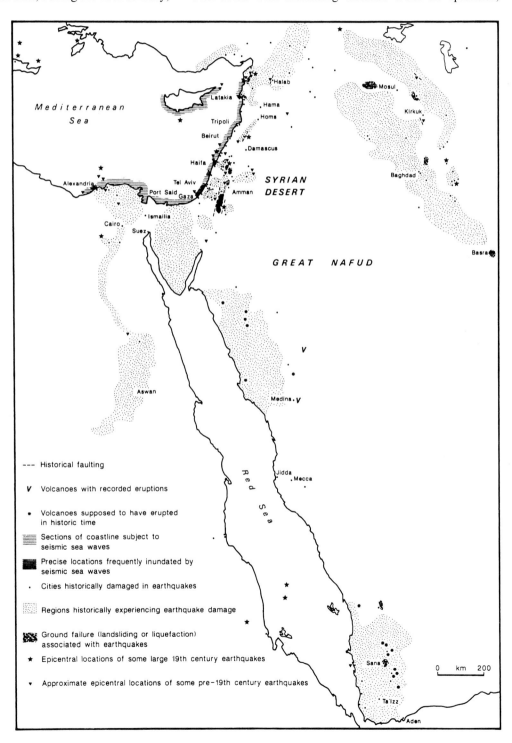

Figure 3 Historical earthquake activity and associated hazards in a part of the Middle East.

leaving motions of predominantly low frequencies of vibration).

On a cautionary note, care has to be taken in interpreting maps (such as Figure 3) which show the distribution of historical earthquake *effects* because the data tend, by their very nature, to be influenced by the historical distribution of human populations. Bearing that in mind, Figures 1, 2 and 3 nevertheless provide a sound basis for a meaningful regional evaluation of the distribution of earthquake hazard in the Middle East, based upon a review of published literature and a compiled record of seismic activity that, for parts of the region, extends back over 4000 years. Such an evaluation is beyond the scope of the present work, but is given in Volume 1 of the Atlas (Degg and Doornkamp, 1989) and in Degg (1990).

Part C of the Atlas: national analyses

The regional evaluations of the atlas serve to set the context within which more localised (e.g. national) hazard assessments may be conducted at a larger scale. These are carried out for areas shown by the regional studies to be exposed to the earthquake threat. The national analyses provide earthquake hazard zonations that quantify this threat in probabilistic terms; an example of one such zonation, for the State of Israel and surrounding areas, is presented in Figure 4.

The zonation uses *intensity* as the zonation parameter. This provides a measure of the effects of an earthquake on humans, structures and the Earth's surface. Intensity was chosen for the atlas in preference to other more quantifiable parameters (e.g. acceleration, velocity) because of its existing widespread use. As Algermissen and Steinbrugge (1984) showed, intensity is the only seismic quantity widely available throughout the world that can be used as a measure of ground motion at a particular site; such a consideration is important in view of the global dimension to the atlas. Intensity is also favoured because of its predominant use in earthquake loss statistics and vulnerability curves (e.g. Sauter and Shah, 1978; Munich Re., 1978, 1988; Swiss Re., 1989: see Table 1) and the relative ease with which it allows historical earthquake data to be incorporated into hazard analyses.

Various intensity scales have been devised, but the one adopted for use in the atlas is the Modified Mercalli scale (MM) of 1956 (see panel). This is a 12-point scale, denoted by Roman numerals, ranging from I (minimum) to XII (maximum). The hazard zonations used in the atlas employ a 50-year return period (see Figure 4) because this bears a relationship to the average life expectancy of a modern building (Smolka and Berz, 1981; Munich Re., 1988), and therefore ensures that the hazard is assessed on a time-scale that is meaningful to a range of potential users.

The zonations of the atlas are based upon the intensity/frequency evaluations produced by the Munich Re. (1988) for seismic areas. These evaluations show expected intensities of shaking for average ground conditions ("average" is defined as firm consolidated sediments). This is an important limitation of the zonations in view of the influence exerted by surface geology and ground conditions upon the severity of earthquake hazard (see Table 2). This influence has been manifest in many

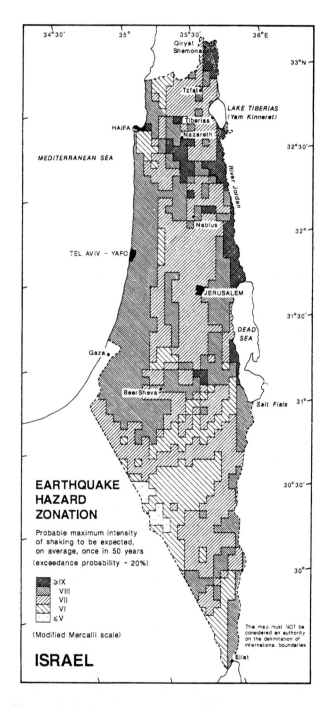

Figure 4 Earthquake hazard map of Israel produced using the ROA scheme.

Table 1 Earthquake loss-susceptibility data for different construction types (MM = Modified Mercalli Scale) (source of data: Sauter and Shah, 1978)

Construction type	Average damage (%) at intensity (MM):				
	VI	VII	VIII	IX	X
1 Adobe	8	22	50	100	100
2 Unreinforced masonry, non-seismic design	3.5	14	40	80	100
3 Reinforced concrete frames, non-seismic design	2.5	11	33	70	100
4 Steel frames, non-seismic design	1.8	6	18	40	60
5 Reinforced masonry, medium quality, non-seismic design	1.5	5.5	16	38	66
6 Reinforced concrete frames, seismic design	0.9	4	13	33	58
7 Shear wall structures, seismic design	0.6	2.3	7	17	30
8 Wooden structures, seismic design	0.5	2.8	8	15	23
9 Steel frames, seismic design	0.4	2	7	20	40
10 Reinforced masonry, high quality seismic design	0.3	1.5	5	13	25

Table 2 Average changes in intensity associated with different types of surface geology (after Munich Re., 1988, p 7)

Subsoil	Average change in intensity
Rock (e.g. granite, gneiss, basalt)	−1
Firm sediments	0
Loose sediments (e.g. sand, alluvial deposits)	+1
Wet sediments, artificially filled ground	+1.5

assigned to a particular square is that which can be expected across the larger part of the unit; to assign a compromise value (i.e. some sort of average) would only lead to misinterpretation.

Earthquake hazard and risk in Israel The hazard zonation of Israel (Figure 4) shows that the maximum intensity of shaking to be expected once in 50 years is IX. It does not attempt to differentiate between intensities greater than IX because such differentiation is extremely difficult, particularly at the macrozonation scale of study (Richter, 1959). Intensities greater than IX will only be experienced during the most severe earthquakes, and will be relatively restricted in extent. Similarly, the zonation does not attempt to distinguish between intensities of less than V. This is because structures that have been reasonably well constructed should

earthquakes, historic and recent (e.g. Celebi, 1987; Degg, 1987, 1989, and this volume; Jarpe and others, 1988; Joyner and others, 1981; Rogers and others, 1984; Tilford and others, 1985). Deviations in ground conditions from the "average", therefore, lead to large variations in exposure to hazard that are not shown in the evaluations of the Munich Re. The atlas zonations are able to overcome some of this limitation by including a consideration of the effects of surface conditions, albeit at the national (macrozonation) scale.

In order to take into account geological effects, the earthquake zonations of the atlas are produced using a grid system similar to that described by Carrara and others (1978) for use in landslide hazard analysis. In the case of the Israel zonation (Figure 4), each grid square represents 25 km². The squares are treated as individual hazard exposure zones to which values for the hazard-controlling elements are assigned and then combined to derive exposure values. In particular, the intensity/frequency data of the Munich Re. are combined with information concerning surface geology and ground conditions in order to arrive at hazard exposure values that more accurately reflect the geological and geomorphological environments of the exposure zones. Problems arise when the surface geology and ground conditions are extremely varied *within* individual grid squares. Under such circumstances, the hazard category

Modified Mercalli Scale

    I Imperceptible

   II Felt on upper floors

  III Objects swing

  IV Felt by everybody, creaking noises

   V Felt by everybody, objects move

  VI Cracks appear in buildings

 VII Fissures in roads, difficult to stand, large bells ring

VIII Steering on cars affected, chimneys and towers fall

  IX Landslides, underground pipes break

   X Most masonry and frame structures destroyed, serious damage to large structures, dams and dykes

  XI Serious damage to large structures and buildings

 XII Total destruction, distortion of the Earth's surface.

99

not show any but the most minor damage when subject to ground motions of this intensity (Richter, 1959). Indeed, the Munich Re. (1976) and Sauter and Shah (1978) demonstrated that significant loss in buildings does not occur until intensity VI-VII is experienced.

By relating the hazard zonation to loss-susceptibility data of the type listed in Table 1, a general indication of the losses to be expected for different construction types can be obtained. In doing so, it should be borne in mind that the hazard zonation is a conservative one with a slight exaggeration of the expected intensity values—such an approach is considered preferable to one that inadvertently underestimates the severity of the hazard.

Figure 5 is a map of population density in Israel. Maps of this type are included in the "National Analyses" section of the atlas in order to allow qualitative evaluations of the distribution of earthquake risk in countries ("earthquake risk" is here defined as the likelihood of loss, whether it be in human or economic terms, due to earthquakes—within an area of high earthquake hazard, earthquake risk can vary according to the distribution, value and vulnerability of elements exposed to the earthquake threat). By comparing Figure 5 with the earthquake hazard zonation (Figure 4) it becomes apparent that earthquake risk in Israel is greatest in the coastal belt; here the combination of a relatively high hazard status and dense concentrations of population (and associated urban and industrial development) give potential for earthquake loss accumulations. The Tel Aviv metropolitan area and the district of Haifa are of particular concern because of their high population densities. In contrast, although the Jordan rift valley (which runs along the eastern margin of the country) and

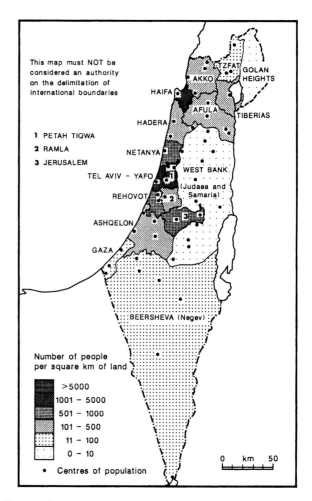

Figure 5 The population density of major districts and sub-districts of Israel (source of data: Central Bureau of Statistics, 1982).

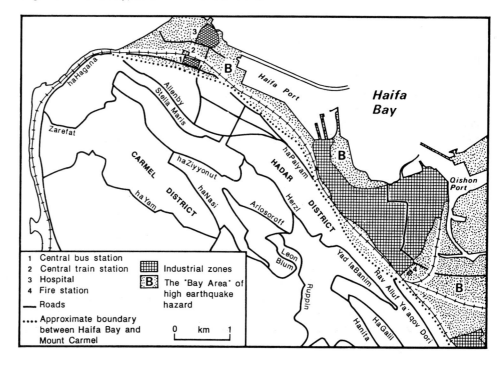

Figure 6 The zone of high earthquake hazard in Haifa.

100

Jezreel valley (which runs in a southeasterly direction from just north of Haifa to the rift valley) are areas of relatively high earthquake hazard, the potential for large loss accumulations in these areas is reduced because they are much less densely populated. However, certain isolated centres of population are of concern, e.g. Tiberias, Eilat and Qiryat Shemona. Earthquake risk in Israel is least in the central southern part of the country where population densities and hazard status are low.

National assessments of this type are invaluable in helping to target hazardous areas within countries where concerted effort is needed to control the escalation of earthquake risk. This will involve the production of large-scale hazard zonations (microzonations) that can be related to existing and proposed developments. A zonation of this type, for the Israeli city of Haifa, is presented in Figure 6. Named streets have been placed on the zonation to enable the boundary between the relatively hazardous bay area and the more secure Carmelite district of the city to be identified with ease. From Figure 4 it can be deduced that the maximum intensity of shaking to be expected (once every 50 years) across the majority of the Mount Carmel region is VI, although earthquake-induced landsliding on some of the steeper slopes could serve to increase this. In the bay area intensities of VIII or greater are to be expected, and there is the added danger of inundation by seismic sea waves across this low-lying part of the city. For example, on 12 October 1856 the area was affected by a seismic sea wave triggered by a high-magnitude earthquake northeast of Crete (Ben-Menahem, 1979). In view of this, it is of concern to note that a number of the essential emergency services upon which Haifa would depend in the event of an earthquake are located in the bay area (see Figure 6). It is to be hoped that the necessary precautions have been taken in the design and construction of these buildings.

In conclusion, the earthquake hazard zonation procedure used in the atlas has a number of limitations, but as a means of producing national hazard zonations (macrozonations) at scales of between 1:500,000 and 1:5,000,000 it has to its advantage the fact that due consideration is paid to surface geology and ground conditions, and the spatial variation that these can introduce into the earthquake hazard. The resultant zonations focus attention upon areas where the hazard is likely to be greatest and where sensible hazard management practice is essential. In order to be truly effective, such management requires the production of refined hazard assessments (e.g. microzonations) at a much larger scale than that of the national analyses, and site-specific evaluations for high-risk projects. The atlas macrozonations nevertheless provide a sound basis for these, and serve to set the national context within which they should be considered.

Part D of the Atlas: historical catalogues

The historical catalogues of the atlas project are produced to provide a basis for the regional hazard evaluations (see Part B above). They are, in part, a response to problems of access and lack of standardisation amongst historical earthquake data. During the present century there has been considerable research concerning aspects of historical seismicity and related hazards for many parts of the world; much of this has involved the analysis of ancient chronicles and religious texts (examples of such work for the Middle East include: Alsinawi and Ghalib, 1975; Ambraseys, 1961, 1962; Ambraseys and Melville, 1982, 1983, 1989; Amiran, 1951, 1952; Ben-Menahem, 1979; Maamoun, 1979; Poirier and Taher, 1980; Poirier and others, 1980; Russell, 1985; Stothers and Rampino, 1983; Willis 1928, 1933). To date, however, the potential of much of the published data has not been exploited to the full, often because of their relative inaccessibility and the time involved in bringing them together. Added to these problems are the confusing discrepancies that exist between some of the published materials, which have served to dissuade individuals from using them, and led to an over-reliance upon 20th century data (which, because they are instrumentally recorded, are often easier to obtain—see Part B above). The atlas project has sought to overcome some of these problems through the production of regional computerised hazard databases, subsequently reproduced in catalogue form in the atlas. Each provides an ordered inventory of the published information concerning pre-20th century seismic activity and associated hazards in a region. The major advantages of the work are:

(a) It has provided a means of introducing homogeneity into the historical data and thereby served to focus attention upon inadequacies and gaps within them;
(b) It has enabled some errors in previous catalogues to be identified and discrepancies corrected;
(c) The regional nature of the databases has allowed the areal extent of some historical earthquakes to be ascertained with greater accuracy, through the amalgamation of reports from different countries within regions;
(d) The databases have provided a sound basis for regional analyses of the earthquake hazard. To date, these analyses have largely concentrated upon spatial aspects of the hazard (e.g. see Figure 3)—the historical data have been used to complement 20th-century seismic data in regional evaluations of the distribution of earthquake activity and related hazards (e.g. see Degg, 1988; Degg and Doornkamp, 1989);
(e) The catalogues in the atlas provide a convenient source of information concerning earthquake effects. This is of considerable value to hazard assessment, particularly at the microzonation scale of study

where data concerning earthquake-related hazards (e.g. landsliding, seismic sea waves, liquefaction) experienced at specific locations may be required.

The databases are of a standard format, with information for each earthquake entry recorded under the following headings:

Date, Epicent(re), **Fault, Intens**(ity), **Mag**(nitude), **Felt in, Details, Hazards, Source, Comment**.

Figure 7 serves to illustrate the types of information recorded—a summary of published facts concerning each event is presented, together with a list of the sources from which the data have been abstracted.

It must be emphasised that the compilation of the databases has not involved the analysis of ancient documents and manuscripts; such work is best left to specialists in regional antiquity. The compilation has sought merely to provide intelligible, standardised summaries of the historical data already published for regions. Ambraseys and others (1983) and Ambraseys and Melville (1988), amongst others, have stressed that these data often contain errors that have been passed on from one catalogue to the next so that they are pervasive throughout the published literature. The errors are frequently in the form of incorrect earthquake dates or duplications of earthquake entries. The use of ordered database structures to assimilate and sort historical data enables many errors of this type to be detected and corrected, but others are virtually impossible to rectify using secondary sources and require recourse to original manuscripts. The effect of such errors upon the type of spatial analysis to which the catalogues in the atlas have been subjected (see Figure 3) is likely to be minimal, but it is to be hoped that the databases can serve a useful purpose in helping to identify areas where confusion exists and where renewed historical research is required.

Conclusion

A problem encountered in earthquake hazard assessment is that whilst the data needed to ensure sensible hazard management often exist, they are not always readily accessible to those who require them and are in a position to encourage sound hazard management practice (e.g. planners, policy makers, direct insurers and reinsurers, engineers, environmental agencies and educationalists). This paper has described the *Earthquake Hazard Atlas* project, a new research initiative funded by reinsurance companies; its main aim is to present standardised data on the earthquake hazard in a format that enables them to be accessed and incorporated into hazard and risk assessments with

DATE	AD 1546 Jan 14
EPICENT.	NEAR DAMIYE, JORDAN (32°N 35.5°E)
FAULT	DEAD SEA
INTENS.	I_0 = X-XI
MAG.	M_L ≥6.8
FELT IN	ISRAEL, JORDAN, SYRIA
DETAILS	AFFECTED JERUSALEM (12 DEAD), GAZA, RAMLE, JERICHO, TIBERIAS, HEBRON (16 DEAD & 70 INJURED), ES-SALT, NABLUS (300-500 DEAD), KARAK, DAMASCUS & JAFFA
HAZARDS	A LANDSLIDE OF THE LISAN MARLS ABOVE JISRED DAMIYE BLOCKED THE RIVER JORDAN FOR 2 DAYS. A FAULT EXTENDED FROM THE VICINITY OF DAMIYE TO THE DEAD SEA & A SEICHE OCCURRED IN THE DEAD SEA. THERE WAS A SEISMIC SEA WAVE AT JAFFA (AND ALONG THE COASTS OF ISRAEL AND LEBANON), PRIOR TO WHICH THE SEA AT JAFFA RETREATED A CONSIDERABLE DISTANCE FROM THE SHORE
SOURCE	6,7,9,12,13,14,17,19,22,23,30
COMMENT	THE MAIN SHOCK WAS FOLLOWED BY A NUMBER OF SMALLER AFTERSHOCKS. IN ADDITION TO NABLUS, (23) CITES SICHEM (SHECHEM?) AS BEING AFFECTED. ALTHOUGH MANY CONTEMPORARY CHRONICLERS EQUATE SHECHEM WITH NABLUS, SHECHEM CORRECTLY = A SMALL SETTLEMENT TO THE E.S.E. OF NABLUS, NOW IDENTIFIED WITH THE ARAB VILLAGE OF 'ASKAR (41)

Figure 7 Format of the earthquake databases.

relative ease, and hopefully ensures their usefulness to a wide range of people.

The structure of the atlas serves to emphasise the value of an appreciation of scale to earthquake hazard assessment: in spatial terms, the importance of understanding the regional context of the hazard before progressing to more refined studies at the national (macrozonation) and local (microzonation) scale; in temporal terms, the importance of incorporating historical data into hazard assessments, thereby ensuring that the hazard is evaluated on a time-scale appropriate to that of the geological processes under analysis.

New atlas parts are currently under research, and it is intended that with time the atlas will build up to provide comprehensive, standardised coverage for all seismic regions of concern.

Acknowledgements

The author is indebted to the Reinsurance Offices Association of London for funding much of the work on which this paper is based. Also to Dr J.C. Doornkamp, and to Dr R.W. Alexander for helpful comments concerning an earlier version of the text.

References

Adams, R.D., and M. Barazangi, 1984. Seismotectonics in the Arab region: a brief summary and future plans. *Bulletin of the Seismological Society of America*, vol 74 (3), p 1011–1030.

Algermissen, S.T., and K.V. Steinbrugge, 1984. Seismic hazard and risk assessment: some case studies. *The Geneva Papers on Risk and Insurance*, vol 9 (30), p 8–26.

Alsinawi, S.A., and H.A.A. Ghalib, 1975. Historical seismicity of Iraq. *Bulletin of the Seismological Society of America*, vol 65 (5), p 541–547.

Ambraseys, N.N., 1961. On the seismicity of south-west Asia. *Revue pour l'étude des calamites*, vol 37 (7), p 18–30.

Ambraseys, N.N., 1962. Data for the investigation of seismic sea waves in the eastern Mediterranean. *Bulletin of the Seismological Society of America*, vol 52 (4), p 895–913.

Ambraseys, N.N., 1971. Value of historical records of earthquakes. *Nature*, vol 232, p 375–379.

Ambraseys, N.N., 1975. Studies in historical seismicity and tectonics. In: *Geodynamics Today: A Review of the Earth's Dynamic Processes*. The Royal Society, London, p 7–16.

Ambraseys, N.N., 1989. Temporary seismic quiescence: SE Turkey. *Geophysical Journal*, vol 96, p 311–331.

Ambraseys, N.N., and M. Barazangi, 1989. The 1759 earthquake in the Bekaa valley: implications for earthquake hazard assessment in the eastern Mediterranean region. *Journal of Geophysical Research*, vol 94 (B4), p 4007–4013.

Ambraseys, N.N., and C.P. Melville, 1982. *A History of Persian Earthquakes*. Cambridge University Press.

Ambraseys, N.N., and C.P. Melville, 1983. Seismicity of Yemen. *Nature*, vol 303, p 321–323.

Ambraseys, N.N., and C.P. Melville, 1988. Analysis of the eastern Mediterranean earthquake of 20 May 1202. In: W.H. Lee, H. Meyers, and K. Shimazaki (eds), *Historical Seismograms and Earthquakes of the World*. Academic Press, San Diego, p 181–200.

Ambraseys, N.N., and C.P. Melville, 1989. Evidence for intraplate earthquakes in northwest Arabia. *Bulletin of the Seismological Society of America*, vol 79 (4), p 1279–1281.

Ambraseys, N.N., and others, 1983. Notes on historical seismicity. *Bulletin of the Seismological Society of America*, vol 73 (6), p 1917–1920.

Amiran, D.H.K., 1951. A revised earthquake catalogue of Palestine. *Israel Exploration Journal*, vol 1 (4), p 223–246.

Amiran, D.H.K., 1952. A revised earthquake catalogue of Palestine. *Israel Exploration Journal*, vol 2, p 48–65.

Arya, A.S., L.S. Srivastava and S.P. Gupta, 1985. Survey of damages during the Dhamar earthquake of 13 December 1982 in the Yemen Arab Republic. *Bulletin of the Seismological Society of America*, vol 75 (2), p 597–610.

Ben-Menahem, A., 1979. Earthquake catalogue for the Middle East (92 BC–1980 AD). *Bollettino di Geofisica Teorica ed Applicata*, vol 21 (84), p 245–310.

Carrara, A., E. Catalano, M. Sorriso Valvo, C. Reali, and I. Osso, 1978. Digital terrain analysis for land evaluation. *Geologia Applicata e Idrogeologia*, vol 13, p 69–127.

Celebi, M., 1987. Topographical and geological amplifications determined from strong-motion and aftershock records of the 3 March 1985 Chile earthquake. *Bulletin of the Seismological Society of America*, vol 77, p 1147–1167.

Central Bureau of Statistics, 1982. *Statistical Abstract of Israel*, No. 33. Hed Press, Jerusalem.

Degg, M.R., 1987. The 1985 Mexican earthquake. *Modern Geology*, vol 11, p 109–131.

Degg, M.R., 1988. *Earthquake Hazard in the Middle East: An Evaluation for Insurance and Reinsurance Purposes*. Unpublished PhD thesis: Department of Geography, University of Nottingham.

Degg, M.R., 1989. Earthquake hazard assessment after Mexico (1985). *Disasters*, vol 13 (3), p 237–246.

Degg, M.R., 1990. A database of historical earthquake activity in the Middle East. *Transactions of the Institute of British Geographers*, vol 15 (3), p 294–307.

Degg, M.R., and J.C. Doornkamp, 1989. *Earthquake Hazard Atlas: 1. Israel*. Reinsurance Offices Association, London.

Degg, M.R., and J.C. Doornkamp, 1990. *Earthquake Hazard Atlas: 2. Egypt*. Reinsurance Offices Association, London.

El-Isa, Z.H., and A. Al Shanti, 1989. Seismicity and tectonics of the Red Sea and western Arabia. *Geophysical Journal*, vol 97 (3), p 449–457.

Jarpe, S.P., C.H. Cramer, B.E. Tucker, and A.F. Shakal, 1988. A comparison of observations of ground response to weak and strong ground motion at Coalinga, California. *Bulletin of the Seismological Society of America*, vol 78 (2), p 421–435.

Joyner, W.B., R.E. Warrick, and T.E. Fumal, 1981. The effect of Quaternary alluvium on strong ground motion in the Coyote Lake, California, earthquake of 1979. *Bulletin of the Seismological Society of America*, vol 71, p 1333–1349.

Kashai, E.L., and P.F. Croker, 1987. Structural geometry and evolution of the Dead Sea-Jordan rift system as deduced from new subsurface data. *Tectonophysics*, vol 141, p 33–60.

Le Pichon, X., and J. Francheteau, 1978. A plate tectonic analysis of the Red Sea-Gulf of Aden area. *Tectonophysics*, vol 46, p 369–406.

Lovelock, P.E.R., 1984. A review of the tectonics of the northern Middle East region. *Geological Magazine*, vol 121 (6), p 577–587.

Lyberis, N., 1988. Tectonic evolution of the Gulf of Suez and the Gulf of Aqaba. *Tectonophysics*, vol 153 (1–4), p 209–220.

Maamoun, M., 1979. Macroseismic observations of principal earthquakes in Egypt. *Bulletin of the Helwan Institute of Astronomy and Geophysics*, vol 183.

Mart, Y., and P.D. Rabinowitz, 1986. The northern Red Sea and the Dead Sea rift. *Tectonophysics*, vol 124, p 85–113.

McKenzie, D.P., 1970. The plate tectonics of the Mediterranean region. *Nature*, vol 226, p 239–243.

McKenzie, D.P., D. Davies, and P. Molnar, 1970. The plate tectonics of the Red Sea and East Africa. *Nature*, vol 226, p 243–248.

Munich Re., 1976. *Guatemala '76—Earthquakes of the Caribbean Plate*. The Munich Reinsurance Company, Munich.

Munich Re., 1978. *World Map of Natural Hazards*. The Munich Reinsurance Company, Munich.

Munich Re., 1988. *World Map of Natural Hazards* (2nd revised edition). The Munich Reinsurance Company, Munich.

Munich Re., 1990. *Windstorm—New Loss Dimensions of a Natural Hazard*. The Munich Reinsurance Company, Munich.

Nowroozi, A.A., 1971. Seismotectonics of the Persian plateau, eastern Turkey, Caucasus and Hindu Kush regions. *Bulletin of the Seismological Society of America*, vol 61, p 317–341.

Nur, N., and Z. Ben-Avraham, 1978. The eastern Mediterranean and the Levant: tectonics of continental collision. *Tectonophysics*, vol 46, p 297–311.

Papazachos, B.C., and P.E. Comninakis, 1978. Deep structure and tectonics of the eastern Mediterranean. *Tectonophysics*, vol 46, p 285–296.

Plafker, G., R. Agar, A.H. Asker, and M. Hanif, 1987. Surface effects and tectonic setting of the 13 December 1982 North Yemen earthquake. *Bulletin of the Seismological Society of America*, vol 77 (6), p 2018–2037.

Poirier, J.P., and M.A. Taher, 1980. Historical seismicity in the Near and Middle East, North Africa and Spain from Arabic documents (VIIth–XVIIIth century). *Bulletin of the Seismological Society of America*, vol 70 (6), p 2185–2201.

Poirier, J.P., B.A. Romanowicz, and M.A. Taher, 1980. Large historical earthquakes and seismic risk in Northwest Syria. *Nature*, vol 285, p 217–220.

Riad, S., and others, 1985. *Seismicity of the Middle East 1900–1983*. Map published by the National Geophysical Data Center and World Data Center A for Solid Earth Geophysics, Boulder, Colorado.

Richter, C.F, 1959. Seismic regionalisation. *Bulletin of the Seismological Society of America*, vol 49 (2), p 123–162.

Rogers, A.M., R.D. Borcherdt, P.A. Covington, and D.M. Perkins, 1984. A comparative ground response study near Los Angeles using recordings of Nevada nuclear tests and the 1971 San Fernando earthquake. *Bulletin of the Seismological Society of America*, vol 74, p 1925–1949.

Rotstein, Y., and Z. Ben-Avraham, 1986. Active tectonics in the eastern Mediterranean: the role of oceanic plateaus and accreted terrains. *Israel Journal of Earth Sciences*, vol 35, p 23–39.

Russell, K.W., 1985. The earthquake chronology of Palestine and Northwest Arabia from the 2nd through the Mid-8th century AD. *Bulletin of the American Schools of Oriental Research*, vol 260, p 37–59.

Sauter, F., and H.C. Shah, 1978. *Estudio de Seguro Contra Terremoto*. Instituto Nacional de Seguros, San Jose (Costa Rica).

Sengor, A.M.C., 1979. The North Anatolian transform fault: its age, offset and tectonic significance. *Journal of the Geological Society of London*, vol 136, p 269–282.

Smolka, A., and G. Berz, 1981. Methodology of hazard mapping—requirements of the insurance industry. *Bulletin of the International Association of Engineering Geologists*, vol 23, p 21–24.

Stothers, B., and M.R. Rampino, 1983. Volcanic eruptions in the Mediterranean before AD 630 from written and archaeological sources. *Journal of Geophysical Research*, vol 88 (B8), p 6357–6371.

Swiss Re., 1978. *Atlas on Seismicity and Volcanism*. The Swiss Reinsurance Company, Zurich.

Swiss Re., 1989. *Natural Hazard and Event Loss*. The Swiss Reinsurance Company, Zurich.

Tilford, N.R., U. Chandra, D.C. Amick, R. Moran, and F. Snider, 1985. Attenuation of intensities and effect of local site conditions on observed intensities during the Corinth, Greece, earthquakes of 24 and 25 February and 4 March 1981. *Bulletin of the Seismological Society of America*, vol 75 (4), p 923–937.

Willis, B., 1928. Earthquakes in the Holy Land. *Bulletin of the Seismological Society of America*, vol 18 (2), p 73–103.

Willis, B., 1933. Earthquakes in the Holy Land—a correction. *Bulletin of the Seismological Society of America*, vol 23, p 88–89.

11 Some implications of the 1985 Mexican earthquake for hazard assessment

M.R. Degg

Abstract The 1985 Mexican earthquake ranks amongst the major earthquake disasters of the twentieth century. One of the few positive aspects of the disaster is that it provided massive quantities of data concerning seismic effects and earthquake damage. Every opportunity should be taken to incorporate the findings from these data into earthquake hazard and risk assessments. The purpose of this paper is to summarise some important lessons from the earthquake, paying particular attention to their implications for seismic regions of the Third World.

The paper arises from field investigations conducted by the author on the behalf of reinsurance companies.

Introduction

On September 19th, 1985, a magnitude (M_s) 8.1 earthquake occurred off the Pacific coastline of Mexico (see Figure 1). It was followed by a $M_s = 7.6$ event on September 21st (Singh and others, 1988). The earthquake and its aftershocks are estimated to have killed at least 10,000 people, injured 50,000 and made 250,000 homeless (see Figure 2). In addition, the event caused US$4000 million worth of damage of which approximately US$275 million (i.e. 7%) was insured, making it one of the most disastrous ever to have affected the insurance industry (Munich Reinsurance Company (Munich Re.), 1986).

Few earthquake disasters have had as profound and far-reaching implications as the Mexican one. It demon-

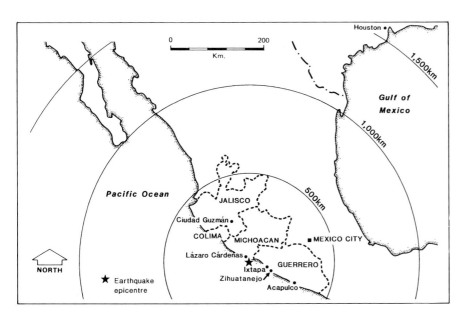

Figure 1 Areas affected by the 1985 Mexican earthquake.

105

strated, probably more convincingly than any previous event of its type, just how vulnerable a modern high-rise city can be to a distant earthquake. The earthquake caused considerable damage in a number of the western coastal states of Mexico (see Figure 1), but by far the greatest losses occurred in Mexico City itself, despite the fact that the city was 370 km from the earthquake epicentre (see Figure 3). Several days after the event, the Institute of Engineering at the National Autonomous University of Mexico (Universidad Nacional Autonoma de Mexico, UNAM) estimated that 65 km² of the city had experienced significant earthquake damage; of this, approximately 23 km² showed a high density of damage (UNAM, 1985). The Munich Re. (1986) determined that 7400 buildings in Mexico City were damaged by the earthquake, of which 770 were total losses and 1665 severely damaged. This represents a very small percentage of the estimated one-million-plus buildings of the city at the time of the earthquake, but nevertheless testifies to the fact that earthquake hazard and risk assessments now need to consider the threat posed to cities many hundreds of kilometres away from the most active seismic belts.

Subsequent to the earthquake, field work in Mexico helped to identify many of the factors that served to control the severity of the earthquake impact. Information of this type, which leads to a greater understanding of earthquakes and their effects, is a primary requirement in the delineation of earthquake hazard and risk. The major findings of the analysis can be

Figure 2 "Disinfected" notice on a collapsed building in Mexico City. Due to the difficulty of extricating bodies from some of the most severe building failures, many were sprayed with disinfectant in the weeks and months following the earthquake (whilst awaiting demolition) in order to prevent disease.

categorised under the following headings, each of which is discussed in further detail below:

(a) the influence of subsoil on the severity of ground motion;
(b) the vulnerability of buildings to damage;
(c) the importance of building-subsoil interaction;
(d) earthquake recurrence; and
(e) escalating vulnerability to catastrophic loss.

Figure 3 November, 1985—severe damage in the part of Mexico City worst affected by the earthquake.

The influence of subsoil on the severity of ground motion

Throughout the areas worst affected by the earthquake (see Figure 1), the influence of surface geology in controlling the severity of ground motion was most apparent. Along the coastal belt, the motions and damage recorded on unconsolidated sediments were greater than those on crystalline bedrock. However, it was in Mexico City that the importance of subsoil was most cruelly and convincingly displayed. Analysis of the distribution of damage in the city showed that it was almost exclusively restricted to saturated clay deposits (the Tacubaya clays) in the western part of an old lake bed (see Figures 4, 5 and 6). These deposits are characterised by low natural frequencies of vibration (Meli and others, 1985), and were therefore excited by the predominantly low-frequency seismic energy experienced at Mexico City during the earthquake —high frequency energy tends to be attenuated rapidly with increasing distance from an earthquake epicentre.

Instrumental recordings show that the clays of the lake zone served to amplify the shock waves by between 8 and 50 times compared to motions recorded on solid rock in adjacent areas (Booth and others, 1986; Singh and others, 1988).

Within the investigation area shown on Figure 4, detailed mapping of the location of damaged buildings revealed that the most severely affected ones (e.g. see Figures 3, 7 and 8) were largely confined to that part of the lake bed where the clay thickness exceeded 37 m (see Figure 9). In addition, a strong correlation was established between clay thickness and damage density (see Figure 10) suggesting that the severity of earthquake ground motion across the western part of the lake bed increased in relation to the thickness of the Tacubaya clays (Degg, 1987).

All in all, the Mexican damage experience demonstrates that within the confines of an individual city exposure to the earthquake hazard can vary considerably, and emphasises the value of detailed analysis of the geological conditions that underlie major cities in seismic zones. Through such analysis it is possible to identify subsoil units that are likely to increase the destructive forces of earthquakes. Data concerning these units should be incorporated into hazard assessments and used as a basis for implementing measures aimed at reducing the severity of earthquake impact (e.g. plan-

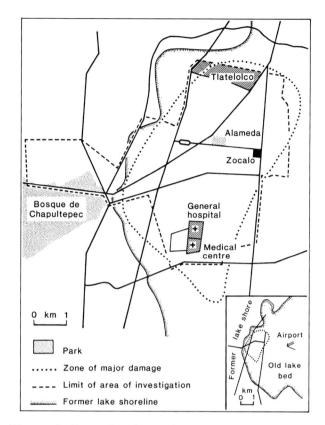

Figure 4 Zone of major earthquake damage in Mexico City.

Figure 5 The *tilting* of an old church (Iglesia de Loreto) —note the angle relative to the lamp post—in the lake zone of Mexico City demonstrates the effects of differential settlement into the lake-bed clays. Settlement of this type probably served to increase the vulnerability of some buildings to the earthquake.

107

Figure 6
Earthquake-induced settlement of a building into the thick, compressible clays of the lake zone.

ning, design and construction regulations). Such work is urgently required in a number of seismic regions of the Third World, where inadequate earth-science data for areas of rapid population growth and urban expansion is a cause of genuine concern.

The vulnerability of buildings to damage

Within the investigation area (see Figure 4) a number of transects were taken across the zone of major damage in order to examine the performance of buildings during

Figure 7 Failure of a reinforced concrete frame building on the lake bed of Mexico City through "top-down" collapse. Medium to high-rise buildings (6–20 storeys) were particularly badly affected.

Figure 8 An example of "pancake" collapse in a reinforced concrete frame building.

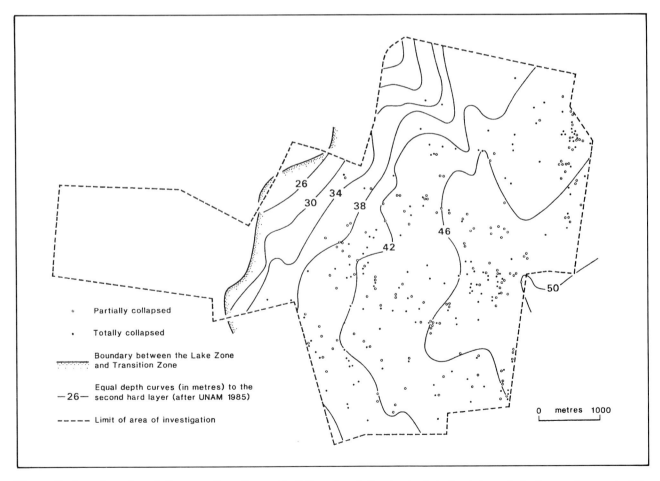

Legend:

○ Partially collapsed

• Totally collapsed

Boundary between the Lake Zone and Transition Zone

—26— Equal depth curves (in metres) to the second hard layer (after UNAM 1985)

- - - - - Limit of area of investigation

0 metres 1000

Figure 9 Location of partially or totally collapsed buildings in relation to depth to hard deposits in the western part of the lake zone of Mexico City.

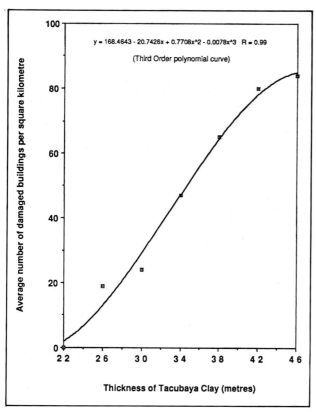

y = 168.4643 - 20.7426x + 0.7708x^2 - 0.0078x^3   R = 0.99

(Third Order polynomial curve)

Figure 10 Density of damage related to clay thickness in the western part of the lake zone of Mexico City.

the earthquake. Along the transects each and every building was examined; types and heights of construction were noted, together with information concerning building damage. The results of the survey are summarised in Figure 11. This shows that on the lake bed, rigid structures (e.g. stone masonry buildings) generally performed better than relatively flexible ones (e.g. many of the reinforced concrete structures).

The greatest single influence on building vulnerability, however, was height of construction (see Figures 7 and 12). Medium to high-rise buildings between 6 and 20 storeys were worst affected, with those between 9 and 11 storeys experiencing the highest incidence of damage. Figure 13 serves to emphasise that the area of major damage on the lake bed was largely confined to that part where the density of medium to high-rise buildings was greatest (i.e. Building Elevation Zones I and II on Figure 13). Damage was considerably restricted in Building Elevation Zone III where 98.5% of the buildings were less than 6 storeys high. Unfortunately no data were available for the remainder of the lake bed, although reconnaissance of the area to the east of Building Elevation Zone III revealed that the vast majority of buildings were low-rise at the time of the earthquake (Degg, 1987).

Medium to high-rise buildings tend to have lower natural frequencies of vibration than low-rise ones. They were therefore much more likely to be sensitive to, and "in-tune" with, the low-frequency ground motions

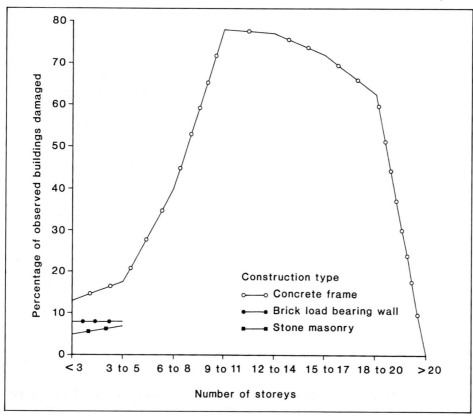

Figure 11 Percentages of buildings damaged according to type and height of construction in the lake zone of Mexico City.

Figure 12 Low-rise buildings to the west of the Zocalo (see Figure 4). Although situated within the zone of major damage, the majority of well constructed and maintained buildings of this type were largely unaffected by the earthquake.

experienced in Mexico City during the earthquake (see Figure 14). This had the effect of causing many of the buildings to resonate, thereby prolonging and reinforcing the vibration within them. This was heightened in the lake zone where the ground motions were amplified to such a considerable extent by the lake-bed clays (Seligman and others, 1989). It is beyond doubt that, had there been no high-rise structures on the lake bed, the amount of damage experienced during the earthquake would have been drastically reduced.

The sensitivity of high-rise buildings to low-frequency seismic energy emanating from a distant earthquake source is something that is becoming of increasing relevance and concern. Not least of all this is because there has been, in recent years, a dramatic increase in the number of high-rise cities, many of which are situated in seismic belts or along their margins. The increased (building) elevation of cities means that some may start to "feel" earthquakes and experience earthquake damage for the first time in their history. Others, such as Mexico City, will find themselves increasingly vulnerable to larger and more frequent losses than those experienced in the past.

The importance of building–subsoil interaction

The resonance coupling experienced in Mexico City between earthquake shock waves, lake-bed clays and

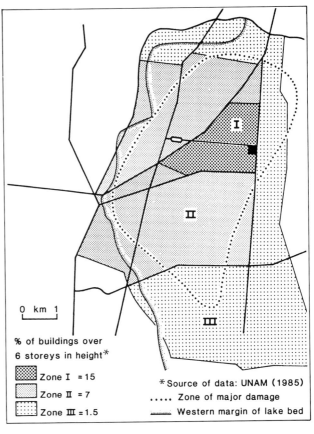

% of buildings over
6 storeys in height*

Zone I = 15
Zone II = 7
Zone III = 1.5

*Source of data: UNAM (1985)
..... Zone of major damage
—·—·— Western margin of lake bed

0 km 1

Figure 13 Zones of building elevation in relation to the distribution of major damage in the lake zone of Mexico City.

Figure 14 Damage caused by buffeting due to inadequate spacing between adjacent tall buildings in the lake zone. The unequal elevation of the two buildings may have given them slightly different natural frequencies of vibration and served to enhance the relative motion between them during the earthquake.

medium to high-rise buildings, highlights the importance in seismic regions of trying to relate the dynamic characteristics of a building to those of the subsoil on which it is situated. The vulnerability of a structure to damage is considerably enhanced if the natural frequency of vibration of the subsoil and that of the structure coincide.

In that context, although the lake-bed clays of Mexico City represent an extreme example, many of the world's major cities stand upon poor subsoil conditions. These often comprise unconsolidated sediments, occasionally water-saturated, associated with river valleys or coastal plains (i.e. areas that have provided the natural resources needed to sustain large sedentary populations). In many cases, however, the possibility of earthquake resonance couplings of the type experienced in Mexico City has not yet been examined.

Analysis of earthquake recurrence

Figure 15 shows the results of an analysis of the distribution of epicentres of major twentieth century earthquakes in Mexico. The Central American Trench delineates a plate boundary, formed where the Cocos Plate (to the south-west) is subducting beneath the much larger North American Plate (to the north-east). The trench forms part of the so-called Pacific "ring of fire", and is one of the most active plate boundaries in the world. Recurrence intervals for large earthquakes along any given section of the trench vary, but are between 32 and 56 years on average (Singh and others, 1981). However, Figure 15 shows that the section that generated the 1985 earthquake was one that had not experienced a major seismic event for a much longer time period. It formed part of a well-known seismic gap, the "Michoacan gap". Gaps of this type in recorded seismic activity can sometimes be taken to indicate a greater risk of large earthquake occurrence, because stresses and strains caused by plate movement have been able to accumulate for a long time without release. Considerable attention in Mexico is now focussed on the Guerrero seismic gap, which lies to the southeast of the 1985 rupture (see Figure 15).

This type of analysis is invaluable in helping to delineate the distribution of earthquake hazard and risk in a region. It often serves to identify spatial and temporal patterns in earthquake activity that can then be used as a basis for predicting, in general terms, the potential for major seismic activity.

Escalating vulnerability to catastrophic loss

A final major lesson from the Mexican earthquake concerns the dangers associated with the over-concentration of people and investment in seismic areas. All other considerations being equal, the larger the number of people and the greater the economic wealth in earthquake regions, the greater the potential for catastrophic loss. These influences are compounded in many Third World regions by a lack of capital to invest in hazard assessment and mitigation measures. Indeed, it is the combination of high hazard exposure, over-concentration of people and economic investment, and the effects of a crippling national debt that serve to make Mexico City the world's archetypal vulnerable city. The city and its metropolitan area accommodate roughly 18 million people (over 20% of the total population of Mexico) in only 0.1% of the nation's land area. Furthermore, the population of the city is continuing to increase at an alarming rate (projected average growth rate for 1985–2000 of 2.56% per annum) and will exceed 24 million by the year 2000 (United Nations, 1989).

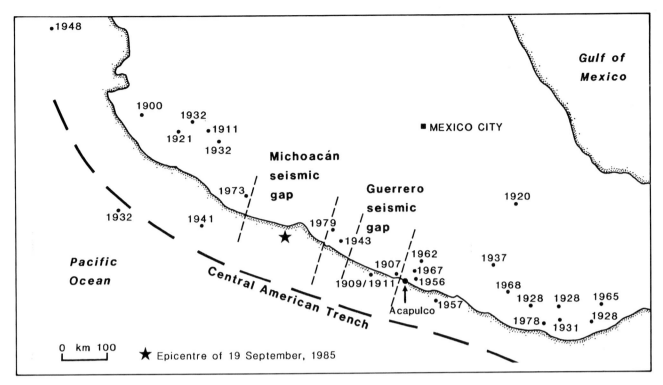

Figure 15 Epicentres of 20th-century Mexican earthquakes with Richter magnitude (M_s) greater than 7.75 (data from Rinehart and others, 1982).

Surrounded by shanty towns and slums, the city serves to symbolise all the dangers of uncontrolled urban growth.

Cities of this type are unfortunately becoming characteristic of many parts of the Third World, where urban primacy on a grand scale is now a severe problem. In the mid-1980s the United Nations listed 34 metropolitan areas with populations greater than 5 million, of which 22 (65%) were in the Third World (Fox and Carroll, 1984). By the year 2025 they estimate that 93 cities will exceed this size, the majority of them (86%) in the Third World. From the work of Bilham (1988) it can be determined that of these rapidly expanding Third World cities, approximately 41% are within 200 km of the location of a major historical earthquake (associated with fatalities in excess of 9000) and/or a plate boundary with the potential to generate magnitude 7.0 earthquakes. Clearly, it is only through concerted international co-operation that the loss-inflicting potential of future earthquakes in these areas can be reduced to levels that are socially and economically acceptable.

Conclusions

Probably the only positive aspect of natural catastrophes such as the 1985 Mexican earthquake is that they provide an opportunity to acquire data that otherwise would be unobtainable. These data should then be incorporated into hazard and risk assessments and used as a basis for implementing measures aimed at reducing the impact of similar events in the future. In that context, major lessons from the Mexican earthquake include the following:

1. The disaster emphasised that high-rise buildings are often sensitive to earthquakes over much greater distances than low-rise ones. This has important implications in view of the ever increasing number of "skyscraper" cities around the world. It suggests that earthquake hazard and risk assessments now need to consider the threat posed to cities many hundreds of kilometres away from the most active seismic belts.
2. The disaster highlighted, once again, the strong influence that subsoil conditions exert on the severity of the earthquake hazard, and emphasised the need for detailed examinations of the geological conditions underlying all major cities in seismic zones.
3. It furnished new information concerning building vulnerability to earthquake ground motions. Vulnerability varied considerably according to type and height of construction, and the nature of the underlying subsoil. The findings emphasise the importance in seismic regions of trying to relate the dynamic characteristics of a building to those of the subsoil on which it is situated. The vulnerability of a structure to damage is considerably enhanced if the natural

frequency of vibration of the subsoil and structure coincide.

4. The earthquake emphasised the value of analysis of the earthquake histories of seismic regions. Analysis of this type may reveal spatial and temporal patterns in earthquake activity that can be used as a basis for evaluating seismic potential.

5. The earthquake served to reiterate the dangers associated with the over-concentration of people and economic investment in areas that are exposed to severe natural hazards. This is of particular relevance to many of the rapidly urbanising countries of the Third World, where concerted international effort is now required to stem escalating vulnerability to loss.

Acknowledgements

At the suggestion of the editors, this paper is a rewritten version of a paper which first appeared in *Disasters* (Degg, 1989). The author and the editors gratefully acknowledge the editorial board of that journal for permission to republish the work.

The author is grateful to the Reinsurance Offices Association of London for funding and facilitating the research, to the many individuals and companies in Mexico City who provided information and assistance in the months immediately following the earthquake, and to Professor J.P. Cole and Dr J.C. Doornkamp for their helpful advice and comments.

References

Bilham, R., 1988. Earthquakes and urban growth. *Nature*, 336 (December 15), p 625–626.

Booth, E.D., J.W. Pappin, J.H. Mills, M.R. Degg, and R.S. Steedman, 1986. *The Mexican Earthquake of 19th September, 1985*. A field report by Earthquake Engineering Field Investigation Team (EEFIT): Society for Earthquakes and Civil Engineering Dynamics, London.

Degg, M.R., 1987. The 1985 Mexican earthquake. *Modern Geology*, 11, p 109–131.

Degg, M.R., 1989. Earthquake hazard assessment after Mexico (1985). *Disasters*, vol 13 (3), p 237–246.

Fox, R.W., and A. Carroll, 1984. The world's urban explosion. *National Geographical Magazine*, vol 166 (2), p 176–185.

Meli, R., and others, 1985. *Efectos de los Sismos de Septiembre de 1985 en las Construcciones de la Ciudad de Mexico: Aspectos Estructurales (Effects of the earthquakes of September 1985 on Buildings in Mexico City: Structural Aspects)*. Second report of the Institute of Engineering, Universidad Nacional Autonoma de Mexico (UNAM), Mexico City.

Munich Re., 1986. *Earthquake Mexico '85*. The Munich Reinsurance Company, Munich.

Rinehart, W., R. Ganse, P. Teik, E. Arnould, C. Stover, and R.W. Smith, 1982. *Seismicity of Middle America*. Map published by the National Geophysical Data Center and National Earthquake Information Service, Boulder, Colorado, USA.

Seligman, T.H., J.M. Alvarez-Tostado, J.L. Mateos, J. Flores, and O. Novaro, 1989. Resonant response models for the Valley of Mexico—I: the elastic inclusion approach. *Geophysical Journal International*, vol 99, p 789–799.

Singh, S.K., L. Astiz and J. Havskov, 1981. Seismic gaps and recurrence periods of large earthquakes along the Mexican subduction zone: A reexamination. *Bulletin of the Seismological Society of America*, vol 71 (3), p 827–843.

Singh, S.K., E. Mena and R. Castro, 1988. Some aspects of source characteristics of the 19 September 1985 Michoacan earthquake and ground motion amplification in and near Mexico City from strong motion data. *Bulletin of the Seismological Society of America*, vol 78 (2), p 451–477.

UNAM, 1985. *Effects of the September 19, 1985, Earthquake in the Buildings of Mexico City*. Preliminary report by the Institute of Engineering (October, 1985), Universidad Nacional Autonoma de Mexico (UNAM), Mexico City.

United Nations, 1989. Prospects of world urbanization, 1988. *Population Studies* No 112, United Nations, New York.

Part Three

Landslide Hazards

12 Landslide hazard assessment in the context of development

D.K.C. Jones

Abstract Landsliding continues to be under-recognised as a major hazard, due to human perception of a phenomenon that is diverse in character and ubiquitous in occurrence but rarely disastrous, so that impacts are frequent, small-scale and undramatic; their cumulative effect is therefore rarely appreciated within the burden of 'natural tax'.

Examples from southern Italy are used to illustrate the potential significance of the landslide hazard. The reasons why landslide hazard assessment should be given more emphasis in development programmes are outlined and the various forms of hazard assessment discussed under the headings direct mapping, indirect mapping, land systems mapping, risk mapping and landslide susceptibility mapping. The conclusions focus on the paradox that satisfactory investigation techniques are available which are both relatively accurate and cost-effective, but remain little used because decision-makers continue to be unclear as to the problem, uncertain of the cause and unconvinced that crisis response is not only easier but less problematic than anticipatory planning. The urgent need is therefore for more education, increased awareness and greater understanding, rather than improved scientific knowledge and further technological advancement.

Introduction

In recent decades it has become fashionable to begin discussions of landslide hazard with dramatic descriptions of the devastation achieved by catastrophic failures involving the collapse of large volumes of rock, usually in mountainous areas (Table 1, Figure 1). The three most quoted catastrophes are: the Vaiont Dam disaster in the Italian Alps (1963), when $250 \times 10^6 \, m^3$ of rock slid into the impounded lake causing huge waves up to 100 m high, to overtop the dam and drown 2600 people in the valley below (Muller, 1964); the Huascarán rock avalanche in the Peruvian Andes (1970), when part of the ice-cap on the northern peak of Nevados Huascarán (6654 m) collapsed, causing 50–$100 \times 10^6 \, m^3$ of snow, rock and ice to descend 2700 m at speeds of up to 400 kph, obliterating two towns and killing 20–25,000 people (Plafker and Ericksen, 1978; Reynolds, this volume); and the huge Mayunmarca rockslide, also in the Peruvian Andes (1974), when $1000 \times 10^6 \, m^3$ of moving material killed 450 (Kojan and Hutchinson, 1978).

The existence of such high-magnitude low-frequency events underlines the potential significance of the land-slide threat to development issues and undoubtedly poses very severe problems in terms of hazard management (see Figures 2 and 3). However, to focus discussion solely on these relatively infrequent, highly conspicuous failures merely diverts attention away from the much more important characteristic of the landslide hazard magnitude-frequency distribution; namely, that there are huge numbers of relatively small-to-medium-sized slope movements which, cumulatively, impose at least as great, if not greater, cost to human society than the rare catastrophic failures. It must be noted that the costs of these smaller scale phenomena are much more widely distributed and both their occurrence and frequency are exacerbated by human activity, so that total losses attributable to landsliding are growing rapidly. There is no cause for undue pessimism, however, for it is often claimed that presently available scientific knowledge enables prediction of such non-catastrophic failures with moderate accuracy, and that their impact can be effectively mitigated by stabilisation measures and planned land management. They therefore represent the most urgent focus of landslide hazard assessments and management strategies, because of the potential for

Table 1 Some major disasters caused by landslides in the 20th century

Place	Date	Type of landslide	Estimated volume (million m^3)	Impact
Java	1919	Debris flow	—	5100 killed, 140 villages destroyed
Kansu, China	16 Dec 1920	Loess flows	—	c200,000 killed
California, USA	31 Dec 1934	Debris flow	—	40 killed, 400 houses destroyed
Kure, Japan	1945	—	—	1154 killed
SW of Tokyo, Japan	1958	—	—	1100 killed
Ranrachirca, Peru	10 June 1962	Ice and rock avalanche	13	3500 + killed
Vaiont, Italy	1963	Rockslide into reservoir	250	about 2600 killed
Aberfan, Wales, UK	21 Oct 1966	Flowslide	0.1	144 killed
Rio de Janeiro, Brazil	1966	—	—	1000 killed
Rio de Janeiro, Brazil	1967	—	—	1700 killed
Virginia, USA	1969	Debris flow	—	150 killed
Japan	1969–72	Various	—	519 died, 7328 houses destroyed
Yungay, Peru	31 May 1970	Earthquake-triggered debris avalanche—debris flow	—	up to 25,000 killed
Chungar	1971	—	—	259 killed
Hong Kong	June 1972	Various	—	138 killed
Kamijima, Japan	1972	—	—	112 killed
Southern Italy	1972–3	Various	—	about 100 villages abandoned
Mayunmarca, Peru	25 Apr 1974	Debris flow	1000	town destroyed, 451 killed
Mantaro Valley, Peru	1974	—	—	450 killed
Mount Semeru	1981	—	—	500 killed
Yacitan, Peru	1983	—	—	233 + killed
Western Nepal	1983	—	—	186 killed
Dongxiang (Salashan) China	1983	—	3	4 villages destroyed, 227 killed
Armero, Colombia	Nov 1985	Lahar	—	about 22,000 killed
Çatak, Turkey	June 1988	—	—	66 killed

achieving significant hazard loss reduction both quickly and cheaply.

Unfortunately, landsliding is a phenomenon that tends to be little understood by the non-specialist, which means that the hazard is frequently poorly perceived and under-estimated by those involved in development programmes. As a consequence, the main problems to be overcome with respect to minimising landsliding costs are:

(a) the need to heighten the perception of development planners as to the nature, scale, distribution and causes of landsliding and the significance and spatial variation of landslide hazard;

(b) the need to increase awareness as to the range of adjustments (both structural and otherwise) that can be adopted to ameliorate the problem;

(c) the need for landslide management to be incorporated as an element of development planning in those areas prone to slope instability; and

(d) the necessity of developing and refining rapid and meaningful landslide hazard evaluation practices for areas which lack the benefit of comprehensive geological and geomorphological base-line information.

These are the topics that will be discussed in this paper.

Landslide phenomena

It is essential to recognise at the outset that the term *landslide* is the most over-used and loosely defined term employed in slope studies. It is merely a convenient short-hand or umbrella term employed to cover a very wide range of gravity-dominated processes that transport relatively dry earth materials (including dumped waste and peat) downslope to lower ground, with displacement achieved by one or more of three main mechanisms: falling, flowing (turbulent motion of material with a water content of less than 21%) and sliding (movement of materials as a coherent body over a basal discontinuity or shear plane). In reality these processes produce a bewildering spectrum of slope failures in terms of form and behaviour: size varies enormously with volumes ranging from 1 m^3 to 1000 million m^3 and displacements measured in metres to tens of kilometres; some movements are rapid, others exceedingly slow; in some cases the majority of displacement is achieved in a single, short-lived event while in other circumstances movement is gradual, cyclical or pulsed; sometimes the displaced material moves in well defined masses to create the familiar irregular terrain of scars, ridges, humps and hollows, while in other cases the earth materials may appear to lose coherence completely and run away from

Figure 1 The Hope Landslide, British Columbia, Canada. This huge rock avalanche occurred in 1965 and involved 47 million m³ of material, thereby ranking as the fourth largest failure in recorded history worldwide. Fortunately, the area is sparsely populated so casualties were limited to an estimated 24 travellers on the road below.

the sites of initial failure like dry sugar or freshly-mixed wet concrete. Occasionally, a single mechanism may operate alone but more usually several mechanisms occur in the same tract of unstable ground (a landslide complex) or the mechanism of movement may change downslope (a complex landslide).

Although there have been many attempts at providing a generally acceptable classification of landsliding (Hutchinson, 1968; Coates, 1977; Varnes, 1978; M.J. Hansen, 1984; Crozier, 1986), the debate continues, for the variable combination of slope-forming materials and agents responsible for movement "opens unlimited vistas for the classification enthusiast" (Terzaghi, 1950). Nevertheless, there is a growing body of scientific opinion in favour of restricting use of the term landslide to cover those situations where coherent masses of material actually move downslope by sliding, and replacing its more general usage by the terms *mass movement* (Hutchinson, 1968) or *slope movements* (Varnes, 1978). However, as neither of these terms communicates the same sense of danger and destruction as does the word

Table 2 Typical causes of landslides

Specific Cause

A. EXTERNAL	B. INTERNAL

1. Geometrical change
    a) Gradient
    b) Height
    c) Slope length
	1. Progressive failure (internal response to unloading, etc)
	    a) Expansion, swelling
	    b) Fissuring
	    c) Strain softening
	    d) Stress concentration

2. Unloading
    a) Natural
    b) Man-induced
	2. Weathering
	    a) Physical property changes
	    —comminution, swelling
	    b) Chemical changes

3. Loading
    a) Natural
    b) Man-induced
	3. Seepage erosion
	    a) Removal of cements
	    b) Removal of fines

4. Shocks and Vibrations
    a) Single
    b) Multiple/continuous
	4. Water regime change
	    a) Saturation
	    b) Rise in water table
	    c) Excess pressures
	    d) Drawdown

Figure 2 The Salashan Landslide, 1983, in the dissected loess uplands of southern Gansu Province, China, was created by the sudden, catastrophic collapse of a section of valley 700 m long and 200 m high. Over 3 million m³ of loess and underlying bedrock were displaced, causing a flowslide to cross the entire width of the valley floor (see Figure 3), obliterating three villages and killing well over 200 people. The photograph was taken in 1988 and shows the rapid recolonisation of the affected area.

"landslide", it is likely that all forms of slope displacement, other than creep and subsidence, will continue to be misleadingly referred to as landslide hazards.

The causes of mass movement phenomena are equally diverse (Table 2) and are well reviewed in Brunsden (1979) and Crozier (1986). Simply stated, all slopes are under stress due to the force of gravity. Should the forces acting on a slope exceed the existing strength of the materials that form the slope, then the slope will fail and movements will occur. The balance of forces can be affected by two main groups of factors: internal changes and external changes. *Internal changes* lead to a decrease in shear resistance and include the weakening of slope-forming materials by physical and chemical weathering, softening and changes in groundwater conditions (pore-water pressure). *External changes* act to increase shear stress and include slope steepening (due to erosion or excavation), slope loading (due to deposition, dumping or construction), vegetation changes and shocks/vibration. These factors may combine to destabilise the slope, first by making the slope increasingly susceptible to failure without actually inducing movement (*preparatory factors*) and second by sufficiently affecting the balance of forces to initiate movement (*triggering factors*).

Both groups of factors can be considerably influenced by human activity, thereby pointing to the fact that although landslides are natural phenomena and are a normal feature of landscapes experiencing dissection, their magnitude, frequency and geographical distribution have been considerably modified in recent centuries by human intervention. Thus assessments of landslide hazard must recognise that slope movements may be wholly natural ('natural' hazards), partly influenced by human activity ('hybrid' or 'quasi-natural' hazards) or due entirely to human activity ('man-made' or 'human-induced' hazards), as in the case with failures in embankments and waste tips.

Figure 3 Vertical air photograph of the Salashan Landslide showing the backscar, collapsed blocks of loess and the 600 m long flowslide tongue extending across the entire width of the valley floor to the stream. This was blocked to form a temporary lake and was eventually re-established 400 m south of its original position. Catastrophic failures of this scale pose severe problems for hazard management because their location is difficult to predict and their timing and size difficult to forecast. In this instance, threats of further failures on adjacent portions of the affected valley flank resulted in the relocation of potentially threatened villages to the opposite side of the valley. However, achieving such major changes in land-use patterns are usually difficult because of socio-political forces, extremely expensive to accomplish, and often seen as unnecessary interference if no catastrophic failures occur in a relatively short period of time. Because of these problems, the most acceptable and cost-effective solutions remain monitoring, emergency action and reconstruction (photograph by courtesy of Wang Jingtai).

Finally, it has to be recognised that the failure of a slope does not necessarily resolve the landslide hazard. Failure at a particular location may be repeated if destabilising forces continue to dominate, especially in those circumstances where erosive processes cause the continued removal of slipped masses and basal under-cutting (e.g. river bluffs, coastal cliffs). In addition, failures may initiate further slope movements upslope (retrogressive) or downslope (progressive), the displaced masses of earth materials may themselves suffer disintegration by further slope movements, and in those cases where displacement has been achieved by slippage over a basal shear surface, further movements (reactivation) may be initiated by loading, unloading or ground-water changes, following periods of stability lasting for hundreds or thousands of years.

Landslide hazard

The diversity of causes and types of mass movement features is matched by their great spectrum of size which ranges from innumerable, insignificant minor failures to occasional very destructive high-magnitude events which can have disastrous impacts on society (Table 1). Excluding the gigantic deformations associated with gravity tectonics, the largest identified subaerial land-slide is the Saidmarreh landslide in the Zagros mountains of Iran, where a mass 15 km wide moved 20 km some 10,000 years ago; the most destructive to date the Huascarán debris avalanche/flowslide of 31st May 1970, which obliterated the Peruvian towns of Yungay and Ranrahirca, killing 25,000 (see Reynolds, this volume).

Such events are clearly of a sufficient scale to merit comparison with other dramatic natural hazards such as hurricanes, earthquakes or volcanic eruptions, and yet global reviews of natural hazards rarely devote much attention to landsliding. There are several reasons for this apparent neglect:

(a) the largest landslides have occurred most frequently in mountainous terrains and have, traditionally, had relatively limited impact on human society;

(b) individual landslides rarely cause sufficiently large death-tolls to claim enough media attention to require political/managerial responses;

(c) appreciation of the significance of landsliding is diminished because widespread landslide impacts are often produced as a secondary consequence of violent geophysical events such as earthquakes, volcanic eruptions, hurricanes or intense rainstorms. As a consequence, the costs of landsliding tend to be subsumed within the effects of these more dramatic and conspicuous hazard events. Thus, while the huge, destructive lahar produced by volcanic activity at Nevado del Ruiz, Colombia in 1985 (Table 1; see also Hall, this volume) is usually recognised as a major landslide impact, the costs resulting from widespread slope movements pro-duced by major earthquakes, such as the ones that occurred in Armenia (1988) and Iran (1990), tend to be attributed to the earthquake hazard. The classic examples of this effect are the 1556 Shensi and 1920 Gansu earthquakes in China, which killed 830,000 and 200,000 people respectively, the majority through shock-induced slope movements.

It can be argued, therefore, that the significance of landsliding has traditionally been underestimated, partly because landslide costs have often been attributed to other hazards and partly because of human percep-tions of landsliding. Although perception of landsliding undoubtedly varies over the globe depending on complex socio-cultural factors as well as the frequency and magni-tude of landslide events, it is generally true that slope movements are not a particularly feared natural hazard. Psychological investigations into perceptions of hazard and risk (e.g. Slovik and others, 1980) have indicated that three measures or dimensions are of importance: familiarity, dread and exposure. Landsliding scores low on all three measures, as most humans consider them-selves familiar with mass-movement features because of their widespread occurrence, are not fearful of being seriously affected by such events because most are small, and consider that very damaging landsliding phenomena tend to be restricted in number and extent and confined to specific geographical locations where the potential dangers are clearly apparent. Thus landsliding tends to be placed at the opposite end of the hazard spectrum from earthquakes or stratospheric ozone depletion where concern is heightened by the fear of unfamiliar, potentially very destructive, spatially extensive threats. Only in those geographic areas where major landslides have occurred in living memory is the perception of hazard significantly raised, at least for a brief period after each event. In most other instances, landsliding tends to be viewed as the cause of inconvenience and irritation; an apparently random, moderately widespread natural phenomenon whose existence has to be accepted and is most efficiently tackled by post-event responses, such as clearance and engineered remedial measures.

Such traditional views of landsliding urgently need to be rejected and replaced by the recognition that land-sliding is far more widespread and costly to societies than previously envisaged. At the outset the popular association between landsliding and high, steep moun-tainous terrains requires modification. Many of the greatest concentrations of slope instability actually occur in coastal situations, where cliff-lines stand as testimony to the significance of mass movement. However, the most dramatic rates of coastal retreat are frequently not associated with high cliff-lines formed in durable rocks, but occur where low cliffs composed of unconsolidated or weakly consolidated materials readily crumble when subject to wave attack. Nor is inland landsliding especially restricted to mountainous areas. Inland slopes developed on clay strata can experience shallow failures at relatively low angles (<15°), while virtually flat ground developed in sensitive clays can be involved in sudden, dramatic displacements if subject to shock.

Further complexity is provided by the influence of climatic change, which may have rendered certain areas more prone to slope failure in the past thus providing a legacy of ancient failures which can be reactivated if disturbed. This is most clearly displayed in certain temperate regions where slopes that appear stable under present climatic conditions are often found to be under-lain by landslides that moved at times in the past when ground conditions were radically altered by the combin-ation of deeply frozen sub-soil (permafrost) overlain by a thawed surface layer saturated with meltwater. The resultant widespread downslope movement of materials by sliding and flowing created landslides and sheets of solifluction deposits underlain by shears. In many cases their features have become camouflaged with the passage of time so that they may attain a similar general appearance to adjacent slopes developed in materials in place, although they are much more prone to failure because of internal weaknesses.

Recognition of the more widespread occurrence of slope instability leads naturally to the conclusion that the costs of slope failure will undoubtedly escalate as a consequence of development pressures unless active steps are taken towards mitigation, and that much more positive measures of hazard management will be required in the future, increasingly focussed on anticipatory planning based on carefully prepared hazard assessments. There are several reasons for this view:

(a) The popular literature on hazards has tended to over-emphasise the significance of death-tolls in assessing the importance of hazard impacts. It is now recognised that death tolls are an extremely imperfect measure of impact, although a large death-toll undoubtedly raises the profile of an event into the attention frame of decision-makers and politicians, thereby increasing the possibility that mitigation measures may be actively considered and eventually implemented. However, death-tolls cannot be used as a surrogate for total impact measured in economic terms resulting from destruction, damage, disruption and delay, because deaths vary dramatically as a function of chance factors;

(b) Economic and technological development has resulted in massive investment in infrastructure, buildings and industry, combined with increasingly complex patterns of commercial activity, all of which indicate growing vulnerability to landslide hazard impacts;

(c) Development pressures in many parts of the World have witnessed the 'opening up' of previously under-developed mountainous regions and highly dissected terrains for power generation, water management, mineral exploitation or 'strategic' reasons, thereby increasingly exposing human activities to landslide hazard;

(d) Urbanisation has caused the rapid expansion of many cities to such an extent in recent decades that most suitable building land has been utilised; so new suburban developments are increasingly located on potentially unstable slopes. This is especially true of many Japanese cities, Rio de Janeiro, Hong Kong and Los Angeles (Cooke, 1984), and has inevitably resulted in rapidly escalating landslide damage costs;

(e) Technological developments and global competition mean that landslide-induced disruption to engineering projects have the potential of being politically embarrassing and damaging to reputations, as well as extremely costly;

(f) Even minor, inconspicuous and relatively slow-moving slope failures can have extremely costly repercussions through the disturbance of structures and infra-structure, especially the dislocation of sub-surface networks; and

(g) There are increasingly numerous examples of landslide impacts due, at least in part, to human activity. Leaking water and sewer pipes can contribute to instability, as was found by Alexander (1983) in Southern Italy following the November 1980 earthquake (see also George, this volume), as can slope loading through construction and slope unloading through excavation. The landslides that affected Hong Kong in 1966 and 1972, killing 64 and 138 respectively (Lumb, 1975) were in part the consequence of new road cuttings, as was the Çatak (Turkey) landslide of 1988 (Figure 4) which left 66 dead (Jones and others, 1989a); while the 1986 failure at Senesi, Basilicata, Southern Italy, which destroyed three blocks of flats (Figure 5), killing eight, is considered to have been triggered by excessive slope remodelling and construction.

Despite the slowly growing awareness of the significance of the landslide hazard, the true costs of mass movement in monetary terms remain unknown. For example, the recently completed survey of reported landslides in Britain undertaken for the Department of the Environment (Geomorphological Services Limited, 1987; Jones and others, 1989b) revealed that, although an unexpectedly large number of landslides had been recorded in publications and on maps (8365), of which 738 were noted as having caused impact/damage, remarkably little was actually known about the costs of these impacts. This is not unusual, as has been shown in a recently published international review by Brabb and Harrod (1989).

The reasons for this are easily identified. Landslides are ubiquitous and many are either so small or occur in such remote locations that their impact does not warrant recording. Damage caused by minor displacements is often attributed to other causes such as settlement, subsidence or defective foundations, or subsumed under the general heading of maintenance, while more serious impacts due to human disturbance are often not publicised because of litigation or commercial interests. In addition, calculations of impact are also beset with problems for, while it is easy to estimate the direct costs of repair and rebuilding, it is extremely difficult to estimate the costs of death, injury, disruption and delay (see Crozier, 1986). Not only are there numerous intangibles that have to be given monetary values, but there are also difficulties of defining the spatial and temporal limits of impact.

Nevertheless, some general estimations of landsliding

Figure 4 View of the Çatak Landslide, Turkey, taken on 1st July 1988, when attempts were still being made to recover bodies. Heavy rains are thought to have initiated failures in the high, steep unprotected cutting faces in colluvium visible on the left-hand side of the photograph. The first small failures occurred at about midnight on 22nd June 1988, causing the road to be blocked, and further landslides probably occurred during the hours of darkness. Travellers congregated in a coffee house which was totally destroyed by a complex catastrophic rock slide at 0800 hours on 23rd June, killing 66 persons (see also Figure 17).

costs have been published that provide an idea of the potential scale of the problem. The annual losses in Italy during the early 1970s were estimated at over US$1140 million (Arnould and Frey, 1977), while the comparative figure for the United States was US$1000 million (Schuster, 1978; US Geological Survey, 1982). Extrapolation of such limited estimations to provide a global figure is clearly not feasible but they indicate that landsliding world-wide undoubtedly results in an annual 'natural tax', measurable in tens of billions of dollars. The significance of landsliding as a natural hazard was clearly highlighted by the study of Alfors and others (1973) for California (Table 3) which estimated potential losses over the period 1970–2000 as US$9850 million, or 25.7% of total anticipated hazard losses, a figure exceeded only by earthquakes.

It is clear that the significance of landslide hazard varies spatially and temporarily in response to physical factors and the character of human activity. Development undoubtedly stimulates mass movements, especially in those cases where transport networks are constructed in difficult, high-energy terrains previously little disturbed by human activity. For example, the construction of the Karakoram Highway in northern Pakistan was fraught with difficulties due to mass movement, and landslides continue to pose costs through disruption and destruction (Jones and others, 1983). Indeed, most roads in the Himalayas suffer severe problems due to landsliding (Hearn, 1987), as is the case in most mountainous regions. However, it is incorrect to envisage that the severest levels of 'natural tax' imposed by landsliding are necessarily to be found in remote, inhospitable and rugged terrains; as has already been mentioned, landslide hazard varies in significance in relation to socio-economic vulnerability and is known to be particularly severe in parts of California (Cooke, 1984) and attains dramatic significance in Southern Italy.

Landsliding in southern Italy

In order to emphasise this point it is informative briefly to consider the situation in Southern Italy, an area particularly prone to landsliding. The accented terrain has been created by the recent, large-scale uplift of tectonically disturbed strata. The erodable lithologies, sparse vegetation, extremes of Mediterranean and mountain climates and seismic events, have all combined to produce circumstances favourable to rapid erosion, floods and landsliding. In addition, long-term human intervention involving forest clearance, cultivation and urbanisation, have exacerbated problems of soil erosion and landsliding, and recent attempts at economic development through the construction of modern infrastructure have frequently experienced costly set-backs. Carrara and Merenda (1976) estimated that damage to roads, railways, aqueducts and houses in Calabria cost US$200 million for 1972–3 alone, and that over the centuries nearly 100 villages have had to be abandoned because of landsliding, involving the displacement of approaching 200,000 people.

The adjacent region of Basilicata experiences equally severe problems. It has recently been estimated (Regione Basilicata, 1987) that 18.5% of the region is affected by landsliding, including 1800 deep landslides (2.6% of the area). Particularly severe problems are being experienced

Figure 5 The Senesi landslide of 1986 (Basilicata, southern Italy), showing one of the collapsed blocks of flats.

Table 3 Projected losses due to geological hazards in California, 1970–2000, indicating the significance of the landslide hazard and the potential for hazard loss reduction (after Alfors and others, 1973). All costs are in US$m at 1970 values

Geological hazard	Projected total losses, without improvements of existing policies and practices US$ million	Possible total loss reduction, if all possible loss-reduction measures are applied		Estimated total cost of applying all feasible measures at current state-of-the-art levels		Benefit/cost ratio if all feasible measures were applied and all possible loss reductions were achieved
		% total loss	US$ million	% total loss	US$ million	
Earthquake shaking	21,035	50	10,517.5	10	2,103.5	5
Landsliding	9,850	90	8,865	10.3	1,018	8.7
Flooding	6,532	52.5	3,432	41.4	2,703	1.3
Erosion activity	565	66	377	45.7	250	1.5
Expansive soils	150	99	145.5	5	7.5	20
Fault displacements	76	17	12.6	10	7.5	1.7
Volcanic hazards	49.38	16.5	8.13	3.5	1.65	4.9
Tsunami hazards	40.8	95	37.76	63	25.7	1.5
Subsidence	26.4	50	13.2	65.1	8.79	1.5
Totals	38,324.58	61	23,411.69	16.0	6,125.64	3.8
Column 1	2	3		4		5

Notes:
Column 2: Estimated total dollar loss, for all of California; about 95% of the loss would be in urban areas. These values are based on the assumptions that the numbers and severity of each type of event occurs as estimated, and that no change is made in the 1970 type, effectiveness, or level of application of preventive and remedial measures.
Column 3: Estimated total loss-reduction in dollars and in per cent of projected loss, for all of California, assuming an aggressive but reasonable degree of improvement in the 1970 type and level of preventive and remedial measures. Conservative improvements in the state-of-the-art application over wider area, and more effective application and follow-up of all known types of loss-reduction measures over the next 30 years, are assumed.
Column 4: Estimated total cost of applying loss-reduction measures of the type, effectiveness, and extent visualised in Column 3, for the period 1970–2000, in dollars and in per cent of projected total loss (Column 4/Column 2).
Column 5: Estimated benefit/cost ratio (Column 3/Column 4) based on the estimated cost (Column 4) of applying the estimated loss-reduction measures to obtain the estimated reductions (Column 3) of the estimated total losses (Column 2).

Figure 6 Calitri, southern Italy: typical southern Italian hill-top town with characteristic threatening landslide, on the Campania/Basilicata border. This town is severely affected by a series of slides that are periodically reactivated by heavy rains or earthquake shocks. This particular site has been described in detail by Hutchinson and Del Prete (1985).

by the relatively ancient settlement pattern (Figure 6), for the majority of the 131 towns (communes) are located either on hilltops (28–21%) or on hill slopes (87–66%). Deforestation, agriculture, urban expansion, leaking infrastructure and earthquake shaking have all conspired progressively to increase mass movement activity on slopes, and to threaten the very fabric of the towns.

The precarious situation of Aliano (Figure 7) merely represents an extreme example of a widespread phenomenon graphically described in a recent official report as follows: "On bleak relief, surrounded by vast areas which are completely naked and affected by Calanchi features and landslides, the towns are being destroyed by an all-round attack" (Regione Basilicata, 1987). No less than 115 out of the 131 communes are experiencing contemporary problems which are of sufficient severity as to require either extensive engineered structural measures (consolidation), relocation (transferal), or a combination of the two (Figure 8, Fulton and others, 1987). The scale of the problem has severe social and

political implications and clearly indicates the need for landslide hazard assessments at scales ranging from the local to the regional.

Basilicata also provides many informative examples of costly and conspicuous structural failures due to inadequate consideration of landslide hazard. The collapse of blocks of flats at Senesi (1986) is thought to have been produced by excessive ground disturbance (Figure 5). The Sinni Highway was completed in the late 1970s but already has three unusuable viaducts with piers that have been displaced by ground movements, and the hill-top town of Craco had to be virtually abandoned in 1971 because of landslide activity exacerbated by human intervention (Figure 9). The Craco landslide is of particular interest and has been written up by Del Prete and Petley (1982). In their view an ancient landslide that had posed few problems over the centuries underwent major reactivation following the dumping of several metres of fill to construct a football pitch in 1954 (Figure 10); no movements were observed until exceptionally heavy rains in November 1959 caused widespread

Figure 7 Aliano, Basilicata, southern Italy, showing well developed "badland" topography (Calanchi) threatening to undermine the town through slope instability.

displacements, destroying the football pitch and causing 20 m uplift of the landslide toe. Further movements in 1963 and 1965 resulted in visible displacements and damage to buildings, and 153 houses were evacuated. A Government Order to evacuate the village totally was subsequently made partial in 1968 and new stabilisation measures were proposed involving the construction of a concrete platform 60 m long by 4 m wide, founded on lines of heavily reinforced piles 800 mm in diameter and 30 m long. This wall was constructed in the autumn of 1970 and in April 1971 "the total collapse of the inhabited area took place" (Del Prete and Petley, 1982) resulting in the virtual abandonment of the site. According to Del Prete and Petley, the stabilisation measures contributed to failure in three ways:

(a) The piles were far too short and failed to reach the basal shear surface estimated to lie at a depth of 60–70 m;
(b) The retaining wall was of massive construction and exerted considerable additional shearing stresses on the landslide body through loading; and

(c) The contiguous pile foundations created a dam effect on the groundwater, thereby raising pore-water pressures.

The abandonment of Craco is an excellent example of the disastrous consequences that can result from failure to take proper consideration of landslide hazard. There are undoubtedly numerous other instances where inadequate geotechnical investigations, cost-cutting exercises and the over-inflated belief in the security of conspicuous engineering have contributed to costly failures. The last-mentioned is especially important, for the post-war emphasis on the power of technology, and the preoccupation with engineered structural solutions involving the use of large volumes of concrete and steel, have tended to breed over-confidence in some quarters, to the extent that prior studies of landslide hazard are seen to be unnecessary 'frills' to standard geotechnical investigations. Such a view is likely to continue to receive support until the true costs of landsliding are identified. Only then will there be a widespread switch from 'reactive engineering' to 'anticipatory planning'

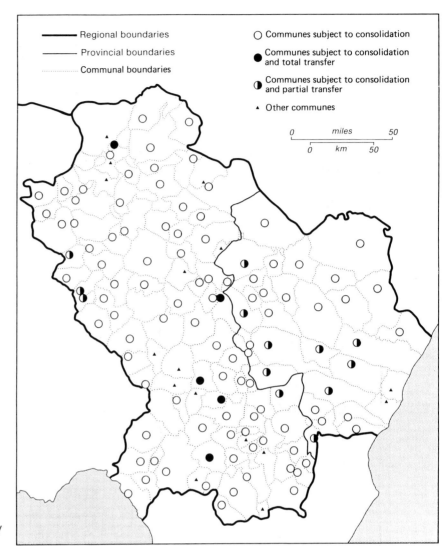

Legend:
——— Regional boundaries
——— Provincial boundaries
·········· Communal boundaries

○ Communes subject to consolidation
● Communes subject to consolidation and total transfer
◐ Communes subject to consolidation and partial transfer
▲ Other communes

Figure 8 Map of Basilicata region, southern Italy, showing communes requiring stabilisation measures (consolidation) or relocation (transferral), largely as a consequence of persistent slope instability. (From *International Geomorphology 1986* by V. Gardiner (ed.), Copyright (1987). Reprinted by permission of John Wiley and Sons, Ltd.).

involving the clear assessment of landslide hazard in the early stages of development programmes.

Landslide hazard assessment

Landsliding is considered to be one of the potentially most predictable of geological hazards (Leighton, 1976), which accounts for the extremely high estimated potential for loss reduction (90%) put forward by Alfors and others (1973) in their study of California (Table 3). Such statements are made with particular reference to small- and medium-scale events, and depend upon the three basic assumptions identified by Varnes (1984):

(a) That uniformitarian principles can be applied to landslide hazard assessment in that the conditions that led to slope instability in the past and the present will apply equally well in the future. Thus, the estimation of future instability can be based on

the assessment of conditions that led to slope failure in the past;

(b) that the main conditions that cause landsliding can be identified; and

(c) where the causes of landsliding can be identified it is usually possible to estimate the relative significance of individual factors. This facilitates assessment of degree of hazard by examining the number of failure-inducing mechanisms present in any area.

Such statements of confident optimism should not be allowed to conceal the existence of some very real problems with landslide hazard assessment. First, while it is indeed true that the main causes of landsliding have been identified, the significance of certain elements, such as chemical changes, have not yet been fully evaluated. Second, uniformitarian principles can only be applied in general terms, as both climatic change and human intervention variably influence the stability conditions of slopes over time. Third, it is not always possible to assign

Figures 9, a and b Two scenes in the town of Craco, Basilicata, seen in 1987, showing the abandoned town following reactivation of a major landslide in 1971.

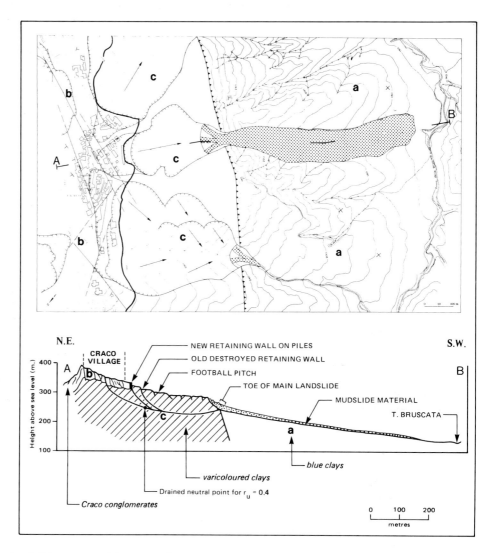

Figure 10 Map of Craco and geological section, showing the major rotational landslide that eventually resulted in the abandonment of the site (after Del Prete and Petley, 1982) (for legend, see opposite page).

relative levels of significance to landslide causes with any degree of confidence, especially in those instances where complex sequences of rocks are involved in failure. Fourth, while it is frequently possible to explain failures adequately by ex-post (post-event) evaluation or hind-casting, it is a very different problem to forecast future events with respect to magnitude, frequency and timing.

Landslide hazard assessment is therefore overwhelmingly concerned with establishing the general likelihood of slope instability (landslide prediction) rather than with specifying the occurrence of specific events (landslide forecasting). Landslide forecasting is focussed on identifying the location, magnitude and timing of specific events, normally based on monitoring procedures, so that emergency actions may be achieved efficiently. It is detailed, site-specific work that is usually undertaken after concern has been raised by some initial movements. Landslide prediction, on the other hand, is spatially extensive and concerned with establishing zones of differ-

ing levels of threat by emphasising the magnitude-frequency characteristics of recorded landsliding and probabilities of slope instability. It is landslide prediction that provides the framework for planning management, engineering and investment decision-making, which explains why it is currently the main goal of landslide hazard studies.

Numerous different approaches to landslide hazard assessment have been developed over the last three decades and these are thoroughly reviewed in A. Hansen (1984) and Geomorphological Services Limited (1986). Hansen identified three main groups of techniques:

(a) *Geotechnical investigations*, involving the detailed analysis of surface and subsurface conditions and ground materials;

(b) *Direct mapping*, involving the analysis of landforms and identification of existing landslides so that areas of past instability can be identified, thereby

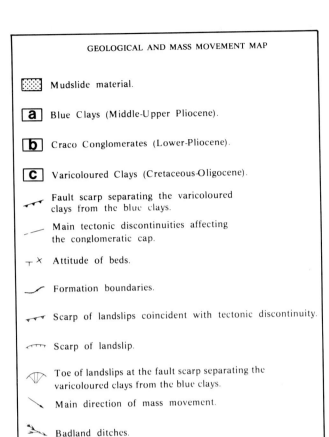

GEOLOGICAL AND MASS MOVEMENT MAP

Mudslide material.

a Blue Clays (Middle-Upper Pliocene).

b Craco Conglomerates (Lower-Pliocene).

c Varicoloured Clays (Cretaceous-Oligocene).

Fault scarp separating the varicoloured clays from the blue clays.

Main tectonic discontinuities affecting the conglomeratic cap.

Attitude of beds.

Formation boundaries.

Scarp of landslips coincident with tectonic discontinuity.

Scarp of landslip.

Toe of landslips at the fault scarp separating the varicoloured clays from the blue clays.

Main direction of mass movement.

Badland ditches.

facilitating extrapolation from areas of recognised past instability to similar situations which may suffer slope failure in the future; and

(c) *Indirect mapping*, which requires the collection of data on the causes and mechanisms of landsliding so that assessments of slope stability can be made by the application of known landslide-inducing parameters.

To these may be added the fourth approach of *Land Systems Mapping* (described below) which lies intermediate between direct and indirect mapping.

Although all four approaches are of relevance in the context of development, geotechnical investigations are site-specific and therefore restricted to particular construction programmes, such as bridges, dams, roads, industrial plant, etc, or situations where stability problems have been identified which require remedial measures. Where hazard assessments are required for the purposes of regional development or general hazard reduction, then the direct, indirect and land-systems mapping approaches are of greatest relevance.

Direct mapping

The simplest form of direct mapping is the *Landslide Inventory Map* which displays the distribution of recognised landslides. Method of portrayal will vary with scale, with dots used for synoptic (1:100,000 scale

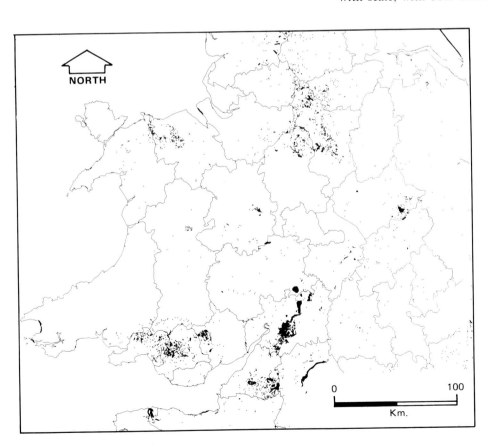

Figure 11 Distribution of reported landslides in the Midlands of England, recorded by a recent survey for the Department of the Environment (Geomorphological Services Limited, 1987).

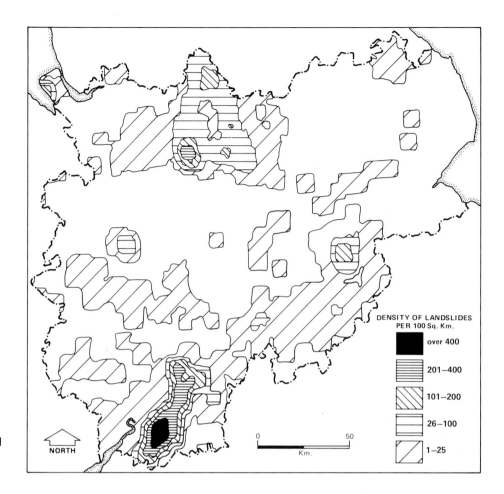

DENSITY OF LANDSLIDES
PER 100 Sq. Km.

over 400

201–400

101–200

26–100

1–25

NORTH

0   50
          Km.

Figure 12 Landslide isopleth map of the Midlands based on data produced by the DoE-sponsored survey. This map should be compared with the dot distribution shown in Figure 11.

and smaller) and medium-scale (1:25,000–1:50,000) maps, but outline shapes and increasing detail on large-scale (1:10,000–1:5000) and detailed (1:2000–1:5000) maps. The information may be obtained by air-photo interpretation, ground survey or literature review (Figure 11), and additional information may be included on the type, size, age or state of activity of the recognised failures.

The obvious next step towards the production of a landslide hazard assessment is to generalise the point data contained in the landslide inventory map over administrative areas, geological outcrops, geomorphological regions or through the application of isopleth or choropleth mapping (Figure 12) to provide *regional landslide distribution maps*.

A different approach, which places landslides in their setting and can indicate the causes of slope failure, is *geomorphological mapping*. The nature and techniques of this have been widely described as have the numerous and varied mapping legends (Demek and Embleton, 1978; Cooke and others, 1982; Cooke and Doornkamp, 1990). Many examples exist of the use of geomorphological mapping to minimise landslide hazard in development projects, a classic case being the mapping

of the Tamar Valley as part of the Dharan-Dhankuta road project in eastern Nepal (Figure 13; Brunsden and others, 1975) which contributed significantly to the identification of a new design philosophy and choice of a more stable alignment. Many other fine examples exist, some beautifully produced in several colours, others utilising the most up-to-date forms of computer drafting. But it has to be recognised that such maps can only portray what is already there and will only show what is identified by the mappers involved in surveying the terrain. Operator variance in terms of the skills, experience and understanding of the personnel involved in map preparation is therefore of the utmost significance in determining the level of information that such maps will portray, and therefore their value to potential users.

Indirect mapping
The causes and mechanisms of landsliding are now thought to be reasonably well understood (Varnes, 1984; Table 2) although it has to be recognised that landsliding in any specific area is almost certainly the result of the complex interaction of a number of causes which will vary in significance over space and time. If causal factors can be identified, recorded, measured and mapped, then

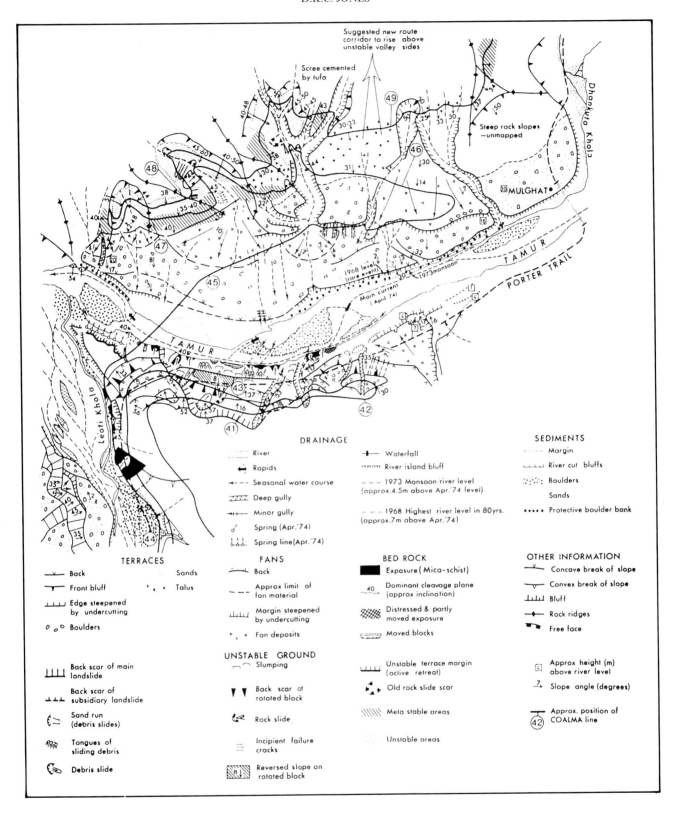

Figure 13 Geomorphological mapping used to assess landslide hazard along a proposed road alignment: the Tamur crossing on the Dharan-Dhankuta Highway, East Nepal. This mapping was instrumental in the rejection of the originally proposed alignment and the development of a new alignment involving hair-pin stacks on relatively stable slopes, including the one shown with a thick arrow on the diagram (Reproduced by permission of the Geological Society from Brunsden and others, 1975. Large scale geomorphological mapping and highway engineering design. *Quarterly Journal of Engineering Geology*, vol 8, p 227–254).

comparison of the distributions of different factors (Table 4) allows the identification of areas with varying potential for landsliding. The resultant element maps or factor maps may either consist of a series of overlays, each of which displays the varying strength of a particular factor, so that the cumulative results can be assessed, or they take the form of a grid or polygon mesh for the purposes of factor scoring (Figure 14). Such approaches essentially entail the collection and assessment of data based on three main groups of analytical techniques:

(a) Graphical methods, mainly involving the use of maps;
(b) The analysis of empirical relationships between landsliding and individual causal factors such as slope steepness, material type, vegetation cover and rainfall intensity; and
(c) Multivariate analysis of the varied causes of landsliding.

A. Hansen (1984) suggested the possibility of a fourth approach, that of *deterministic modelling*, where the detailed, site-specific results obtained through geotechnical investigations may be extrapolated spatially using probability theory.

Table 4 List of element (factor) maps typically produced in landslide hazard studies

Geological structure
Chronostratigraphy of rocks
Lithostratigraphy of rocks
Lithostratigraphy of soils
Rockhead contours
Geotechnical properties of soils
Geotechnical properties of rocks
Hydrogeological conditions
Geomorphological processes
Geomorphological history, palaeodeposits, surfaces or
 residual conditions
Seismic activity
Hydrological conditions
Climate (including precipitation)
Ground morphology, especially slope height, length and
 angle
Land use
Vegetation, type, cover, root density and strength
Pedological soils, type and thickness of regoliths

Some schemes also include:
 Landslide deposits
 Landslide morphometry

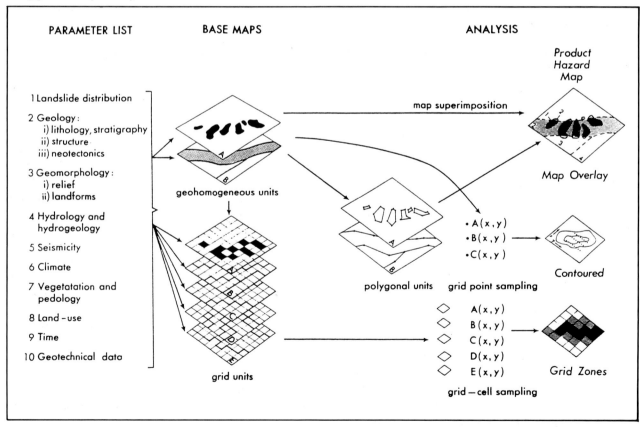

Figure 14 Graphical methods of data collection, presentation and analysis (from *Slope Instability* by D. Brunsden and D.B. Prior (eds), copyright (1984). Reprinted by permission of John Wiley and Sons, Ltd.).

Land systems mapping

The land systems mapping approach occupies an intermediate position between direct and indirect mapping. Land systems mapping involves the definition of areas within which occur certain predictable combinations of surface forms, soils, vegetation and surface processes. The result is a hierarchical classification of terrain into *land systems* (large-scale divisions) composed of *land facets/land units* which are further divisible into *land elements*. The technique was first developed in Australia where there was an urgent need for maps to serve as a basis for rural planning (Christian and Stewart, 1952) but has subsequently been widely applied elsewhere for it "produces an extremely simple, cost-effective and versatile method for rapidly classifying large areas of relatively unknown territory, and providing a regional framework for environmental data collection and storage" (Cooke and Doornkamp 1990, p 25). Good descriptions of the technique are to be found in Mitchell (1973) and Ollier (1977) and recent practice is well described in King (1987).

The use of the land systems approach for the analysis of landslide hazard was initially undertaken in the United States, where Wilhusan (1979) described the distribution of different types of mass movement in Pennsylvania by reference to terrain provinces. Other early examples include the national study undertaken by Radbruch-Hall and others (1976) and for the Franconia area of Virginia by Froelich and others (1978). All of these involved the preparation of small-scale maps (1:1 million–1:7.5 million) but more recent developments in Hong Kong have focussed on systematic terrain classification at scales of 1:20,000 and 1:2500 for engineering appraisal, planning and land-use management purposes (Styles and others, 1982, 1984; Burnett and others, 1985).

Landslide hazard and susceptibility maps

Landslide Hazard Maps may be derived from direct, indirect or land systems mapping approaches. Some confusion exists over the terms "hazard", "susceptibility" and "risk", which are often used interchangeably in the literature. There are some who argue that maps showing the distribution of landslides (*landslide inventory maps*) or landslide deposits, should be referred to as *hazard maps*; however, the term *hazard* only has meaning when viewed from a human perspective because it is culturally defined—if there are no humans present, or nothing of human value, then strictly speaking there is no hazard. Thus past events are only of significance if they are likely to be repeated or reactivated in the future. True hazard maps display the extent of hazardous events (*hazard zone*) together with an internal division (*hazard zonation*) reflecting the magnitude-frequency distribution of the hazardous events (*hazard rating*). Such maps must be based on evidence of past failures together with judgements as to the likelihood of future events. It is these maps that have often been referred to as *landslide susceptibility maps* or *landslide potential maps* and which are of greatest value to development planning as they attempt to assess spatial dimensions of the potential threat from slope instability.

Risk maps are significantly different from hazard maps because they involve the assessment of the potential losses that may be incurred by society through hazardous impacts. There are numerous and diverse definitions of risk, but the most useful with respect to landsliding is:

$$R_s = E \times H \times V$$

where R_s = Specific Risk associated with a hazard of particular magnitude;

E = Elements at Risk or the total value of population, properties and economic activities within the area under consideration;

H = the probability of occurrence of the specified magnitude of hazard event within the area; and

V = vulnerability, or the proportion of E affected by the specified magnitude of hazardous event.

The most important feature of landslide susceptibility maps and landslide risk maps is the spatial division of the earth's surface into areas of different levels of threat (zonation and micro-zonation). It is these divisions that provide essential frameworks for land-use planning, building regulations and engineering practices.

Numerous examples of landslide susceptibility maps exist and have been thoroughly reviewed by Brabb (1984), Carrara (1984), A. Hansen (1984), Varnes (1984) and Geomorphological Services Limited (1986). Initially such maps were relatively simple and generalised; for example, Radbruch and Wentworth (1971) employed three factors (slope steepness, rainfall, ground materials) to produce a six-fold zonation of the San Francisco region of California at a scale of 1:170,000. More detail was provided by Brabb and others (1972), in a study of San Mateo County, California, where a specially commissioned slope category map was used to produce a 1:63,360 scale map with a seven-fold zonation (Figure 15) by means of the following steps:

(a) Measurement of each rock outcrop;
(b) Area of landsliding within each rock outcrop measured;
(c) Rock types ranked in order of percentage affected by landsliding;

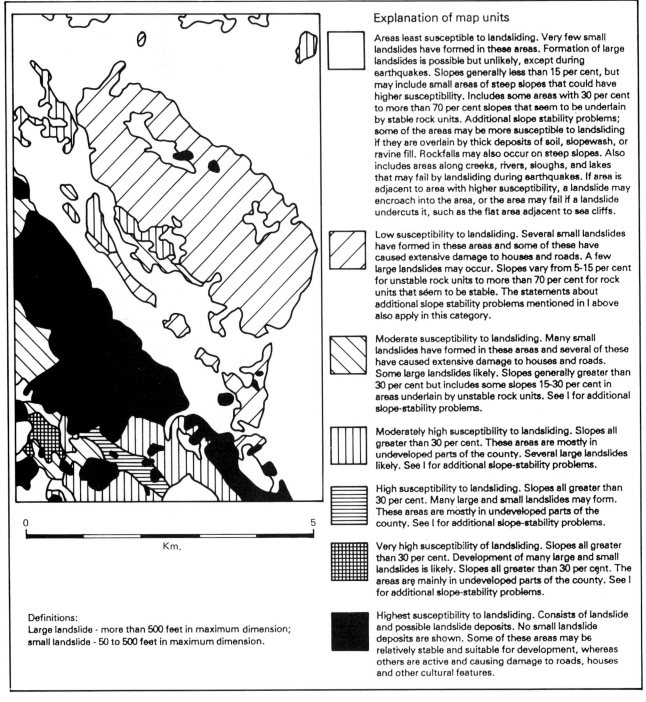

Explanation of map units

Areas least susceptible to landsliding. Very few small landslides have formed in these areas. Formation of large landslides is possible but unlikely, except during earthquakes. Slopes generally less than 15 per cent, but may include small areas of steep slopes that could have higher susceptibility. Includes some areas with 30 per cent to more than 70 per cent slopes that seem to be underlain by stable rock units. Additional slope stability problems; some of the areas may be more susceptible to landsliding if they are overlain by thick deposits of soil, slopewash, or ravine fill. Rockfalls may also occur on steep slopes. Also includes areas along creeks, rivers, sloughs, and lakes that may fail by landsliding during earthquakes. If area is adjacent to area with higher susceptibility, a landslide may encroach into the area, or the area may fail if a landslide undercuts it, such as the flat area adjacent to sea cliffs.

Low susceptibility to landsliding. Several small landslides have formed in these areas and some of these have caused extensive damage to houses and roads. A few large landslides may occur. Slopes vary from 5-15 per cent for unstable rock units to more than 70 per cent for rock units that seem to be stable. The statements about additional slope stability problems mentioned in I above also apply in this category.

Moderate susceptibility to landsliding. Many small landslides have formed in these areas and several of these have caused extensive damage to houses and roads. Some large landslides likely. Slopes generally greater than 30 per cent but includes some slopes 15-30 per cent in areas underlain by unstable rock units. See I for additional slope-stability problems.

Moderately high susceptibility to landsliding. Slopes all greater than 30 per cent. These areas are mostly in undeveloped parts of the county. Several large landslides likely. See I for additional slope-stability problems.

High susceptibility to landsliding. Slopes all greater than 30 per cent. Many large and small landslides may form. These areas are mostly in undeveloped parts of the county. See I for additional slope-stability problems.

Very high susceptibility of landsliding. Slopes all greater than 30 per cent. Development of many large and small landslides is likely. Slopes all greater than 30 per cent. The areas are mainly in undeveloped parts of the county. See I for additional slope-stability problems.

Highest susceptibility to landsliding. Consists of landslide and possible landslide deposits. No small landslide deposits are shown. Some of these areas may be relatively stable and suitable for development, whereas others are active and causing damage to roads, houses and other cultural features.

0 ————— 5
Km.

Definitions:
Large landslide - more than 500 feet in maximum dimension;
small landslide - 50 to 500 feet in maximum dimension.

Figure 15 Landslide susceptibility classification used in a study of San Mateo County, California (Brabb and others, 1972).

(d) Map of slope steepness (six divisions) superimposed on geology map and landslide distribution map to determine associations;
(e) Definition of seven landslide susceptibility classes and delimitation on maps.

Perhaps the best known hazard assessment based on geomorphological mapping was that carried out by

Kienholz (1977, 1978) in Grindelwald, Switzerland. An extensive programme of field mapping resulted in the identification of 4000 geomorphological units which were then hazard-rated on a scale of 0 to 3 with respect to five main causative hazards (ice-avalanche, avalanche, landslip, rockfall and water) to yield 1:10,000 scale maps of combined hazard. Mapping is also the basis of the extensive hazard-assessment programme in France

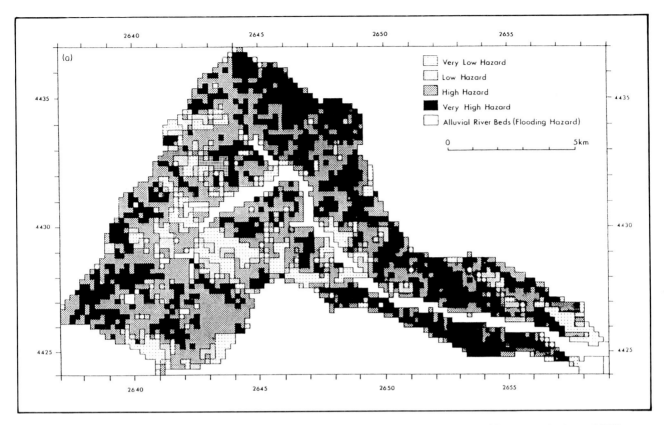

Figure 16 Landslide hazard and erosion zones, Ferro River basin, Calabria, southern Italy (Carrara and others, 1978).

(ZERMOS) which has to date resulted in the production of over 30 maps showing seven hazard categories (Laudrey, 1979; Varnes, 1984).

Indirect mapping techniques are resulting in the production of increasingly complex hazard assessments. One of the most comprehensive is that undertaken in Calabria, Southern Italy (Carrara and others, 1977, 1978) where landslide registration, field mapping, air-photo interpretation and the incorporation of existing geological, geomorphological and hydrological data, yielded information which was analysed within a framework of cells 200 m square to produce a landslide hazard classification (Figure 16). Recent developments, including computerisation and application to planning, are well described in Einstein (1988) and Siddle and others (1989).

Discussion

Hazard assessments are undertaken for three main purposes (Jones, 1983):

(a) To assess the reasons for widespread or repeated losses affecting a broad range of interests within society;

(b) To assess hazard potential with regard to specific proposed developments so that future losses can be

minimised by relocation (hazard avoidance) or the adoption of protection measures, and

(c) Evaluation of the likelihood of hazardous events occurring as a consequence of proposed developments (*environmental impact assessments*).

Clearly the techniques that can be used do not vary with the purpose of the study, but rather with regard to other considerations including the spatial extent of the area under investigation, the time available, the range and quality of accessible background (base-line) information, facilities, man-power skills and finance. Thus, while ground-based geomorphological mapping may prove highly successful and cost-effective in problem evaluation regarding the assessment of alternative road alignments, as was the case in the 60 km Dharan-Dhankuta Highway in Eastern Nepal (Brunsden and others, 1975; Jones, 1983; Hearn, 1987), such an approach would not prove so successful if used to assess landslide hazard throughout a region.

If landsliding can be considered a predictable hazard (Leighton, 1976) with good potential for hazard loss reduction measures that have favourable cost-benefit ratios, especially where high value developments are threatened (Alfors and others, 1973), then it is necessary to question why landslide hazard assessments have not been more widely applied. As has already been shown, a

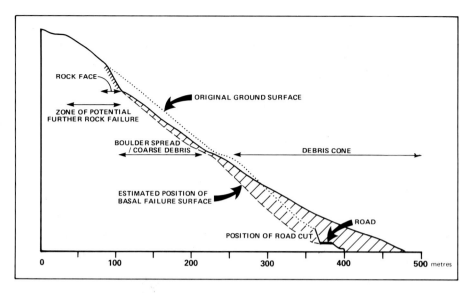

Figure 17 Section through the Çatak landslide, Turkey, showing position of road cutting (after Jones and others, 1989a).

variety of techniques exist which should prove suitable to a wide range of situations, and although many have yet to be vigorously tested by ex-post evaluation (hindsight review), it seems reasonable to assume that satisfactory results could be obtained. Thus the applicability of landslide hazard assessments seems clear but has yet to be matched by application.

Several reasons can be put forward to explain the limited application of landslide hazard assessment techniques:

General ignorance as to the nature, significance and causes of landsliding As was discussed earlier, perceptions of landsliding are generally very imperfect. To many, the causes of landsliding are obscure and often attributed to some overwhelming force or higher authority (an "Act of God", etc). As a consequence, a fatalistic attitude tends to prevail in many areas, especially within the Developing World, which naturally negates the need for assessments of landslide potential. Secondly, in other instances, ignorance as to the nature and causes of slope failure results in underestimation of the potential role of human activity in stimulating slope instability. This, coupled with an over-confidence in engineering following the introduction of massive machinery and reinforced concrete, often resulted in excessive earthworks with limited stabilisation measures. For example, investigations at the site of the Çatak landslide disaster (Turkey, 23rd June 1988) revealed that a road had been cut into the base of the slope in 1984, involving the creation of cutting faces 25 m high in both rock and colluvium (Figures 4 and 17), the latter standing at 55° but lacking support except for a 3 m-high mortared masonary wall at the base (Jones and others, 1989a). Thirdly, the true costs of ground movements are seldom calculated and even more rarely publicised. As a consequence, the frequency of landslides tends to be grossly underestimated and their significance unappreciated, except during the immediate aftermath of a serious impact.

Clearly, the first step in overcoming these problems is the initiation of regional/national surveys of landsliding, which seek to establish the distribution of past landslides, together with their causes, consequences and costs. Such censuses are an essential basis for determining the varying scale of the problem, trends in impacts, the needs for further research and the desirability of policy responses. This need was recognised by the International Ad Hoc Group of Experts on the International Decade for Natural Disaster Reduction (United Nations, 1989, p 41) when they stated that one project that could contribute to reducing the impact of landslides was "the establishment of a world landslide centre concerned with data gathering and dissemination, information transfer, mapping and training of land-use specialists and structural engineers".

General ignorance as to the potential for landslide assessment Because landsliding has for long been viewed as a randomly distributed phenomenon, the traditional response to the threat of landsliding has focussed on post-event clearance, remedial measures and reactive engineering rather than anticipatory planning involving some degree of prediction. There has been relatively little general interest in landslide hazard assessments because emergency action and response measures are widely deemed to be cheaper and, therefore, more cost-effective. Only in the case of the engineering profession has there been a widespread shift to landslide evaluation over the past four decades, largely due to the growth of geotechnical investigations and the development of soil mechanics. But even in these instances, studies largely remain site-specific in focus and heavily dependent on the results obtained from the detailed examination of boreholes and trial pits, with few attempts at extrapolation involving mapping.

The incorporation of landslide hazard assessments in land planning policies has been extremely limited to date, for a number of reasons. Ignorance as to the nature and potential of landslide susceptibility studies is clearly a major factor. Thus, while Cooke and Doornkamp (1990, p 28) were undoubtedly correct when they stated "the main problem lies in transferring a landslide hazard awareness from the geomorphologist to the planner and legislator", it is also necessary to add *"and to make them aware of the existence of a range of reasonably reliable predictive techniques"*. However, further problems undoubtedly exist regarding the disparity between timescales applicable to geomorphological processes and the time frames of politicians and policy-makers, and it is here that landslide susceptibility studies face their major challenge. It is still only possible to predict slope failure in the most general of terms and virtually impossible to forecast the location, magnitude and timing of specific future events. It is this inexactitude that has caused disinterest amongst planners and politicians. Faced with the choice of blanket controls based on what are seen to be "vague and uncertain predictions" as against undertaking reactive responses to specific impacts, it is hardly surprising that the majority choose the latter where actions and expenditure are clearly driven by event-generated needs.

Problems with landslide hazard assessment application
Landslide susceptibility studies result in the identification of tracts of land with different levels of hazard potential. Such hazard zones provide an ideal framework for land-use planning, development control, the application of building codes/ordinances, guidelines for engineering practice and the establishment of insurance-premium levels. Where the impact of landsliding has been significant in the past, high levels of public awareness have ensured that such zoning policies have been applied with relative success. The best example is the case of the Los Angeles area where increasingly rigorous laws have been adopted since October 1952 (Fleming and others, 1979) which are claimed to have resulted in a significant reduction in landslide losses (Cooke, 1984). Similarly, in Tasmania a landslide hazard scheme has been established based on mapping at scales of 1:5000 and 1:10,000. In those locations where marginal stability conditions are identified, decisions are made either to warn intending builders and purchasers or to restrict development under the Local Government Act of 1973 (Stevenson and Sloane, 1980).

While the desirability of the above approaches is undeniable, major problems are encountered with respect to implementation and operation. In many areas, especially the poorer parts of the Developing World, the cost of landslide susceptibility studies may appear excessive. This is especially true where there are greatly competing calls for finance, and where the benefits accruing from landslide hazard assessment may be seen to be long-term and uncertain. In addition, the implementation of hazard zonation requires the existence of a planning system capable of enforcing development controls, curbing speculative building and limiting shanty developments produced by rural-urban migrations. Finally, any hazard zonation applied to an already populated region faces the problem of landowner hostility, for land values are often unfavourably affected by increased hazard rating. As a consequence, commercial and political pressures may be brought to bear with a view to limiting the use of zoning, and litigation can result.

Thus severe problems exist regarding how hazard zones can be depicted and described, and what measures of control will prove acceptable to the local communities. In many instances it still proves simpler and easier to make individual developers and property owners responsible for ensuring their own safety, thereby avoiding the need for any broad-scale landslide hazard assessment.

Conclusions

Because slope stability is a function of the interplay of several factors which vary in space and time, slope movement assessment must remain vague and hedged with probabilities. Herein lies a fundamental dichotomy: decision-makers naturally limit expenditure on landslide hazard evaluation to the very minimum because of perceptions regarding need and usefulness, with the result that the potential costs of failure may be several orders of magnitude greater than the costs of investigation. It is natural, therefore, that proponents of landslide susceptibility studies are cautious regarding further developments without clarification of the legal status of the hazard assessments so produced.

Great strides have been made over the last four decades with respect to the range and accuracy of landslide hazard assessments, although certain groups of decision-makers may still find them vague and consider that they have the potential to cause more anxiety than they resolve. However, the identification and delimitation of hazard zones undoubtedly provides an extremely valuable framework for decision-makers.

Future development of landslide hazard assessments are, therefore, not so much dependent on further refinement and improvement of the techniques of hazard assessment, valuable though these will be, but rather on the increased awareness of the need for hazard evaluation. For demand to be consumer driven, it is necessary that engineers, planners and developers become aware of the potential scale of the problem, and this will only be achieved by the production of systematic surveys and censuses.

References

Alexander, D., 1983. Earthquakes and the continuing problem of landslides in Southern Italy. *Disasters*, vol 7, p 88–90.

Alfors, J.T., J.L. Burnett and T.E. Gay, 1973. Urban geology masterplan for California. *California Division of Mines and Geology Bulletin*, no 198.

Arnould, M., and P. Frey, 1977. *Analyse detaillée des réponses à l'enquête internationale sur les glissements de terrain.* UNESCO, Paris.

Brabb, E.E., 1984. Innovative approaches to landslide hazard and risk mapping. *Proceedings of the 4th International Symposium on Landslides*, Toronto, vol 1, p 307–324.

Brabb, E.E., and B.L. Harrod (eds), 1989. *Landslides: Extent and Economic Significance*. A.A. Balkema, Rotterdam.

Brabb, E.E., E.H. Pampeyan and M.G. Bonilla, 1972. Landslide susceptibility in San Mateo County, California. *US Geological Survey Miscellaneous Field Studies Map*, MF 360.

Brunsden, D., 1979. Mass Movements. In: C. Embleton and J.B. Thornes (eds) *Processes in Geomorphology*, Arnold, London, p 130–186.

Brunsden, D., J.C. Doornkamp, P.G. Fookes, D.K.C. Jones, and J.M.H. Kelly, 1975. Large-scale geomorphological mapping and highway engineering design. *Quarterly Journal of Engineering Geology*, vol 8, p 227–253.

Burnett, A.D., E.W. Brand and K.A. Styles, 1985. Terrain classification mapping for a landslide inventory of Hong Kong. *Proceedings of the 4th International Conference and Field Workshop on Landslides*, Tokyo, p 63–68.

Carrara, A., 1984. Landslide hazard mapping: aims and methods. In: J.-C. Flageollet (ed.), *Mouvements des Terrains*, Serie Documents du BRGM, no 83, p 141–51.

Carrara, A., and L. Merenda, 1976. Landslide inventory in northern Calabria, Southern Italy. *Bulletin of the Geological Society of America*, vol 87, p 1153–62.

Carrara, A., E. Pugliese-Carratelli and L. Merenda, 1977. Computer-based data bank and statistical analyses of slope instability phenomena. *Zeitschrift für Geomorphologie*, vol 2l, p 187–222.

Carrara, A., E. Catalone, M. Sorriso Valvo, C. Realli and I. Ossi, 1978. Digital terrain analysis for land evaluation. *Geologia Applicata e Idrogeologia*, vol 9, p 237–255.

Christian, C.S., and G.A. Stewart, 1952. *Summary of General Report on Survey of Katherine-Darwin Region, 1946*. CSIRO, Land Research Series, no 1.

Coates, D.R., 1977. Landslide perspectives. In: D.R. Coates (ed.) *Landslides*, Geological Society of America, p 3–28.

Cooke, R.U., 1984. *Geomorphological Hazards in Los Angeles*. Allen and Unwin, London.

Cooke, R.U., and J.C. Doornkamp, 1990. *Geomorphology in Environmental Management* (2nd Edition). Clarendon Press, Oxford.

Cooke, R.U., D. Brunsden, J.C. Doornkamp and D.K.C. Jones, 1982. *Urban Geomorphology in Drylands*. Clarendon Press, Oxford.

Crozier, M.J., 1986. *Landslides: Causes, Consequences and Environment*. Croom Helm, London.

Del Prete, M., and D.J. Petley, 1982. Case history of the main landslide at Craco, Basilicata, South Italy. *Geologia Applicata e Idrogeologia*, vol 17, p 291–304.

Demek, J., and C. Embleton (eds.), 1978. *Guide to Medium-Scale Geomorphological Mapping*. IGU, Stuttgart.

Einstein, H.H, 1988. Landslide risk assessment procedure. In: C. Bonnard (ed.), *Proceedings of the 5th International Symposium on Landslides*, Lausanne, vol 2, p 1075–1090.

Fleming, R.W., D.J. Varnes and R.L. Schuster, 1979. Landslide hazards and their reduction. *American Planning Association Journal*, vol 45, p 428–39.

Froelich, A.J., A.D. Garnas and J.N. van Driel, 1978. Franconia area, Fairfax County, Virginia. Planning a new community in urban setting: Lehigh. In: G.D. Robinson and A. Spieker (eds.), *Nature to be Commanded . . . Science Maps Applied to Land and Water Management*. US Geological Survey Professional Paper no 950, p 69–89.

Fulton, A.R.G., D.K.C. Jones and S. Lazzari, 1987. The role of geomorphology in post-disaster reconstruction: the case of Basilicata, Southern Italy. In: V. Gardiner (ed.), *International Geomorphology 1986*, Wiley, Chichester, p 24l–62.

Geomorphological Services Limited, 1986. *Landslide Hazard Assessment*. Report to the UK Department of the Environment.

Geomorphological Services Limited, 1987. *Review of Research into Landsliding in Great Britain: National Summary and Recommendations*. Report to the UK Department of the Environment.

Hansen, A., 1984. Landslide hazard analysis. In: D. Brunsden and D.B. Prior (eds) *Slope Instability*, John Wiley & Sons, p 523–602.

Hansen, M.J., 1984. Strategies for classification of landslides. In: D. Brunsden and D.B. Prior (eds.) *Slope Instability*, John Wiley & Sons, p 1–25.

Hearn, G.J., 1987. An evaluation of geomorphological contributions to mountain highway design with particular references to the Lower Himalaya. PhD thesis, University of London.

Hutchinson, J.N., 1968. Mass movement. In: R.W. Fairbridge (ed.), *Encyclopaedia of Earth Sciences*, Reinhold, New York, p 688–695.

Hutchinson, J.N., and M. Del Prete, 1985. Landslides at Calitri, Southern Appenines, reactivated by the earthquake of 23rd November 1980. *Geologia Applicata e Idrogeologia*, vol 20, p 9–38.

Jones, D.K.C., 1983. Environments of concern. *Transactions of the Institute of British Geographers*, New Series, p 429–57.

Jones, D.K.C., D. Brunsden and A.S. Goudie, 1983. A preliminary geomorphological assessment of part of the Karakoram Highway. *Quarterly Journal of Engineering Geology*, vol 16, p 331–55.

Jones, D.K.C., E.M. Lee, G.J. Hearn and S. Genc, 1989a. The Çatak landslide disaster, Trabzon Province, Turkey. *Terra Nova*, vol 1, p 84–90.

Jones, D.K.C., D. Brook and D. Brunsden, 1989b. Grounds for improvement. *Geographical Magazine*, vol 16 (8) (October), p 38–43.

Leighton, F.B., 1976. Urban landslides: targets for land-use planning in California. *Geological Society of America Special Paper* no 173, p 89–96.

Kienholz, H., 1977. Kombinierte geomorphologische gefahrenkarte 1:10,000 von Grindelwald. *Catena*, vol 3, p 265–294.

Kienkolz, H., 1978. Maps of geomorphology and natural hazards in Grindelwald, Switzerland, scale 1:10,000. *Arctic and Alpine Research*, vol 10, p 169–84.

King, R.B., 1987. Review of geomorphic description and classification in land resource surveys. In: V. Gardiner (ed.), *International Geomorphology 1986*, Part II, Wiley, Chichester, p 384–403.

Kojan, E., and J.N. Hutchinson, 1978. Myanmarca rockslide and debris flow in Peru. In: B. Voight (ed.), *Rockslides and Avalanches, I: Natural Phenomena*, Elsevier Science Publishers, New York, p 315–361.

Laudrey, J., 1979. *Zones exposées a des risques lies ceux mouvements du sols et des sous-sol. Region de Lons-le-Sauviera Polijniy (Jura) Orleans.* Bureau de Recherches Géologiques et Minières, 1:21,000 scale.

Lumb, P., 1975. Slope failure in Hong Kong. *Quarterly Journal of Engineering Geology*, vol 8, p 31–65.

Mitchell, C.W., 1973. *Terrain Evaluation*, Longman, London.

Muller, L., 1964. The rock slide in the Vaiont Valley. *Rock Mechanics and Engineering Geology*, vol 2, p 148–212.

Ollier, C.D., 1977. Terrain classification, principles and applications. In: J.R. Hails (ed), *Applied Geomorphology*, Elsevier, Amsterdam, p 277–316.

Plafker, G., and G.E. Ericksen, 1978. Nevado Huascarán avalanches, Peru. In: B. Voigt (ed.) *Rockslides and Avalanches, I: Natural Phenomena*, Elsevier Science Publishers, New York, p 277–314.

Radbruch, D.H., and C.M. Wentworth, 1971. *Estimated Relative Abundance of Landslides in the San Francisco Bay Region, California.* US Department of the Interior, Geological Survey.

Radbruch-Hall, D.R., R.B. Colton, W.E. Davis, B.A. Skipp, I. Lucchitta and D.J. Varnes, 1976. *Preliminary Landslide Overview Map of the Continental United States.* US Geological Survey Miscellaneous Field Studies Map, MF 771.

Regione Basilicata, 1987. *Gli Interventi per Senise, per il Consolidamento e il Transferimento di Insedimenti Abitati in Basilicata.*

Schuster, R.L., 1978. Introduction. In: R.L. Schuster and R.J. Krizek (eds.), *Landslides: Analysis and Control.* Transport Research Board Special Report no 176, National Academy of Sciences, Washington.

Siddle, H.J., M.D. Turner and S.P. Bentley, 1989. Computer aided landslip potential mapping and its application to land use planning and development control. *International Conference on Computers in Urban Planning and Urban Management, Hong Kong, Proceedings*, p 1–11.

Slovik, P.L., B. Fischhoff and S. Lichtenstein, 1980. Perceived risk. In: R.C. Schwing and W.A. Albers (eds.), *Societal Risk Assessment: How Safe is Safe Enough*, Plenum Press, New York.

Stevenson, P.C., and D.J. Sloane, 1980. The evolution of a risk-zoning system for landslide areas in Tasmania, Australia. *Proceedings of the Australian and New Zealand Geomechanics Conference*, Wellington.

Styles, K.A., A.D. Burnett and D.C. Cox, 1982. Geotechnical assessment of terrain for land management and planning purposes in Hong Kong. *Proceedings of the 1st International Symposium on Soil, Geology and Landforms—Impact of Land Use Planning in Developing Countries*, Bangkok, vol 16, p 1–16.

Styles, K.A., A. Hansen, M.J. Dale and A.D. Burnett, 1984. Terrain classification methods for development planning and geotechnical appraisal: a Hong Kong case. *Proceedings of the 4th International Symposium on Landslides*, Toronto, 1984, vol 2, p 561–568.

Terzaghi, K., 1950. Mechanisms of landslides. *Geological Society of America, Berkey Volume*, p 83–123.

United Nations, 1989. *Report of the International Ad Hoc Group of Experts on the International Decade for Natural Disaster Reduction.* General Assembly, Economic and Social Council, A/44/322/Add.1.

United States Geological Survey, 1982. Goals and tasks of the landslide part of the ground-failure hazard reduction program. *US Geological Survey Circular 880.*

Varnes, D.J., 1978. Slope movements and types and processes. In: E.B. Eckel (ed.), *Landslides and Engineering Practice*, Highway Research Board, National Academy of Sciences, Washington, Special Report no 176, p 11–33.

Varnes, D.J., 1984. *Landslide Hazard Zonation*, UNESCO, Paris.

Wilhusen, J.P., 1979. Geologic hazards in Pennsylvania. *Pennsylvania Geological Survey, Educational Series*, no 9.

13 The identification and mitigation of glacier-related hazards: examples from the Cordillera Blanca, Peru

J.M. Reynolds

Abstract Glaciers and snowfields can form potential hazards which may be significant on a short term (minutes-days) time scale, such as ice and snow avalanches, jökulhlaups, aluviones and iceberg impacts. Glacier-related hazards can be equally serious but less obvious when considered on a much longer time scale (months-years-decades), such as glacier surges and ice front fluctuations.

While most, if not all, of these phenomena occur quite commonly, they only form a hazard when human lives and/or property are at risk. As populations increase, there is greater pressure to inhabit more marginal terrain, thereby exposing more people to risks associated with glacierised mountainous regions, such as in the Andes and the Himalayas.

In the Cordillera Blanca of Peru, some 32,000 people have died in this century alone (20,000 in one incident) as a direct consequence of catastrophic glacier-related events, and many more people are still at risk. Apart from seismically-induced rock and ice avalanches, deaths have resulted from catastrophic floods caused by the sudden mechanical failure of moraine dams. There are at least four possible mechanisms for the breaching of moraines which dam high-altitude glacial lakes: (1) the melting of ice cores within a moraine; (2) overtopping of the moraine freeboard by displacement waves caused by rock/ice avalanches; (3) settlement and piping within the moraine, or piping alone, as a result of seismic activity; and (4) catastrophic sub-glacial drainage of local glacier tongues.

Successful engineering works have been undertaken to mitigate further aluviones at Laguna Paron and, more recently, at Hualcán.

Introduction

There are two objectives of this paper—first, to provide a brief outline of glacier-related and snow hazards (Section 1) and, secondly, to present examples of hazard assessment and mitigation from the Cordillera Blanca, Peru (Section 2).

Section 1: Glacier-related and snow hazards

When a glacier or glacier-related feature adversely affects human activities, directly or indirectly, it is considered to be a hazard. This general statement covers the range of human activities from hill farmers in Peru or Switzerland, for example, having their livelihoods (and lives) threatened by avalanches, through to the environmental effects of iceberg impacts on seabed oil installations off Labrador; from catastrophic glacier outbursts destroying Icelandic road communications, to glacier recessions resulting in increased storage of water behind fragile terminal morainic dams at high altitudes such as in the Andes in Peru, the Himalayas, etc. What is critical, therefore, is not only the *type* of glacial phenomenon but the *rate* at which an event takes place (Table 1).

The types of hazard can be sub-divided into two groups, those which are a *direct* action of ice or snow (e.g. an avalanche, iceberg impact) and those which give rise to *indirect* hazards (e.g. aluvión, glacier outburst, flooding, etc.). The former are much simpler to identify than the latter, which tend to be much more complex as there are many more factors which can interact to form

Table 1 Types of snow and ice hazard

Time Scale	Hazard	Description
Minutes	Avalanche	Slide or fall of large mass of snow, ice and/or rock
Hours	Glacier outburst	Catastrophic discharge of water under pressure from a glacier;
	Jökulhlaup	Outburst which may be associated with volcanic activity (Icelandic);
	Débâcle	Outburst but from a proglacial lake (French);
	Aluvión	Catastrophic flood of liquid mud, irrespective of its cause, generally transporting large boulders (Spanish; plural: aluviones)
Days-weeks	Flood	Areal coverage mostly by water
Months-years	Glacier surge	Rapid increase in rate of glacier flow
Years-decades	Glacier fluctuations	Variations in ice front positions due to climatic changes, etc.

sums of money, not only in terms of actual damage, but also indirectly due to lost production, delays in transportation, remedial action, etc., all of which can easily total many hundreds of millions of dollars, comparable to the financial costs of many major earthquakes in industrialised areas. Although governments are still slow to recognise the importance of pre-disaster planning and preventative measures, the insurance industry is now taking an increasing role in encouraging hazard assessment and mitigation research.

Evaluating the time scale of the various phenomena is important, as it provides a means of strategic planning. If a region is known to suffer from frequent avalanches, appropriate remedial measures can be instituted to reduce the damage and delays that may have otherwise occurred. Similarly, bridge and motorway designs in eastern Iceland need to take into consideration the probability, frequency and magnitude of jökulhlaups to minimise the recurrent costs of rebuilding after each event. These are obvious examples. However, where there is a long-term trend such as glacier recession, the hazards may develop unobtrusively until it is almost too late to do anything, or the problem is so great that a solution is either not affordable, practical or both. An example of this will be described in more detail later.

The strategic assessment of hazards and the way they change in style as a function of time is of utmost importance.

multiple hazards, as will be described in more detail later.

Glacier-related hazards are usually neither as dramatic nor catastrophic in a single event as a major earthquake or volcanic eruption, but the cumulative annual cost in those areas where such events are common can be equally significant. For example, flooding in the Reuss River valley in Switzerland in August 1987, which occurred as a consequence of deforestation of the local hillslopes with high avalanche rates, resulted in about US$400 million of damage (Mehr, 1989). Strategically, key communications routes can pass through areas prone to avalanche risk, such as the Gotthard-Reuss area of Switzerland so, without remedial action, avalanche activity can cost vast

Outline of snow and glacier hazards

Avalanches

The term *avalanche* covers a very wide range of types of mass movement from the slight sloughing of snow down a small incline through to the violent extreme which can result in total devastation and major loss of life. A qualitative scale of snow and ice avalanches is given in Table 2; the frequency of occurrence is an estimated figure. In remote areas, many avalanches may occur but are not reported. In central Europe, there are on average about 40 casualties (fatalities and injured) per year, although the actual numbers are increasing as more people are attracted

Table 2. Scale of snow and ice avalanches (after Perla, 1980)

Size	Potential effects	Order of magnitude estimates		
		Vertical descent	Volume (m^3)	Frequency of occurrence
Sluffs	Harmless	~10 m	1–10	10^4/year
Small	Could bury, injure or kill a human	10–10^2 m	10–10^2	10^3/year
Medium	Could destroy a timber-frame house or car (localised damage)	10^2 m	10^3–10^4	10^2/year
Large	Could destroy a village or forest (general damage)	1 km	10^5–10^6	one per decade
Extreme	Could gouge the landscape (widespread damage)	1–5 km	10^6–10^9	two per century

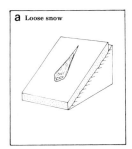

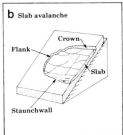

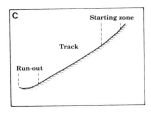

Figure 1 (a) Loose-snow and (b) slab avalanches with (c) the starting zone, track and run-out zone in profile.

to participate in recreational activities in the mountains; however, care must be taken when examining the statistics on the number of avalanches which occur each year, as an apparent increase may be a result of improved reporting techniques.

Avalanche paths comprise three parts: the starting zone, the track and the runout-deposition zone. The style of avalanche is based on the type of initiation mechanism and starting-zone failure patterns and falls into two categories, the loose-snow avalanche and the slab avalanche (Figure 1), the latter of which tends to be the more dangerous.

Loose-snow avalanches usually occur in cohesionless surface layers of dry snow, or wet snow containing liquid water, and in many cases start when the angle of repose is exceeded resulting in rotational failure. Initial failure occurs almost at a single point and the avalanche fans out in a triangular shape. Dangerous avalanches, particularly to mountaineers, also occur when overhanging cornices break off and the track is very steep (Figure 2).

In contrast, a slab avalanche occurs when a cohesive snow slab is released over a plane of weakness. As a general rule, the average slope angle of the track is 15° and often as much as 20–25°. Dry avalanches tend to travel in a straight line whereas wet-snow avalanches hug the contours more and become channelled by gullies. Unconfined tracks, such as open planar surfaces, are generally shorter than channelled tracks. In attempting to determine the extent of the run-out (e.g. Lied and Toppe, 1989), allowance must be made of an additional 100 m to account for the accompanying airblast.

Figure 2 Avalanche starting zones, Nevado Pyramide, Cordillera Blanca. Vertical height shown is about 600 m.

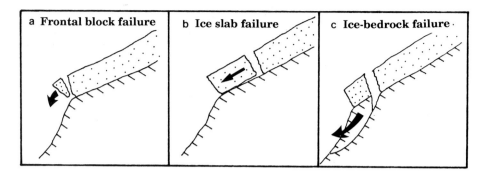

Figure 3 Three styles of ice avalanches (after Perla, 1980).

Ice avalanches (as opposed to those of snow) have three modes of release: frontal block failure, ice slab failure and ice-bedrock failure (see Figure 3). One notable ice avalanche occurred from the Allalin Glacier at Mattmark, Switzerland, when 1 million m³ of ice crashed down on accommodation huts, killing many of the local workmen employed in the construction of the Mattmark dam (Röthlisberger, 1977; Tufnell, 1984). Other major snow/ice/rock avalanches happened at Huascarán, Cordillera Blanca, Peru in 1962 and again in 1970; more details of these will be given later. Ice avalanches are common in the Cordillera Blanca.

For much more detailed descriptions and discussion of avalanches, the reader is referred to Perla (1980) and the *Avalanche Handbook* by Perla and Martinelli (1978). More specialised aspects of avalanches and avalanche control can also be found in the proceedings of the symposia on 'Snow in Motion' (*Journal of Glaciology*, volume 26, 1980) and 'Snow and Glacier Research relating to Human Living Conditions', Lom, Norway, 4–9 September 1988 (*Annals of Glaciology*, volume 13, 1989). Popular and spectacularly illustrated accounts of avalanches have been written by Cupp (1982) and Mehr (1989).

Glacier floods
In Table 1, four terms are listed—*glacier outburst,*

jökulhlaup, débâcle and *aluvión,* but these refer to only two main types of phenomena known collectively as *glacier floods.* Glacier outburst and jökulhlaup both refer to a rapid discharge of water under pressure from within a glacier. Although the term jökulhlaup was used initially to describe the type of glacier outburst which occurs as a consequence of volcanic eruption underneath a glacier, as has occurred for example at Grimsvatn, Vatnajökull, Iceland, the term is now commonly used as a synonym of glacier outburst.

There are three main mechanisms by which glacier outbursts occur: rupture of an internal water pocket, progressive enlargement of internal drainage channels, and catastrophic buoyancy with sub-glacial discharge. Each has a distinctive hazard character.

Water within the glacier's internal plumbing may build up pressure to such a point that the hydrostatic pressure exceeds the constraining cryostatic (ice-overburden) pressure and the water is able to rupture through the glacier ice. Internal water pockets may burst into existing sub-glacial drainage, increasing its discharge to produce a characteristic hydrograph (Figure 4a): the peak discharge forms a sudden frontal wave which may be many metres high and which, because of its unpredictability and rapid onset (only minutes), is extremely dangerous. This type of event accounts for all the known fatalities in Switzerland caused by glacier outbursts (Haeberli, 1983).

Figure 4 Schematic hydrographs for two common types of glacier outbursts: (a) temporary blockage and rupture of sub-glacial channel, and (b) progressive enlargements of sub- and englacial channels (after Haeberli, 1983).

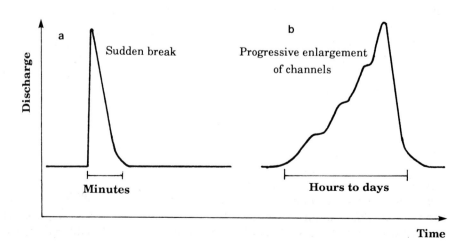

In the second mechanism, progressive enlargement of intra- and sub-glacial channels produces a build up of the discharge followed by a rapid fall-off in rate (Figure 4b). This form of hydrograph is formed by smaller channels rupturing into adjacent larger ones thereby adding to their discharge. The time period over which this process takes place can range from several hours through to days. When there is a more gradual build up, there is some time to take limited preventative measures. In some cases, water from the outburst can fountain upwards through its newly ruptured opening and can transport sub- and/or englacial debris to the ice surface. This has happened, for example, at Austre Okstindbreen in northern Norway in 1976 (Figure 5a). The resulting discharge flooded an adjacent proglacial lake whose moraine dam failed, so releasing the excess water downstream (Figure 5b,c). This deluge of water resulted in a nearby gravel roadway being washed away (Figure 5d). This has happened several times since then (Theakstone and Knudsen, 1989) and each time the little-used road has been repaired by use of a mechanical digger, this being far more cost-effective than building a more permanent rupture-proof road.

The sudden sub-glacial discharge of water beneath an ice dam, such as part of a glacier tongue damming a river, may result in larger and more devastating glacier outbursts. Water ponding behind an ice dam may cause the ice to detach from the substrate by its buoyancy. The ponded water then drains catastrophically beneath the ice tongue until the head of water drops sufficiently to bring the glacier back in contact with the ground, so terminating the outburst. The rupture of the Orba Glacier (Italy) in 1935 provides an example of the scale of discharges possible: about 22.5 million m³ of water and discharge rates of up to 13,000 m³/sec have been estimated for this event (Haeberli, 1983). Even peak discharges of only 10 m³/sec can cause damage. It has also been found that the great majority of glacier outbursts in Europe and North America occur in summer months (June to September). Seasonality and morphological factors affecting glacier outbursts were discussed more fully by Tufnell (1984) and Haeberli (1983).

Débâcle and *aluvión* are French and Spanish terms respectively for a rapid discharge of water, typically from a proglacial lake which is moraine dammed. Lliboutry and others (1977) added that, for an aluvión, the rapid discharge is of liquid mud in which large boulders are transported, irrespective of the cause (glacial or otherwise). There have been 60–70 such outbursts in Switzerland which have been well documented (see Haeberli (1983) for further references, for example). Elsewhere, such as in the Himalayas and the Andes, where populations are quite dispersed and communications more difficult, far fewer have been analysed. Indeed, there is no published 'typical' hydrograph of a débâcle/aluvión, as the recording stations along those rivers which have been instrumented have themselves been destroyed by the event!

Figure 6 illustrates the form of debris left from a typical Peruvian aluvión, which occurred in 1951 with discharges of 1.13 and 3.52 million m³ in two separate events within three months of each other. Artesanraju, the glacier at the head of the rock cliff shown in the figure, has also been notable for its ice avalanches.

Glacier fluctuations

The advance of a glacier tongue into a valley can cause two types of hazard. Should a local river be dammed by the extended ice tongue dam, there is a danger of some form of glacier outburst, as previously described. The second form of hazard is the inundation of land which is far less dangerous in terms of risk to human life, though the cost of lost revenue to farmers, repairs to damaged buildings, roads, bridges and railways, etc., can be very great indeed. In the 17th and 18th centuries, and again in the early 19th century, deterioration of the climate in Norway resulted in widespread famine due to failure of cereal crops and loss of livestock (Hoel and Worenskiold, 1962). Similar events are recorded as affecting Alpine communities at the same time, and direct damage was also caused by the advance of glaciers onto farmland.

The problem, however, is not restricted to past times: Bossons glacier in the Chamonix Valley, which began a new advance in 1953 which reached and partly destroyed woodlands, was still in progress in the early 1980s (Tufnell, 1984). A recent account of similar detrimental effects on the local land use caused by the advance of Bualtar and Barpu glaciers in the Karakoram Himalaya in northern Pakistan was given by MacDonald (1989). Another example is that of the modelling of the behaviour of Griesgletcher, Switzerland, in response to climatic changes which has been described by Bindschadler (1983). Given the continuation of the present climate, the model predicts that the glacier terminus will retreat 200 m until around the year 2023 and then advance some 150 m over several centuries. A nearby hydro-electric dam is not likely to be affected by small changes in the ice-front position in response to short-term climatic changes; however, the position of the ice front and its associated calving would threaten the dam if the long-term mass-balance, the vertical accumulation of material, increases by 0.19 m per year. This situation underlines the importance of obtaining accurate climatic models with sufficient resolution in both space and time so that potential problems can be anticipated in sufficient time so that remedial action can be taken.

Glacier fluctuations have adversely affected mining operations at, for example, the Kennecott Copper Mines in Alaska around 1911. Another example is that of the Isua mine in Greenland which is at the edge of the main

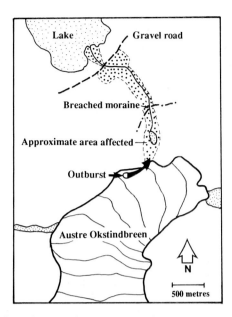

Figure 5 (a) Map of the location of the glacier outburst of Austre Okstindbreen, 5th August 1977; (b) Ice blocks and sub-glacial debris strewn on glacier surface (photographed several days after outburst); (c) outburst descending after moraine dam had ruptured; (d) (opposite page) failure of gravel road—12 hours after being deluged; the river level had abated sufficiently so as to be quite fordable on foot (photographs: M. Alexander, Department of Geography, Durham University).

ice sheet, where the economic resource comprises an iron-ore deposit 2 km long and 250–450 m wide but lying largely under 200 m of moving ice. A study of the glacier ice flow around the site of the mine was undertaken by Colbeck (1974); as it was anticipated that the economic life of the mine could exceed 30 years, the fluctuations of the ice sheet and associated processes were considered likely to change with time and the operation of the mine would have to take these into consideration, not only for economic reasons, but also for the safety of the staff working there.

Glacier surges

Rapid changes in ice front positions caused by glacier surges (anomalously fast flow rates for periods of several months) can lead to major potential hazards. A recent example of this, which also caught the attention of the world media, was that of the advance of the Valerie lobe of Hubbard Glacier in Alaska (e.g. *New Scientist*, 1986, volume 111, numbers 1519 and 1526). The surge advance, which occurred in May 1986, blocked off Russell Fiord which then overfilled with 5.4 km³ of fresh water. On 8 October 1988, the ice dam failed and discharged about 380 million m³ of water in one hour with a colossal peak discharge rate of 108,000 m³/sec. Further advances and repeated blocking off of this fiord are anticipated, which will cause consequent flooding and erosion of forest land, fish habitats, fishing camps, archaeological sites and roads (Mayo, 1989). In addition to the substantial physical disturbance of the area, the change from salt to fresh water within Russell Fiord caused severe environmental stress on marine wildlife.

Icebergs

Icebergs, made infamous by the tragedy of the sinking of the SS *Titanic* in 1912, pose a serious threat to shipping

and to oil/gas installations, particularly off the Labrador coast. Monitoring iceberg drift trajectories is important in order to determine whether, given a particular size and movement of an iceberg, it is better (more economic) to move a rig out of the way of an approaching iceberg or to attempt to divert its route around a drilling rig, for instance. Special sea-floor well heads have been designed with low profiles so as to reduce the possibility of the submerged base of an iceberg impacting them. In the Antarctic, small icebergs of several tens of thousands of tonnes have been successfully towed by a service vessel out of the way of a Deep Sea Drilling Ship which was on station.

Radar monitoring of icebergs is essential in poor visibility so as to avoid possible collisions. Even a minor brush with an iceberg can cause costly damage to a ship. With the increasing popularity of tourists visiting polar regions by ship, the risks of iceberg impact are growing. Furthermore, should the utilisation of Antarctic icebergs as a freshwater resource become feasible and economic, icebergs within major shipping lanes could prove to be very problematic (Reynolds, 1979).

Figure 6 Residual debris fan from the two aluviones in 1951 from Artesancocha. The ice cliff of Artesanraju can be clearly seen (arrowed).

Section 2: Case histories from the Cordillera Blanca, Peru

Introduction

The Cordillera Blanca in the Peruvian Andes (Figure 7) consists of granodioritic peaks up to 6,500 m with deeply dissected U-shaped valleys draining into the Callejón de Huaylas, a fertile valley corridor through which the Rio Santa flows northwards and ultimately westwards to the Pacific Ocean. The region is seismically very active, both with many earth tremors associated with the Cordillera Blanca fault (e.g. over 150 tremors with magnitudes up to Richter 3.2 were recorded in 34 days in 1987 by Deverchère and others, 1989) as well as major quakes associated with subduction of the Nazca Plate in the Peru-Chile Trench. It was one such major earthquake with an epicentre some 75 km west of Chimbote (Figure 7) which occurred in May 1970 with the loss of 80,000 lives. Of these deaths, about 20,000 were due to the catastrophic rock/ice avalanche from Huascarán Norte, details of which were given by Plafker and Erickson (1978) and summarised by Reynolds (1990).

Many of the steep-sided valleys contain lakes which have developed behind dams formed by moraines, by avalanche cones or by alluvial fans from hanging valleys at higher altitudes. Frequent avalanches of ice and/or rock result from the steepness of the slopes and the nature of the rock fractures, in conjunction with over-hanging ice cliffs of the precarious high-altitude glaciers. Occasionally, these develop into exceedingly rapid debris flows known as *sturzstroms* (Hsü, 1978, after Heim, 1882) such as happened at Huascarán in 1970 (Figure 8).

Corporación Peruana del Santa (CPS) was established in order to evaluate the safety of the mountain lakes and to use the naturally-stored water in the generation of hydro-electric power. In the mid 1940s, what is now the second largest power station in Peru was constructed at Huallanca with its water intake 9 km upstream at Cañon del Pato. This intake had to be rebuilt after it was very badly damaged by an aluvión on 20 October 1950.

CPS later became known as ElectroPeru, of which Hidrandina S.A. is a subsidiary, based in Huaraz, with responsibility for the Ancash Department, amongst others. Hidrandina S.A. has conflicting interests; on one hand it has to provide more water storage in order to maintain and, if possible, increase the water flow through the Canon del Pato intake to raise the power output still further from 150 MW, while on the other hand, it is responsible for making the high mountain lakes safe against avalanches and resultant flooding. Often this has involved the complete drainage of lakes or the lowering of water levels. Their work is further

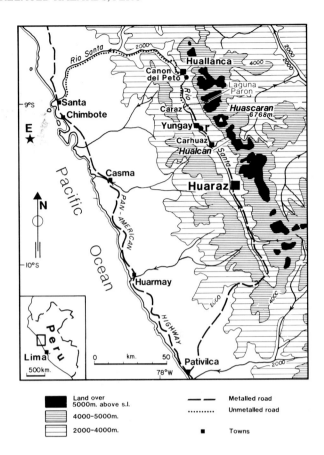

Figure 7 Map of the western central Peru. The Callejón de Huaylas is in the part of the Rio Santa valley between Huaraz and Huallanca. E indicates the epicentre of the main shock of the earthquake on 31 May 1970. Other key locations are also marked.

hampered by the extreme political unrest and poor economic climate within Peru, which can ill afford the engineering works necessary to meet both objectives. Nevertheless, the staff at Hidrandina S.A. at Huaraz have been remarkably successful in what they have achieved so far.

History of aluviones in the Callejón de Huaylas

The earliest record of an aluvión in this area is that of 4 March 1702, which was noted in the Chronicles of Father Bertrano although there was no record of fatalities or damage. This and other aluviones in the area are listed in Table 3 which is derived from Zapata (1977) and Lliboutry and others (1977).

Huaraz has suffered particularly badly from aluviones. Those who cannot afford the higher-priced, brick-built houses in safe areas, construct shacks out of whatever they can find but on land prone to such inundation which, as a consequence, is cheap (Figure 9).

150

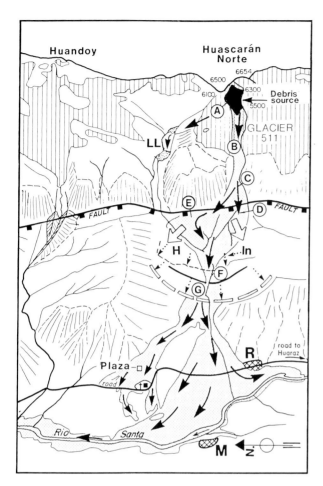

Figure 8 Sketch based on an oblique aerial photograph taken by Servício Aerofotografíco Nacional de Peru, 13 June 1970. Letters **M**, **R** and **Y** mark the positions of the towns of Matacoto, Ranrahirca and Yungay respectively; **LL**, Lagunas Llanganuco; **H** and **In**, the hamlets of Huashua and Incayoc. The general direction of movement of the main avalanche is indicated by arrows with solid tails, the limits of airborne rocks by line F and of mud spatter by line G.

Figure 9 Shanty suburb of Huaraz built along the banks of the Rio Quilcay which has been inundated each time there is an aluvión upstream.

Types of glacier hazard

One of the problems faced by Hidrandina S.A. is the recognition and identification of potential hazards. In addition to the risks of ice avalanches, which are readily identifiable, and rock avalanches, which are recognisable by appropriately trained staff, the main hazard is that of the potential failure of moraine-dammed lakes. Four possible natural mechanisms by which natural dams can fail are listed in Table 4 and illustrated in Figure 10; there have been two cases where aluviones have been triggered accidentally by engineering works being undertaken in an attempt to reduce the chances of such aluviones happening. One or more of these mechanisms can occur simultaneously, compounding both the complexity of the hazard and the possible engineering solutions, as will be described briefly later.

Melting ice cores
The impervious ice core within a moraine dam melts, so lowering the effective height of the dam: this allows lake water to drain over the residual ice core (Figure 10a). The discharge increases as the ice core melts, and as greater amounts of water filter through the moraine taking out fine material. The resulting regressive erosion of the moraine dam ultimately leads to its failure.

Overtopping by displacement waves
Lake water is displaced by the sudden influx of rock and/or ice avalanche debris (Figure 10b). The resultant

151

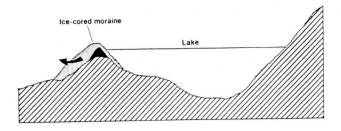

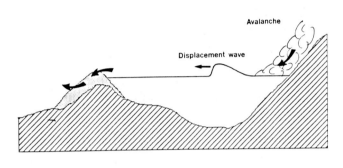

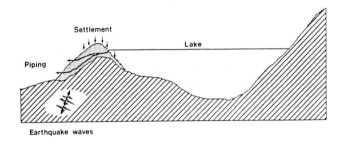

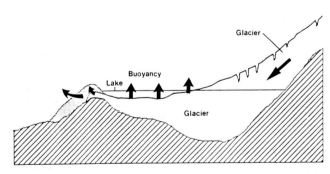

Figure 10 Four mechanisms for moraine dam failure: (a) melting ice core; (b) displacement wave from rock/ice avalanche into a dammed lake; (c) settlement and/or piping within a moraine dam (resulting from seismic activity); (d) catastrophic sub-glacial drainage.

Table 3 Historical aluviones in the Callejón de Huaylas

1702 Mar 04	No specific details
1725 Jan 06	1,500–2,000 dead (Ancash town)
1869 Feb 10	Destroyed several houses
1883 Jun 24	Much damage and many deaths
1911 Feb 10	Destroyed a small village
1932 Mar 14	No details of damage or casualties
1938 Jan 20	Local damage only, no deaths; destroyed 32 houses and 13 bridges
1941 Apr 20	Damage only to agricultural land
1941 Dec 13	6,000+ dead; discharge 4 million m^3 (Huaraz)
1945 Jan 17	Destroyed the historic ruins at Chavin; many deaths
1950 Oct 20	Severe damage to water intake, Cañon del Pato; 200–500 deaths following hazard mitigation works
1951 Jul 16/17	Rupture of Artesancocha; discharge 1.13 million m^3
1951 Oct 29	Rupture of Artesancocha; discharge 3.52 million m^3
1952 Nov 06	Damage to agricultural land
1953 Mar 01	Destruction of three hamlets and one town; many deaths
1962 Jan 10	4000 deaths; destruction of Ranrahirca and nine hamlets destroyed following an ice avalanche from Huascarán Norte; other aluviones on eastern Cordillera Blanca, of which one flooded the Pompei Mine
1965 Jan 19	Few deaths and localised damage to buildings
1970 May 31	Major earthquake and sturzstrom; 20,000 killed by sturzstrom; 60,000 killed by earthquake; total devastation.

Table 4 Possible mechanisms for the failure of moraine dams

- Melting ice-cores within the moraine dam
- Rock and/or ice avalanches into a dammed lake
- Settlement and/or piping within the moraine dam
- Sub-glacial drainage
- Engineering works

waves overtop the freeboard of the dam causing regressive erosion and eventual failure.

Settlement and/or piping
Earthquake shocks can cause settlement of the moraine, reducing the dam freeboard to a point that lake water drains over the moraine causing regressive erosion and eventual failure (Figure 10c).

Sub-glacial drainage
A receding glacier with a terminus grounded within a proglacial lake can have its volume reduced without its ice front receding up-valley. This comes about by the volume of meltwater within the lake increasing to a point

that the formerly-grounded glacier floats, thus allowing instantaneous sub-glacial drainage (Figure 10d). Such drainage can destroy any moraine dam, allowing the lake to discharge until the glacier loses its buoyancy and grounds again.

Problems with hazard-mitigation engineering works

One of the main difficulties in changing water levels or dam structures artificially, to reduce hazard, is knowing what effects such actions will have on the cause of the original hazard. Commonly, having identified a glacial hazard, there is insufficient time or resources, or both, to identify the full extent of the problem. In two cases in the Cordillera Blanca, remedial engineering works have unintentionally actually triggered catastrophic discharge events.

Lliboutry and others (1977) gave an account of two examples in Peru where engineering works complicated matters. The first happened in 1953 during the artificial lowering of the water level within Laguna Tullparaju. An earth slide caused a 12-m high displacement wave which poured into a trench, excavated as part of the engineering works, and almost led to the total failure of the moraine dam. The lake level fell by 18 m but the leakage stopped due to the presence of resistant black glacial clay within the moraine and to a temporary dam formed by sandbags thrown in by local workmen.

In the second case, Glacier Kogan in 1949–1950 generated frequent ice avalanches into Laguna Jancarurish which was dammed by a moraine with only a 10-m freeboard. As this was regarded as too little to protect the dam from regressive erosion from displacement waves, a trench through the moraine was dug to lower the lake water level in careful stages. Unfortunately, the regenerated glacier, which lay beneath the cliffs down which the ice avalanches fell, had been supported by its natural buoyancy within the lake (Figure 11); with this support removed, the toe of the regenerated glacier collapsed into the lake causing its own displacement wave,

'Glacier Kogan'

Figure 11 The failure at Laguna Jancarurish in Quebrada Los Cedros, Peru (after Lliboutry and others, 1977).

destroying the trench and sluice gates. Rapid regressive erosion took place around a waterfall around the remains of the sluice gates, and eventually led to the failure of the moraine dam. The lake level fell 21 m and the volume of water discharged was 6–10 million m³. The resulting aluvión resulted in up to 500 deaths along the Quebrada Los Cedros and severe damage to the water intake at Canon del Pato. The Huallanca power station was out of operation for over a year as a consequence.

Another graphic example of the destructiveness of displacement waves was given by Plafker and Eyzaguirre (1978). On 18 March 1971 a rock avalanche slid into Laguna Yanahuin, about 200 km north east of Lima, and the resulting displacement wave destroyed a mining camp killing 400–600 people.

Laguna Paron: a successful example

One successful example where major engineering works have been completed is that of Laguna Paron (Figure 12). Prior to remedial works, the lake covered an area of about 1.6 km² and had a volume of around 75 million m³. The lake is dammed by Hatunraju, a spectacular glacier-filled moraine some 250 m high. Water from the lake drains through the moraine and discharges by way of a series of small springs on the down-valley side. The freeboard was less than 2 m at the toe of Hatunraju until, in 1952, it was raised 2.6 m by the construction of an earth dam. Since then, a tunnel has been excavated through the granodiorite bedrock on the northern side of the lake. The tunnel discharges water into the existing water course with the consequent lowering of the lake level by 20 m.

Drilling through Hatunraju moraine failed to reach bedrock (Huaman, 1983) and even subsequent geophysical surveys carried out by the author in 1988 failed to provide an unambiguous estimate of the thickness of the moraine dam (Reynolds, 1989). To date, the exact structure of Hatunraju moraine dam remains unclear and, as long as this uncertainty exists, there is no way of knowing how stable the moraine dam is. Should a cataclysmic failure occur, an aluvión of some 50 million m³ could occur which would be sufficiently large to destroy the town of Caraz 16 km downstream.

The freeboard at Hatunraju came perilously close to being overtopped in 1951. On the night of 16/17 July, an ice avalanche involving 4–5 million m³ of ice fell from the sheer ice-front of Artesanraju into Artesancocha, which resulted in 1.13 million m³ of water escaping over the lower moraine which dams the lake (see Figure 6). The water level in Artesancocha fell by 7 m but raised Laguna Paron by 0.7 m. The leak in the dam created a 22-m wide waterfall, causing regressive erosion leading to the total failure of the moraine three months later. In the second aluvión, 3.52 million m³ of water discharged into Laguna Paron raising its level by 2 m. Had the failures happened

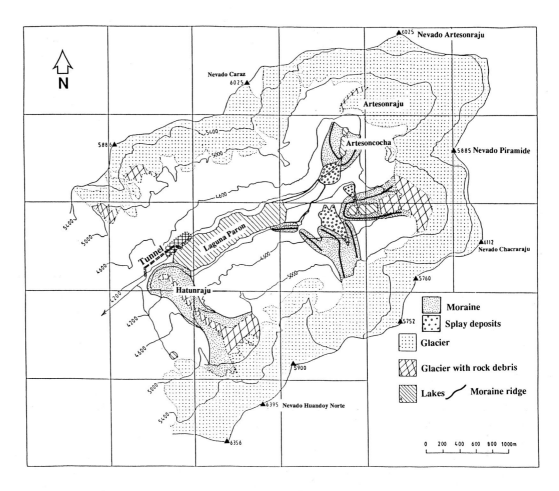

Figure 12 Map of the area around Laguna Paron.

during the rainy season, Hatunraju would have also failed and Caraz would have been flooded, possibly causing thousands of deaths. Artesancocha and other lakes in the Paron catchment area still represent potential hazards to Laguna Paron. While the freeboard is as large as it is at present (some 20+ m) and as long as Hatunraju remains intact, Caraz is safe. However, Laguna Paron is silting up; eventually the drainage outlet to the tunnel will be covered and the discharge stopped. The lake level will then rise and create a major hazard. Although water levels in Hatunraju have been at their lowest levels since the tunnel was constructed, piping is known to occur within the moraine (Reynolds, 1989). If the water level rises, so too will the water pressure on the moraine and the resulting increased discharges through the internal piping could undermine the structural stability of the entire moraine. It is thus imperative that the water level of Laguna Paron does not rise much above its present level.

Furthermore, parts of the drainage tunnel which are associated with a major fault are collapsing. If no remedial work is done to counteract this internal spalling of the tunnel, complete blockage could occur. Further details of the hazards have been given in a report to Hidrandina S.A. by the author (Reynolds, 1989).

Hualcán: a brief case history

In August 1988, the author was asked to investigate a lake at Hualcán, at the head of a valley upstream of the town of Carhuaz (Figure 13). Hidrandina S.A. had previously surveyed the area in 1985 (Huaman, 1985) and, on a visit in early 1988, had discovered that the moraine dam had lowered by about 4 m. After studying the area briefly, it appeared that the moraine dam was ice-cored and was lowering at an average rate of 11 cm per month. Water from the lake was already filtering over the ice core within the moraine (Figure 14) and draining via two springs on the down-valley side. If nothing had been done, it was anticipated that the moraine dam would have failed within two months, inundating Carhuaz in an aluvión and probably resulting in many casualties. The ice core was in fact part of the grounded snout of the two glaciers which drain into the lake (Figure 15).

The solution to the immediate problem was to lower the water level by 8 m, preferably 12 m, to below the level of a natural bedrock rim beneath the moraine. However, by September 1988, no remedial work had been put into effect, as apparently there was no money to pay for

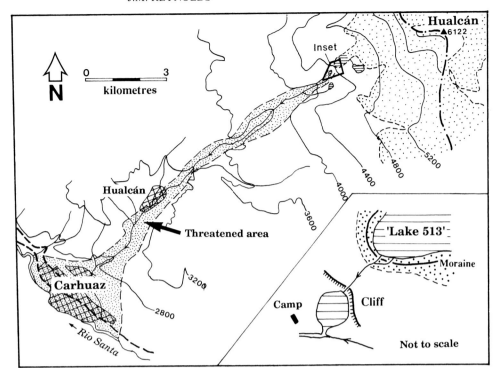

Figure 13 Map of the Hualcán-Carhuaz area showing the location of the problematical 'Lake 513'.

it. Following media attention initiated by the author, ElectroPeru found sufficient money to pay for one drainage syphon which was installed by 23 October, the syphon discharging at a rate of 190 l/s. By Christmas 1988, the lake level had not changed even though about 1 million m³ had been discharged, a volume probably equivalent to the rainy-season precipitation recharge within the lake catchment. Following intervention by the author to the British Embassy in Lima, funds for a

second, larger diameter pipe were made available. The new pipe was installed in January 1989 and the two pipes discharged at a rate of about 500 l/s and the lake level began to go down; by 31 March 1989, the level had fallen 2 m having discharged an additional 480,000 m³ (Quiroz, 1989) and by June 1989 it had fallen a further 2 m. By June 1990, the lake level was down a total of 5 m and a drainage channel had been excavated through the moraine to ensure that the lake level could not rise

Figure 14 Moraine dam at Hualcán prior to remedial works.

Figure 15 The two source glaciers which drain into 'Lake 513' at Hualcán.

beyond this level (C. Portocarrero, personal communication; see also Reynolds, 1990).

Further monitoring of the lake-glacier system has ceased due to the lack of funds. Whether the lake at Hualcán is now safe remains to be seen. The lack of monitoring means that important practical information about the response of the local glaciers to induced drainage may not be obtained. Even though funding could be made available through international aid, the presence of Maoist terrorists in the area has made it highly unlikely that any further remedial action can be taken until the political problems have been resolved.

Conclusions

Glacier hazards and related phenomena constitute major hazards in the Cordillera Blanca, Peru, and in similarly glacierised regions of the world. In addition to seismic activity, the most significant hazard in the Peruvian region is through the mechanical failure of natural dams of high-altitude lakes. Given adequate remote-sensing mapping techniques, it should be possible to identify areas with high risk and advise the local authorities accordingly. The example of remedial work at Hualcán demonstrates that effective measures can be instituted at costs of only a few thousand dollars and, if carried out quickly enough, can prove effective to mitigate against potentially devastating aluviones.

The success of future remedial hazard work is now dependent upon resolving internal political troubles, especially with regard to the Maoist terrorists. The combination of earthquake and aluvión could result in not just one lethal event but many, culminating in widespread destruction including that of the second largest power plant in Peru. Hazard assessment in the Cordillera Blanca is not just a matter of draining a few lakes, but rather the establishment of a hazard assessment strategy so that local authorities can plan for the safe generation of electricity and for the overall safety of the local population well into the next century.

Acknowledgements

The author is extremely grateful to colleagues at Hidrandina S.A. for providing the opportunity to study these problems. Thanks are due to Ing Cesar Portocarrero, Ing Alcides Ames and Ing Marcos Zapata for their assistance and encouragement during the field work, and particularly to Ing Portocarrero for continuing to supply reports on the progress of the work at Hualcán. The author is also grateful to the Royal Society for financial support of this work.

References

Bindschadler, R., 1983. The predicted behaviour of Griesgletcher, Wallis, Switzerland, and its possible threat to a nearby dam. *Annals of Glaciology*, vol 4, p 295.

Colbeck, S., 1974. A study of glacier flow for an open-pit mine: an exercise in applied glaciology. *Journal of Glaciology*, vol 13 (69), p 401–414.

Cupp, D., 1982. Avalanche! *National Geographic Magazine*, vol 162 (3), p 280–305.

Deverchère, J., C. Dorbath and L. Dorbath, 1989. Extension related to high topography: results from a microearthquake survey in the Andes of Peru and tectonic implications. *Geophysical Journal International*, vol 98 (2), p 281–292.

Hoel, A., and W. Worenskiold, 1962. Glaciers and snowfields in Norway. *Norsk Polarinstitutts Skrifter*, Number 126.

Haeberli, W., 1983. Frequency and characteristics of glacier floods in the Swiss Alps. *Annals of Glaciology*, vol 4, p 85–90.

Heim, A., 1882. Der Bergsturz von Elm. *Zeitschrift der Deutchen Geologischen Gesellschaft*, vol 34, p 74–115.

Hsü, K.J., 1978. Albert Heim: observations of landslides and relevance to modern interpretations. In: Voight, B. (ed.), *Rockslides and Avalanches. 1: Natural Phenomena*, Elsevier, Amsterdam, p 71–93.

Huaman, A., 1983. Descripcion—pruebas de permeabilidad y colocacion de piezometros de las perforaciones S-100, S-101, S-102, S-103, P-120 y G-110 efectuadas en el dique de la Laguna Paron y lectura de la variacion del espejo de agua. Electroperu, Glaciologia y Seguridad de Lagunas, Huaraz, August 1983 (unpublished).

Huaman, A., 1985. La laguna en formation en el glaciar 513A (Hualcán). Hidrandina S.A., Huaraz (unpublished).

Lied, K., and R. Toppe, 1989. Calculation of maximum snow-avalanche run-out distance by use of digital terrain models. *Annals of Glaciology*, vol 13, p 164–169.

Llliboutry, L., B. Morales, A. Pautre and B. Schneider, B., 1977. Glaciological problems set by the control of dangerous lakes in Cordillera Blanca, Peru. I: Historical failures of morainic dams, their causes and prevention. *Journal of Glaciology*, vol 18 (79), p 239–254.

MacDonald, K.I., 1989. Impacts of glacier-related landslides on the settlement at Hopar, Karakorum Himalaya. *Annals of Glaciology*, vol 13, p 185–188.

Mayo, L.R., 1989. Advance of Hubbard Glacier and 1986 outburst of Russell Fiord, Alaska, USA. *Annals of Glaciology*, vol 13, p 189–194.

Mehr, C., 1989. Are the Swiss forests in peril? *National Geographic Magazine*, vol 175 (5), p 637–651.

Perla, R.I., 1980. Avalanche release, motion and impact. In: Colbeck, S.C. (ed.), *Dynamics of Snow and Ice Masses*, Academic Press, New York, p 397–462.

Perla, R.I., and M. Martinelli, 1978. *Avalanche Handbook*. US Department of Agriculture, Forest Service, Agriculture Handbook 489. US Government Printing Office, Washington, DC, USA.

Plafker, G., and G.E. Ericksen, 1978. Nevados Huascarán avalanches, Peru. In: Voight, B. (ed.), *Rockslides and Avalanches. 1: Natural Phenomena*, Elsevier, Amsterdam, p 277–314.

Plafker, G., and V.R. Eyzaguirre, 1978. Rock avalanche and wave at Chungar, Peru. In: Voight, B. (ed.), *Rockslides and Avalanches. 2: Engineering Sites*, Elsevier, Amsterdam, p 269–279.

Quiroz, J., 1989. Proyecto de desague laguna en formacion No 513. Hidrandina S.A., Huaraz, Glaciologia e Hidrologia. Internal report (unpublished).

Reynolds, J.M., 1979. Icebergs are a frozen asset. *The Geographical Magazine*, vol 52 (3), p 177–185.

Reynolds, J.M., 1989. Hazard assessment in the Callejón de Huaylas, Cordillera Blanca, Peru. Hidrandina S.A. Glaciologia e Hidrologia, Huaraz, Peru (unpublished).

Reynolds, J.M., 1990. Geological hazards in the Cordillera Blanca, Peru. *AGID News*, No 61/62, p 31–33.

Röthlisberger, H., 1977. Ice avalanches. *Journal of Glaciology*, vol 19 (81), p 669–671.

Theakstone, W.H., and N.T. Knudsen, 1989. Temporal changes of glacier hydrological systems indicated by isotopic and related observations at Austre Okstindbreen, Okstindan, Norway, 1976–87. *Annals of Glaciology*, vol 13, p 252–261.

Tufnell, L., 1984. *Glacier Hazards*. London, Longman.

Zapata, M., 1977. Origen y evolucion de las lagunas. Internal report (unpublished), Instituto de Geologia y Mineria, Glaciologia y Seguridad, Huaraz, Peru.

Part Four

The 'Quiet' Hazards

14 Sea-level changes in China—past and future: their impact and countermeasures

Wang Sijing and Zhao Xitao

Abstract With the rapidly increasing concentration of carbon dioxide in the atmosphere and the resulting "greenhouse effect", the global climate of the next century will get warmer and there will be dramatic environmental changes. One effect that human beings must face seriously is the rapidly rising sea level, especially in low-lying coastal plain and delta regions with dense populations and highly advanced industry and agriculture.

China has not only an extensive land area, but also a long, sinuous and wandering coastline characterised by a large number of large and small islands; big cities and industrial regions such as Shanghai, Tianjin and Guangzhou are located on coastal plains and deltas. Therefore, the history, characteristics and tendency for sea-level changes in China, and the impact on environment and engineering, should all be studied carefully and the general and specific policies discussed and implemented strictly; Chinese scientists, government officals and departmental personnel responsible for developing strategies and policies must pay much attention to the prospect of sea-level rise.

Assessment of the history and characteristics of Quaternary and present-day sea-level changes in China allows a preliminary estimate to be made on the possible tendency of future sea-level changes and potential impacts; account is also taken of crustal movement and land subsidence.

History and characteristics of sea-level changes in China

Quaternary sea-level changes

Along the coasts of mainland China and her islands, reliable outcrops of Pleistocene marine strata are found only on the coasts of Taiwan and the northwestern coast of Hainan Island. Some reported outcrops of possibly marine Pleistocene strata on the mainland, Hainan Island and Xisha Islands, such as the "Old Red Sand" and "Lufeng Formation" (on the northern coast of Shandong Peninsula) and the "Shidao Formation" (in the Xisha Islands) are subject to controversy and are gradually being disproved as marine Pleistocene by more and more researchers.

Generally, information on the Pleistocene sea-level changes in China is chiefly obtained from cores from the coastal plain and the adjacent shelf. Information from these cores includes biostratigraphy, sedimentology, magnetostratigraphy and other data such as ^{14}C dating, and division and comparison of the marine to transitional-continental strata. The results obtained from these studies establish a general trend of sea-level changes for China in the Pleistocene.

Taking 2.5 Ma as the lower boundary of the Quaternary, between nine and eleven marine beds of Pleistocene age have been found in China during the last decade. The characteristics of the transgressive marine layers are shown in Table 1. As shown in the table,

(a) The number and distribution of Quaternary transgressive marine layers in China are closely related to geographical location, i.e. the number of layers reduces from the western coasts of the Yellow Sea, Bohai Sea and East China Sea southwards to the Taiwan Strait and the northern coast of the South China Sea and to the Shandong and Liaoning Peninsulas; in this direction also the ranges of transgression reduce and the degree of marine character becomes less pronounced. A contrary trend is

Table 1 Divisions and correlations of Quaternary marine transgressive layers along the coast of China and the adjacent continental shelf (after Lin Hemao and Zhu Xionghua, 1989, and others)

Epoch	Distribution	Environment and facies	Age* (ka BP)
Holocene			
	1. Whole seas and coasts in China	Neritic, nearshore, littoral and transitional	10–0
Late Pleistocene			
Late	2. Bohai Sea and its coasts, Yellow Sea, East China Sea and Taiwan Strait and their west coasts, South China Sea and some of its coasts	Neritic, nearshore, littoral and transitional	29–21
Middle	3. Bohai and Yellow Seas, Shanghai	Very weak marine	69–53.5
Early	4. Middle of Bohai Sea and its north and west coasts, Yellow Sea, East China Sea and their west coasts	Littoral, neritic, nearshore bay	130–70
Middle Pleistocene			
Late	5. North and west coasts of Bohai Sea, southern Yellow Sea and East China Sea and their west coasts	Nearshore or transitional	—
Early	6. North and west coasts of Bohai Sea, southern Yellow Sea and its coasts, Shanghai	Nearshore or transitional	—
Early Pleistocene			
Late	7. Southern Yellow sea, Hebei Plain and Shanghai	Littoral or transitional	970–730
Middle	8. Southern Yellow Sea	Littoral	1700–1650
Early	9. Beijing Plain	Pelagic or stenohaline	2260

* [14]C or palaeomagnetic dating

evident on Hainan and Taiwan Islands, however, where some Pleistocene marine strata have occurred at various heights from 20 m to 300 m above sea level.

(b) Even in the same region, the marine transgressive layers of different dates are different in their ranges of distribution, elevation and marine character.

These characteristics show that fluctuations of sea level in China occurred at least nine times since the beginning of the Quaternary, and it is possible there were

eleven. Each fluctuation is, however, different along the coast of China and the adjacent continental shelf.

Crustal movements with different rates, amplitudes and directions, have had significant effects on changes of relative sea level which are manifested in the number, thickness, distribution, degree of marine character and depth of burial or elevation of the transgressive layers.

Holocene transgressive sequences, because of their good condition—not being deeply buried, good outcrop, moderate thickness and availability of samples for ^{14}C dating—have been subject to much more detailed study in relation to sea-level changes. These transgressive layers (No 1 in Table 1) extend over a wide area of China's coastline, an area greater than that covered by the Pleistocene transgressions. Not only did they reach those regions where the Pleistocene trangressions had occurred in the past, such as the coasts of the Bohai Sea, Yellow Sea and East China Sea, and the Zhujiang (Pearl) River, Hanjiang and Minjiang deltas, but also the bays and deltas of the Liadong and Shandong Peninsulas and the rocky coasts of Guangxi and Hainan provinces. Furthermore, the marine character of the Holocene transgression was most strongly developed in the Quaternary. Hundreds of ^{14}C dates have demonstrated that the rate of transgression kept pace throughout the Holocene, covering the range of dates from 8000 years BP to the present; the maximum transgression occurred during the period 7000–3000 years BP with dates varying in different regions, though most fall within the range of 6000–5000 years BP.

For the last few thousand years since the Holocene marine transgressive maximum, the coastline retreated seaward step by step, leaving behind a series of sedimentary, morphological and biological features closely indicative of ancient coastlines and sea levels. These features have abundant materials for ^{14}C dating.

In China, widely distributed indicators of former sea levels, which have been relatively well studied, are chiefly as follows (Figure 1):

(a) *cheniers*, chiefly occurring in the west and south coasts of the Bohai Gulf, the west coast of Laizhou Bay, the northern Jiangsu Plain, the southern Yangtze Delta, some small bays and deltas in Liaoning, southern Fujian and eastern Guangdong;

(b) *coral reefs*, chiefly distributed around the islands and northern coast of the South China Sea, Taiwan and Hainan;

(c) *beachrock deposits*, chiefly occurring in the same areas as the coral reefs.

A series of sea-level change curves has been obtained for the Holocene and the latest Quaternary since the 1970s with respect to location, sediment thickness and ^{14}C dates of transgressive layers or ancient sea-level

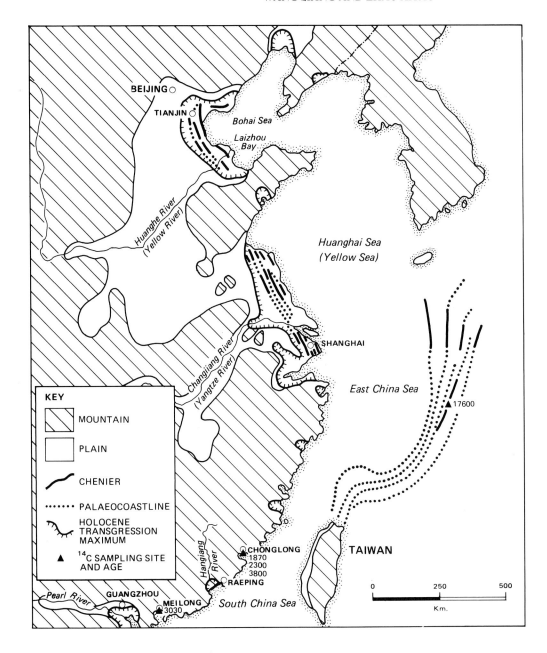

Figure 1 Distribution of cheniers and palaeocoastlines along the coasts and on the adjacent shelf of China.

indicators (Zhao Xitao and others, 1982; Zhao Xitao and Zhang Jingwen, 1982; Yang Huiaren and Xie Zhiren, 1984; Huang Zhenguo and others, 1987). Zhao Xitao and others pointed out that Holocene sea-level changes (Figure 2) can be divided into three periods: the rapid rise before 6000 years BP, the highest sea level between 6000 and 5000 years BP, and a slight descent during the last 5000 years. They also found that the Holocene sea level had fluctuated during that period, with six peaks at 8500–7800, 7300–6700, 6000–5000, 4600–4000, 3800–3100 and 2500–1500 years BP. The last four of these were higher than present-day sea level.

According to records recently obtained from the northern Jiangsu Plain, the Haihe Estuary and the Hangjiahu Plain, the existence of high oscillatory

Holocene sea levels is not only verified but an additional oscillation has been demonstrated at about 9000 years BP (Zhao Xitao and others, 1990; Li Yuangfang and others, 1989; Hong Xueqing, 1989).

Recent sea-level changes

Although a few hundred tidal stations have been established along the coasts of China during the last decades for the purposes of navigation, only a very few of these have long-term tidal records available for studying sea-level changes. Moreover, the reliability of the data is affected by many factors, notably two basic geological problems which affect data from tidal stations when sea-level changes and crustal movements are discussed:

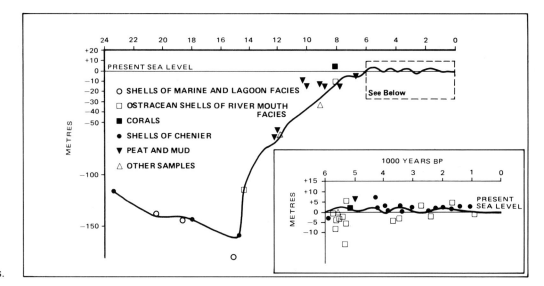

Figure 2 Sea-level changes curve of eastern China during the past 20,000 years.

(a) changes of the gauge datum as a result of crustal movements and land-surface subsidence are not usually considered;

(b) differences of regional crustal movements with different signs, amplitudes and rates, and of land subsidence due to non-tectonic causes, are not considered in regional or national studies.

On the other hand, bench marks on land are defined according to mean sea level, assuming that the land on which the stations stand is stable relative to sea level.

To solve this problem, the authors have suggested that a common datum must be set up for sea-level changes and crustal movements. With the advantage of a long and wandering coastline, and taking into account the known crustal movements with signs, amplitudes and rate, an isostatic datum has been developed in China by Huang Liren and others (1990) to measure vertical movements of the crust, and work begun on some representative tidal stations. All the work follows the concepts of Zhao Xitao and others (1982), and is based upon data obtained from the precise relevelling of China from the 1950s to the 1980s.

Based on the isostatic datum, the recent vertical movement of the land along the coasts of China has been obtained in the form of curves (Figure 3). Using the results of the federal levelling survey, corrections for crustal movements and land subsidence for gauge zeros for some tidal stations have also been made, and so the rates of recent sea-level changes around these stations have been obtained (Figure 4, Table 2).

Based on Figures 3 and 4, and Table 2, and all the data mentioned above, it is evident that:

(a) The vertical movements of the crust in different parts of the coasts of China show regional features

which are evident in the succession of Quaternary crustal movements, and the net change in eustasy derived by removing the effect of crustal movement from the total change in sea level, is seen to be of an oscillatory character;

(b) after correction, the sea-level changes around the 13 stations in Table 2 generally show a rise at a rate of 0.65 mm/year.

This rate of rise is in step with global tendencies which have rates of 1.0–1.5 mm/year for the last century of records, though the data for China indicate a lower value: the cause of this difference will be discussed later.

Future sea-level change tendencies

Data from instrumental observations suggest that from 1880 to 1980, the average temperature of the Northern Hemisphere had risen about 0.4°C (Hanson and others, 1981) and the sea level about 10–15 cm, which was explained as the result of the thermal expansion of the upper layers of the ocean and the melting of ice caps (Gornitz and others, 1982; Barnett, 1983). The data also show that the concentration of carbon dioxide had increased from 315 parts per million (ppm) in 1958 to 343 ppm in 1984, a rise of a quarter over the value for the 1860s of 270 ± 10 ppm. In considering the use of fossil fuels, which constitute about 80% of existing energy systems, together with their expected growth in the energy systems of the future, scientists have estimated various scenarios for future sea-level rise, as shown in Table 3 (see also Warrick and Oerlemans, 1990).

Forecasting by the US Environmental Protection Agency (EPA) is in general agreement with the details for East China, though the values are slightly higher.

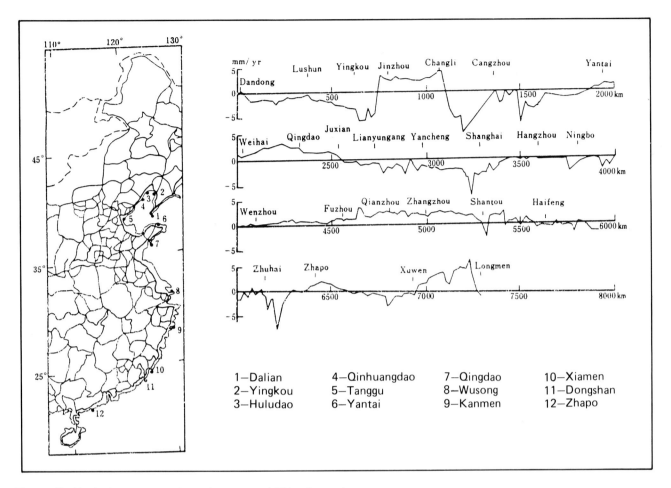

Figure 3 Vertical movement along the coasts of China in mm/yr.

Thus, the EPA researchers take 1.0–1.5 mm/year as the rate of rise in the "eustatic" sea level (ESL) in the past century and as a basis for forecasting. In fact, what the tide-gauge records have demonstrated is the rise of sea level *relative to the continents* (RSL); supposing that tidal stations had been positioned all over the oceans, in subsiding and uplifting regions in the same numbers, the average rise would be the eustatic value. Unfortunately,

Table 2 Sea-level changes according to the data from tide-gauge stations along the coasts of China relative to the isostatic datum

No	Tide-gauge station	Sea-level changes relative to bench mark (mm/year)	Correction due to the isostatic datum (mm/year)	Sea-level changes relative to the isostatic datum
1	Qinhuangdao	−3.3±2.2	4.2	0.9
2	Tanngu A	1.7±3.8	−1.9	−0.2
3	Tanggu B	2.0±3.6	−1.9	0.1
4	Longkou	0.2±2.9	−0.9	−0.7
5	Yantai	−2.3±1.5	2.1	−0.2
6	Rushan	−3.2±2.2	2.6	−0.6
7	Qingdao	−1.4±1.5	1.6	0.2
8	Wusong	2.0±3.4	−1.5	0.5
9	Kanmen	1.5±3.3	−2.2	−0.6
10	Xiamen	1.1±3.3	1.5	2.6
11	Dongshan	−1.6±5.5	1.6	0.0
12	Zhapo	0.3±4.0	2.5	2.8
13	Beihai	3.3±3.4	0.4	3.7
	Mean	0.03	0.62	0.65

165

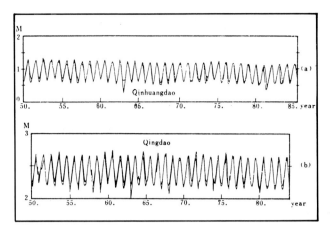

Figure 4 Monthly average curves from the tide-gauge, observed and fitted, at Qinhuangdao (Hebei) and Qingdao (Shandong).

Table 3 Scenarios for future sea-level rises (expressed in centimetres) (Titus, 1986)

Year	2000	2025	2050	2075	2085	2100
NAS (1983)	—	—	—	—	70	—
EPA (1983)						
Low	4.8	13	23	38	—	56
Mid-range low	8.8	26	53	91	—	144
Mid-range high	13.2	39	79	137	—	217
High	17.1	55	117	212	—	345
Hoffman and others (1986)						
Low	3.5	10	20	36	44	57
High	5.5	21	55	191	258	368

in actuality stations are poorly distributed, especially the long-term stations being mainly on the east and west coasts of the Atlantic Ocean in the middle latitudes of the Northern Hemisphere where a glacio-isostatic depression zone has been active in post-glacial times. So the rate of "eustatic" sea-level change deduced by general statistical methods is not the real ESL, in which the effects of vertical crustal movement and relative sea-level rise due to thermal effects and melting of glaciers and icecaps are the chief contributors (Fairbridge, 1989; Healy, 1989; Warrick and Oerlemans, 1990).

Based on the mean annual sea-level data selected from 162 stations around the globe, Xie Zhiren estimated that the average rate of sea-level rise from 1900 to 1987 was 0.7–1.2 mm/year, which is close to that of China, i.e. 0.65 mm/yr in recent years. Taking into account the crustal and land subsidence during the last century, the predicted value of ESL derived by EPA and others from determinations of past average rates should be corrected and, after correction, its value is close to the predictions of other workers. Due to the distortion of the geiod by the melting of the Antarctic and Greenland ice sheets, the value of ESL in East China is 5–15% higher than that of the world's high scenario (Table 3).

Taking account of all the crustal movement and land subsidence factors mentioned above (Figure 5), the ESL prediction scenarios for East China and the Yangtze Delta are shown in Table 4, these figures being regarded as a case study for future ESL rise and its impacts.

Impact of sea-level rise on the environment and engineering countermeasures

Possible impacts of sea-level rise on the environment

A future rise in sea level of several centimetres will have serious impacts on China's coasts, especially on the low-lying coastal plains and deltas where there is rapid subsidence of the land at rates higher than those of the sea-level rise. As a case study, some impacts of this rise on the environment and engineering constructions of the Yangtze Delta area will be dealt with here.

Inundation of wetlands Along the coasts of China, wetlands are mainly located on sandy or muddy coasts of the Yellow Sea, the East China Sea and the Bohai Sea. The coastal wetlands in the Yangtze Delta, with an area of 1000 km², are chiefly distributed on the stable sectors, the

Table 4 Scenarios for future eustatic sea-level rise in East China and of the future relative sea level in the Yangtze Delta area (cm)

Year	2000 high	2000 medium	2000 low	2025 high	2025 medium	2025 low	2050 high	2050 medium	2050 low
East China	8.1	5.4	3.0	29.2	19.6	11.7	71.2	46.0	26.1
Northern North Jiangsu littoral plain	10	7	4	37	25	16	83	55	33
Southern North Jiangsu littoral plain	13	8	5	45	30	17	98	64	35
Eastern Yangtze Delta (Xie Zhiren, 1989)	16	11	7	56	40	26	116	81	51
Southern Hangjiahu Plain	10	7	4	39	25	16	83	55	33
Shanghai	17	11	6	61	40	22	124	81	44

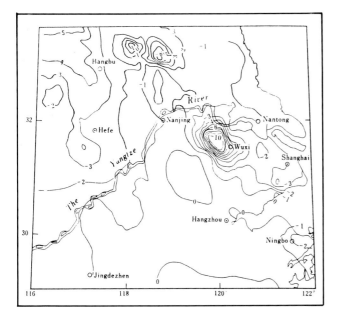

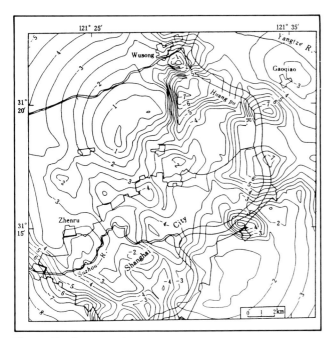

Figure 5 Crustal deformation (a) in the Yangtze Delta and (b) in Shanghai, in mm/yr.

agricultural lands, of which two-thirds extends across the area from Sheyang Mouth to Dongzao Harbour. The coastal wetland has very high agricultural productivity and supports large human and animal populations. With a rate of sediment supply equal to the rise in sea level, the wetlands would be continuously aggraded and keep pace with the rise, and with a greater rate new wetlands would be created. Thus the area of wetland would be maintained or even expanded slightly. If the rate of rise in sea level is greater than the rate of sedimentation, wetland would be reduced. According to observed data

of the rate of sedimentation along the North Jiangsu coast, there would be no loss of coastal wetland area, whereas in stable areas such as the Dongyuan coast in Gidong county the loss due to the low scenario would be 30% by 2050.

Erosion of coasts and effects on sea walls Because sea walls along the coasts of the Yangtze Delta have been built up, the erosive effects of sea-level rise are increased sea-wall erosion and damage, countering the beneficial effects of beaches banked up against the sea walls.

The main effects on sea walls are the increase in overwash due to rising storm levels and consequent destruction. While overwash is a function of the present heights of the walls, the degree of damage is a function of future storm levels and their return periods. The sea walls in this area are now designed to meet the highest tidal level from past records plus wave run-up expected in a typhoon of scale 10 with a freeboard (additional height above static high tide level) of 2.0–2.5 m. From now on, heights of sea walls are being designed to take into account the highest expected 100-year tidal level plus the wave run-up height and freeboard, a standard that is slightly higher than the original one. The effectiveness of such sea-wall design also depends on the strength or weakness of the slope-protection measures.

According to the low and mid-range scenarios for 2050, the heights of new sea walls should be 1.0–0.6 m or 0.9–0.5 m higher than at present.

A numerical method of assessing the levels of protection needed for the coasts of this area has been devised, using the following four factors:

(a) the differences in height between the new and present sea walls,
(b) the rate of retreat of the shoreline,
(c) the width of the intertidal zone, and
(d) the density of economic activity in the zone being protected.

The coasts of the area can be divided into four categories of those affected most seriously, seriously, moderately and slightly (Figure 6).

Storm disasters One of the principal disasters to affect the coasts of China is storm, especially in the case of cities on large estuaries. Because of the interaction between run-off and tidal currents, a combination of typhoon with floods and exceptional high astronomical tides can lead to serious disaster. Sea-level rise will aggravate the situation in the future. For example, Shanghai City (Zhu Jiwen and others, 1988a and b) has an elevation of 4 m according to the Wusong datum but the height of the average high tide for the majority of its history has been 5 m; so industry and agriculture as well

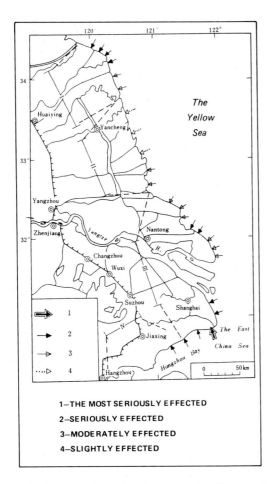

1—THE MOST SERIOUSLY EFFECTED

2—SERIOUSLY EFFECTED

3—MODERATELY EFFECTED

4—SLIGHTLY EFFECTED

Figure 6 Impact of sea-level rise on the Yangtze Delta area: 1, the most seriously affected; 2, seriously affected; 3, moderately affected; 4, slightly affected.

Sea-water intrusion and the effect on low-lying coastal land Low-lying land located in the area of Taihu Lake and eastwards, including the lowlands of northern Kunshan, Qinsong, Taipu River, the western Punanyun, Deqing and Yuhang, mainly stands at an elevation of 3.0–4.5 m (by the Wusong datum), though parts are under 2.0 m. Even at present this area is often much influenced by floods. For example, the 5-year return-period flood of 1982 covered most of the area from May to July. The water level in Taihu Lake is 4.4 m (Mao Rui and Xu Pengzhu, 1989) and when the five sluice gates in Yangcheng Lake were opened the volume of water discharged was 2,150 m³/second but this was not sufficient to drain the flood in this area. Jiaxing and Suzhou cities alone accounted for a total inundated area of 2430 km², which is 21% of the total area of the region, and the economic loss was put at one billion Yuan Renminbi. If the sea level rises, the height of low tides outside the gates will rise too, and the discharge ability will be reduced accordingly: a rise of 1 m in sea level and of 0.8 m in low tide level will reduce this to 80%, and most of the floods in a year of floods will accumulate in this area also. If the low tide level rises 1 m in the low-water season, equivalent to the gate heights along the river, the drainage ability will decrease 2–3 times more, and there will also be the possibility of sea water entering.

Countermeasures for mitigating the hazards of rising sea-level

Rising sea level is a phenomenon of world-wide scope and is a common tendency in the coastal areas of many countries. However, uplift and subsidence vary from place to place, so the hazards due to the rise are different in their extent and character. In coastal areas where the crust is rising or where there is bedrock erosion, the rise in sea level will cause no serious hazard; but where crustal or land subsidence is occurring this hazard may threaten human activity. Therefore, the basic task in planning measures for hazard mitigation is to analyse the characteristics of neo-tectonic movement in the coastal zone and to contour the area and locate the centre of subsidence. Close attention should be paid to land subsidence due to extraction of groundwater for industrial and agricultural water supply or due to mine drainage.

On the basis of hydro- and engineering-geological studies, measurement of levelling profiles would be useful to determine the area and rate of subsidence. This is a basic measure for the prediction of hazards caused by rising sea level.

The following countermeasures should be taken to mitigate the hazards in coastal zones:

Control of land subsidence due to human activity For an area with industrial and agricultural water supply, in general, the management of underground water on the

as the security of the population of the area depends on protection by sea walls. At present the heights of sea walls in the Shanghai area vary from 6.0 m to 9.5 m in different districts, and their ability to defend against storm is accordingly variable. The present penetration-proof wall design is worked out according to the height needed for defence against a 5.3 m tidal level but, to fit the standard of protection for 2020, the present walls should be reinforced with about 40 cm of height added; even so, of the 186 km length of sea wall in Shanghai, only 20% meets the original standard set in 1974.

Under the hypothesis that the high tide level in the Huangpu River will be 2–3 m higher than the land surface in Shanghai City, the city would be inundated if the levees broke up; flooded to a depth of 1 m, the centre of the city would have 60 km² flooded and, if industry were brought to a halt for ten days, the loss would be more than 600 billion Yuan Renminbi (US$125 billion). To increase the height of the present levees against a one-metre sea-level rise will take 2,550,820 m³ of earth and stone for each 70,000 m³ centimetre of rise.

basis of water-table levels is necessary to prevent a fall in the groundwater level and the land subsidence process that may accompany it. In urban areas of large cities or other areas of large groundwater abstraction, the back-injection of water into the aquifer can be adopted to control land subsidence, provided measures are taken to protect the water quality of the aquifer. This method has proved effective in Shanghai, for instance.

Another important measure is to control the water table to prevent sea-water intrusion into aquifers under land, which results in deterioration of groundwater quality. For this purpose it is necessary to ensure that the water table is higher than that of the sea, though this will become increasingly difficult as sea level rises.

Sea water intrusion into mine chambers due to mining extraction and drainage is a serious hazard in the coastal zone of Shandong Peninsula. A procedure of filling cavities and water-tight grouting is used in many cases.

Prevention of sea-water seepage and windstorm tides In many areas, the coastal embankment plays a vital role, particularly in the case of sea-level rise. To check the safety of the embankment and to maintain and strengthen it are important countermeasures. Prevention of wave erosion is necessary to protect coastal embankments. In especially vulnerable areas, a water-tight screen formed by grouting a series of drill holes is an effective measure.

Drainage of low-lying land and land improvement In addition to protecting embankments and water-tight grouting, the drainage of low-lying land, levelling and creation of drainage canals and ditches are necessary. In some areas, the salinisation of groundwater and soil takes place due to sea-water intrusion. Creation of surface or underground water reservoirs may provide a suitable source for soil improvement and drinking-water supply. This is especially important for the urban areas of large cities. For example, Qingdao city often suffers a shortage of drinking water in the summer, seriously affecting peoples' lives in the area.

Taking into account sea-level rise in planning new urban areas When new towns or industrial areas are planned, the rise in sea level must be taken into account. It is important to distinguish high flood-risk areas from those with only slight risk, for it is essential to locate the new towns in safe areas. In some areas, hazard mitigation measures as described above should be taken during the process of urbanization.

References

Barnett, T.P., 1983. Recent changes in sea level and their possible causes. *Climate Changes*, vol 5 (1), p 15–38.

Fairbridge, R.W., 1989. Crescendo events in sea-level changes. *Journal of Coastal Research*, vol 5 (1).

Gornitz, V, L. Lebedeff and J. Hassen, 1982. Global sea level trend in the past century. *Science*, vol 215 (4540), p 1011–1614.

Hanson, J.E., D. Johnson, A. Lacis, S. Lebedeff, D. Rind and G. Russell, 1981. Climate impact of increasing atmosphere carbon dioxide. *Science*, vol 213, p 957–966.

Healy, T., 1989. Report of meetings. *Journal of Coastal Research*, vol 5 (1), p 161–162.

Hong Xueqing, 1989. Low temperature climates and sea-level changes during Holocene. In: Yang Zigeng and Lin Hemao (eds), *Quaternary Processes and Events in China Offshore and Onshore*, China Ocean Press, Beijing, p 111–116 (in Chinese, with English abstract).

Huang Liren, Yang Guohua, Hu Huimeng and Ma Qing, 1990. Isostatic datum and its application in sea level change.

Huang Zhengguo, Li Pingri, Zhang Zhongying and Zong Yongqiang, 1987. Sea level changes along the coastal area of south China since the late Pleistocene. In: Qin Yunshan and Zhao Songling (eds), *Late Quaternary Sea Level Changes*, China Ocean Press, Beijing, p 142–154.

Li Yuangfang, Niu Xiujun and Li Qingchun, 1989. Environment and stratigraphy in the mouth region of Hihe River during the Holocene Epoch. *Acta Geographica Sinica*, vol 44 (3), p 363–375 (in Chinese, with English abstract).

Lin Hemao and Zhu Xionghua, 1989. Correlation of Quaternary marine transgressions in southern Huanghai (Yellow) Sea and in China coastal areas. In: Yang Zigeng and Lin Hemao (eds), *Quaternary Processes and Events in China Offshore and Onshore*, China Ocean Press, Beijing, p 33–45 (in Chinese, with English abstract).

Mao Rui and Xu Pengzhu, 1989. Effects of the future sea level rise on the tidal level of the Yangtze Estuary and on the flood defence in the low lands east of Taihu Lake.

Titus, J.G., 1986. Causes and results of sea level rise. In: J.G. Titus (ed.), *Effects of Changes in Stratospheric Ozone and Global Climate*, EPA, vol 1, p 219–248.

Warrick, R.A., and H. Oerlemans, 1990. Sea level rise. In: J.J. Houghton, G.J. Jenkins and J.J. Ephraums (eds), *Climatic Change—the IPCC Assessment*, Cambridge University Press, p 257–281.

Xie Zhiren, 1989. A preliminary forecast on the future tendency of sea level changes in the Yangtze Delta area (abstract).

Yang Huairen and Xie Zhiren, 1984. Sea level changes along the east coast of China over the last 20,000 years. *Oceanologia et Limnologia Sinica*, vol 15 (1), p 1–12 (in Chinese, with English abstract).

Zhao Xitao, 1989. Cheniers in China: an overview. *Marine Geology*, vol 90, p 311–320.

Zhao Xitao and Zhang Jingwen, 1982. Basic characteristics of the Holocene sea level changes along the coastal areas in China. In: *Quaternary Geology and Environment in China*, China Ocean Press, Beijing, p 155–160.

Zhao Xitao, Gen Xiushan and Zhang Jingwen, 1982. Sea level changes in eastern China during the past 20,000 years. *Acta Oceanologica Sinica*, vol 1 (2), p 248–258.

Zhao Xitao, Lu Gangyi, Wang Shaohong, Wu Xuezhong, Zhang Jingwen, Xie Zhiren, Yang Dayun, Zhang Changsu and Ren Jiangzhang, 1990. A preliminary study on the Holocene stratigraphy and changes of environment and sea level. *Kexue Tongbao*, vol 35 (4), p 285–288 (in Chinese).

Zhu Jiwen, Shi Yafeng, Ji Zhixiu, Jiang Zhixuan and Xie Zhiren, 1988a. A preliminary study of impacts of sea level rise on the Yangtze Delta and the littoral Northern Jiangsu Plain.

Zhu Jiwen, Jiang Zhixuan, Ji Zhixiu and Yang Guishan, 1988b. Trend of sea level rise and its impact on the storm disasters in Shanghai City.

15 Rising groundwater: a problem of development in some urban areas of the Middle East

D.J. George

Abstract Rising groundwater has caused considerable, widespread and costly damage to structures and services in many urban areas of the Middle East, and represents a significant hazard to public health. The principal cause is artificial groundwater recharge from potable supplies, sewer systems and irrigation returns which may greatly exceed both the natural rate of recharge and the capacity of the natural subsurface and surface drainage systems to receive them.

Much damage has occurred because the potential for rising groundwater and associated problems were not recognised prior to development, the effect of a higher water table not often being considered in the design of deep basement buildings and buried services. Site-specific remediation is generally costly as considerable damage has often been done prior to implementation. It may also adversely affect adjacent structures or services. Regional groundwater control may be more cost-effective, but problems of settlement due to removal of soluble salts, suspended sediments or rock matrix are often difficult to predict and may have severe consequences.

With hindsight it is possible, from the Middle East experience, to identify mitigation measures which could reduce the impact of rising groundwater upon future developments under similar circumstances. These include (1) water-use control and conservation measures to reduce the rate of artificial recharge, which could be seen as part of an overall water management plan, and (2) the adoption of appropriate planning and construction regulations to enable structures and services to be appropriately designed. It is envisaged that the latter strategy would involve the hydrogeological and geotechnical investigation of the proposed development areas, together with the formulation of a groundwater budget to enable susceptible areas to be identified. Appropriate development, which may include limiting basement depths or giving priority to foul or stormwater sewer systems in susceptible areas, may greatly reduce damage as a result of rising groundwater.

Introduction

Rising groundwater has affected many urban areas in the Middle East from large cities to individual development sites, resulting in widespread and costly damage to property, and dangers to public health. The prior identification of areas at risk from potential rising groundwater problems and the adoption of appropriate planning and development strategies could have helped to prevent such damage occurring.

This paper draws upon relevant publications and the author's experience gained during the completion of a number of rising groundwater projects and studies in the Middle East. The aim is to provide a broad understanding of the causes and problems associated with rising groundwater and to present some considerations appropriate to the design of remedial schemes. With hindsight it has been possible to identify mitigative measures which, based upon the Middle East experience, could have reduced the scale and impact of rising groundwater problems now being experienced. Much can be learned from the Middle East experiences, and the adoption of appropriate strategies for future urban developments in similar circumstances could greatly reduce the economic and human cost of future rising groundwater problems.

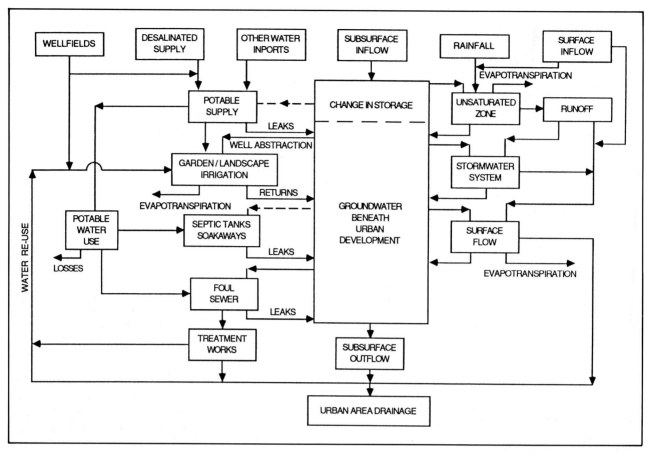

Figure 1 Typical simplified water budget for groundwater beneath a Middle East urban development.

The scale of the problem

Rising groundwater beneath urban developments in the Middle East has especially given cause for concern over the last decade. Problems have affected developments on the scale from major cities to individual isolated developments. Rising groundwater problems have been reported from Kuwait (Kuwait City), Qatar (Doha), Egypt (Cairo) and Saudi Arabia (Riyadh, Jeddah, Jizzan, Tabuk, Buraidah, Madinah and Jubail)—and this list is almost certainly incomplete.

Measured in terms of economic cost and loss and injury to human life, the problem is one of major proportion. The cost of studies related to rising groundwater, such as those completed by the Kuwait Institute for Scientific Research for the Ministry of Electricity and Water in Kuwait, Arriyadh Development Authority in Riyadh, Saudi Arabia, and the Ministry of Electricity and Water in Qatar, now runs into millions of dollars. The cost of damage caused by rising groundwater is large but, due to the scale of the problem, difficult to estimate. The cost of known site-specific remedial measures runs to hundreds of millions of dollars (Salih and Swann, 1989) and this estimate is thought to represent only a small proportion of the real total. The cost to human life or health is much more difficult to assess, because direct

identification of rising groundwater as a cause is difficult to prove; however, there have been reports of loss of life in Qatar (Anon, 1983a) and many more potential public health problems have been identified.

Rising groundwater problems generally occur most acutely where water tables were typically low and the rise in groundwater was not foreseen prior to development. In areas where water tables were typically closer to the surface, buildings and structures have often been designed to resist the effects of a high water table and therefore damage has been less acute but problems arise where water tables were initially lower.

The causes of rising groundwater in urban areas of the Middle East

Urban developments, by their very nature, will result in changes in the amount of groundwater recharge reaching the water table beneath developed areas (Price and Reed, 1989). Paved areas may increase direct run-off, whereas irrigation returns and leaks from water supplies and sewer systems may provide direct recharge to the water table. In the Middle East, where rainfall is commonly low, potential evaporation high and recharge

Figure 2 Dry weather flows from part of Riyadh's storm-water sewer system now exceeds 100,000 m³/day.

to the water table small and sporadic, the net effect of urban development is often to increase the contribution to the water table; in some cases the urban recharge rate may exceed estimated naturally occurring rates by an order of magnitude.

Figure 1 illustrates the components of a typical simplified water budget for groundwater beneath a Middle East urban development. Only the major elements have been identified, with their principal interactions shown by arrowed lines; dotted lines show where the interaction is small or uncertain. The inputs can be numerous and composed of both natural and man-made sources. There can be many cycles within the system, such as the abstraction of groundwater from wells and re-entry as irrigation returns. The principal outputs from the entire system are evapotranspiration and surface and subsurface drainage. It is often the lack, or overloading, of these systems which contributes to the rising groundwater problems.

The scale of the drainage problem is illustrated by observations completed to the south of Riyadh, in Wadi Hanifah, during 1988. Dry weather flows in the wadi, immediately to the south of the Wadi Batha outfall, were estimated to be in excess of 3 m³ per second. The water feeds a series of lakes and ponds which continue down Wadi Hanifah for over 60 km south of Riyadh. The drainage water was derived from dry weather flows from the stormwater system (Figure 2) and effluent from the sewage treatment plant, although groundwater may also have made a contribution.

Table 1, based upon work published by Salih and Swann (1989), provides an estimate of the importance of some of the principal components of the groundwater budget for the City of Riyadh for 1987, the components

being divided into man-made and natural. From these values it can be shown that man-made inputs to the system exceed the natural input by a factor of eight or more. Hassan (1989) reported that in 1989 the daily water supply to Riyadh was more than 900,000 m³, of which 622,000 m³ enters the groundwater; of this 38% originates from cesspools, infiltration pits and leaks

Table 1 Estimated groundwater budget for the main inputs and outputs, 1987, Main Drainage Area, Ar Riyadh; based on data from Salih and Swann (1989)

Inputs and Outputs of Groundwater	Value in m³/day
Main inputs to the groundwater system	
Man-made	
Potable system leaks	111,000
Septic tanks/soakaways	57,000
Irrigation returns	134,000
Total	302,000
Natural	
Rainfall infiltration	14,000
Subsurface inflow	23,000
Total	37,000
Main outputs from the groundwater system	
Man-made	
Net flow to stormwater system	112,000
Well abstractions	70,000
Net flow to foul drainage	70,000
Total	252,000
Natural	
Subsurface flow	50,000
Total	50,000

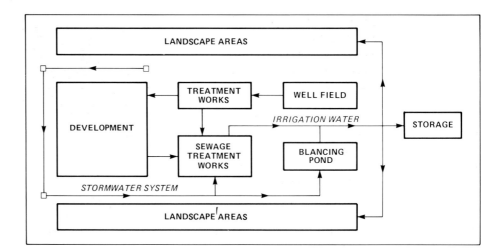

Figure 3 Schematic layout of typical single site development which has suffered from rising groundwater problems in the Middle East where subsurface and surface drainage are poor.

from sewers, 30% comes from potable system leaks, 23% from irrigation water returns, and 9% from natural sources.

Results of other studies have attributed the cause of rising groundwater to various elements of the water budget. In Doha (Anon, 1986) and Kuwait (Anon, 1987), irrigation returns, leaks from septic tanks and sewers and potable system leaks were identified as the main cause. In Jeddah (Abu Rizaiza, 1987) the lack of a centralised sewage disposal facility and low permeability soils were blamed.

Smaller-scale rising groundwater problems show a similarity to those of large urban developments in that they stem from a lack of natural drainage and a plentiful supply of water. A schematic layout of a typical single-site development, which has experienced rising ground-water problems on flat, poorly drained strata, is shown in Figure 3. The only major output from such a system is evapotranspiration; this inevitably results in a water table close to the surface and the accumulation of salts in root and soil zones in the landscape and surrounding areas.

The principal source of water fuelling the rising groundwater is therefore, in most cases, the import of water into the urban area from wellfields and desalination plants for potable and irrigation purposes. Where this is combined with a lack of surface or subsurface drainage, contributions to the groundwater budget often exceed losses and the groundwater levels will rise.

In some cases the groundwater has risen to the surface, where it forms springs and contributes to surface flows; in other areas the rise has been constrained by regulating mechanisms. These include flow to sewer systems (as in Kuwait and Riyadh), increased subsurface flow due to higher gradients, and loss of water through evapo-transpiration. The rate of rise may also be influenced by the abandonment of existing groundwater abstractions. In Riyadh rates of groundwater rise of 1 m per year were common, although up to 3 m per year occurred in some

areas. In Kuwait, rates of rise were lower and generally less than 0.5 m per year.

Hurst and Wilkinson (1986) described the causes of rising water and damage to structures and services in European cities including London and Birmingham. These showed some remarkable similarities with situations in the Middle East, although on a longer time scale. The principal cause of rising groundwater in London and Birmingham has been attributed to the reduction in groundwater abstractions due to the decline in manufacturing industries since the 1960s. The main cause of damage was attributed to the assumption that the water levels reduced by abstractions were considered "stable" and consequently the effect of a higher water table was often not considered in the geotechnical design of deep basement buildings and buried services.

In some Middle East cities the abandonment of existing groundwater supplies has also contributed to the rising groundwater problem. In Riyadh, for example, the simplified sequence of events leading to the present rising groundwater problems can be summarised as follows:

Pre-1950	Riyadh undeveloped: hand dug wells, water abstracted by animal power for irrigation and potable use.
1950–1953	Tubewells constructed into shallow limestone; some decline in groundwater levels.
Late 1950s	Deep tubewells sunk to the sandstone aquifer to supplement shallow wells.
1960s	Development of city commencing: abandonment of shallow wells due to contamination and large drawdowns.
1970s	Rapid development: water imported into city from external wellfields.
Early 1980s	Desalinated water from Jubail and Wasia wellfield.
Early 1980s	Rising groundwater problems encountered.

Mid 1980s   Site-specific remediation of rising
            groundwater problems.
1987   Arriyadh Development Authority
            commenced regional study of rising
            groundwater.

As with some European cities, the period of rapid development of large buildings with deep basements occurred when groundwater levels were depressed following abstraction, and the effects of higher water tables were not always considered in the design of such buildings.

Problems associated with rising groundwater in some urban areas of the Middle East

There are a wide range of problems associated with rising groundwater, depending upon the particular geotechnical and hydrogeological conditions, and the construction practices employed at the locality concerned. For the purpose of this paper they have been divided into the following categories:

(a) Damage to structures;
(b) Damage to services and roads;
(c) Overloading of sewer systems and treatment plants;
(d) Salting and waterlogging of soils; and
(e) Public health hazards.

Many of the problems caused by rising groundwater in urban areas stem from the failure to identify it as a potential problem and to plan, design and construct so as to minimise the potential impact, both on a local and regional scale.

Damage to structures
Damage to structures in the Middle East is widespread and has resulted in severe damage both to large developments with deep basements (Figure 4) and small residential developments with shallow foundations. The principal causes of damage are chemical attack of materials, settlement and heave of soils and fills, and mechanical damage due to erosion and hydrostatic forces. Settlement due to solution of rock matrix and construction materials is a particular problem in some areas.

Geotechnical problems due to fluctuations in the water table or flowing groundwater have resulted in structural damage in many areas, notably Jizzan, Jeddah, Kuwait and Riyadh. The main problem areas identified are:

(a) Heave and swelling of cohesive soils;
(b) Settlement and collapse of cohesive and granular soils;
(c) Consolidation of soils on dewatering;

Figure 4   Damage to basement wall and water seepage, Kuwait.

(d) Heave and settlement of poorly consolidated backfill (Figure 5);
(e) Solution of rock matrix;
(f) Wash-out of fines (Figure 6); and
(g) Loss of pile or foundation bearing capacity.

Some of the geotechnical problems associated with rising groundwater in Riyadh are described by Salih and Swann (1989), who reported many cases of rising groundwater causing substantial structural damage; settlement and collapse of both cohesive and granular soils has also been reported. Heave and settlement problems were found to be particularly common in backfill materials. The presence of deep cohesive and granular soils contained within steep sided valleys cut into the underlying limestone also provides opportunity for the differential settlement of buildings which straddle these features.

Engineering structures founded on or in soluble rock are at risk if the water moving through them is unsaturated in respect to the soluble rock (James and

Figure 5 Subsidence damage due to consolidation of poorly compacted fill is a common problem.

Edworthy, 1985). In Kuwait, removal of fines and the solution of gypsum from the underlying sediments is a particular problem, especially when associated with inappropriately designed drainage systems. In Jizzan, solution of halite and anhydrite by recharged water has resulted in substantial damage to buildings and services (R. King, personal communication, 1989).

Intensively loaded foundations on soluble soils are vulnerable when only very small settlements can be tolerated. Such problems have been met in Iraq in relatively shallow gypsiferous soils with a fluctuating water table (Anon, 1983b).

Flooding of basements has occurred in many urban developments; often the basements have not been designed to resist the hydrostatic forces or chemical attack imposed by the rising groundwater. In Kuwait and Riyadh it is common to find basement floor slabs lifted and a central longitudinal crack developed. Salih and Swann (1989) described vertical movements of basement slabs of up to 9 mm; the cracking may result in a flow of water into the basement, bringing with it suspended sediment and resulting in sub-slab voids.

Figure 6 Inappropriate solutions to dewatering problems can lead to loss of fines and soil bearing capacity.

Figure 7 Damage to structural concrete by aggressive groundwater is a problem in some basements.

Cracking also exposes the reinforcement to corrosion. Chemical damage of reinforced concrete in basements (Figure 7) is a common occurrence as a result of the moderate to high chloride and sulphate concentrations in some Middle East urban groundwaters, as illustrated in Table 2. Salts have been observed to accumulate on the surface and within the concrete just above the level of the groundwater outside the basement; the likely mechanism is illustrated in Figure 8. Damage is enhanced where the concrete has been poorly specified or constructed, and spalling of the concrete with subsequent corrosion of the reinforcement is commonly observed.

Where groundwater rises close to the surface, capillary rise and evaporation may take place, resulting in the accumulation of salts at the surface. This has resulted in salina conditions, as described by Fookes and Collis (1975); these conditions are commonly found where the urban developments are situated on flat impermeable soils and there is a lack of foul or stormwater drainage. The engineering problems encountered under such conditions were described by Fookes and others (1985) and

need no further amplification here. A particular problem is the chemical attack of materials and fills, leading to accumulation of salts and damage to structures just above ground level.

Damage to roads and services

The geotechnical problems associated with rising groundwater, as previously described, also result in damage to roads and services. Salih and Swann (1989) gave examples of service-trench collapse and damage to services. In Doha, the aggressive groundwater was found to destroy asbestos cement pipes used for the potable water supply (Eccleston, 1986). A further potential problem identified in Doha was the effect of hydrogen sulphide, generated as the result of anaerobic decomposition of sewage effluent, upon concrete and metal services above the water table.

The effect of groundwater close to the surface upon roads which have not been appropriately designed has been severe. The principal damage mechanisms observed are mechanical erosion and salt accumulation, both within the sub-base and beneath the bitumen; a review of damage to thin bituminous roads and preventive and remedial measures was given by Obika and others (1989). The damage to aggregates comprising the pavement was described by Fookes and French (1977); once the road has been damaged by salt accumulation, the pavement quickly deteriorates under traffic use. Flowing surface water in Riyadh has resulted in the erosion of pavements, leaving channelled surfaces and undermining the surface layers (Figure 9).

Overloading of sewer systems and treatment plants

In Kuwait, Riyadh and Doha, water levels have risen above the invert levels of the foul or stormwater drainage

Table 2 Typical range of concentrations of selected analytes for groundwater samples from beneath Kuwait and Riyadh

	Central Riyadh	Kuwait
Total dissolved solids	190–2700	5000–8500
pH value (pH limits)	6.9–7.7	7.8–8.3
Chloride (Cl)	30–400	350–400
Sulphate (SO$_4$)	73–1000	2400–4300
Nitrate (NO$_3$)	35–100	—

All values in mg/litre except for pH

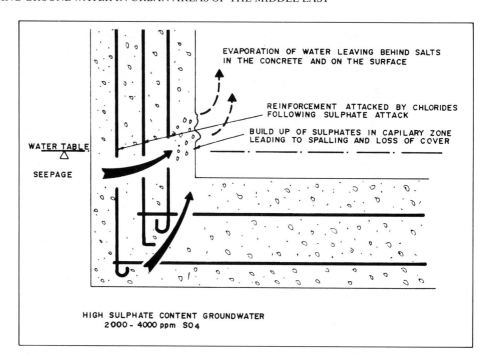

EVAPORATION OF WATER LEAVING BEHIND SALTS
IN THE CONCRETE AND ON THE SURFACE

REINFORCEMENT ATTACKED BY CHLORIDES
FOLLOWING SULPHATE ATTACK

BUILD UP OF SULPHATES IN CAPILARY ZONE
LEADING TO SPALLING AND LOSS OF COVER

WATER TABLE

SEEPAGE

HIGH SULPHATE CONTENT GROUNDWATER
2000 - 4000 ppm SO4

Figure 8 Chemical attack of reinforced concrete basement structures by sulphate and chloride.

systems. Leaks from the groundwater to these systems act as regulating mechanisms, reducing the rate of further rise; as a result, flows to sewage works are increased, and dry weather flows occur within the stormwater systems. In Riyadh the net flow to the foul sewer system in 1987 was estimated at 70,000 m³ per day (Salih and Swann, 1989), and the dry-weather flow to the stormwater system some 112,000 m³ per day, some of which was derived from discharge by site-specific dewatering systems within the city (Figure 10).

Increased flows to the treatment plants result in increased operating costs and may require the installation of additional capacity. An increase in the salt content of treated water intended for after-use may also be a consequence.

Salting and waterlogging of soils

Where the water table rises close to the surface, accumulation of salts may occur due to capillary rise and evapotranspiration, and surface deposits of salt are a common feature in many undrained urban areas. The accumulation of salt and waterlogging has resulted in

Figure 9 Groundwater may rise to the surface resulting in springs and damage to road pavements.

Figure 10 Dewatering requirements for deep basements in highly transmissive aquifers can be large and result in overloading of sewer systems.

Figure 11 Surface flows of polluted groundwater present obvious potential hazards to public health.

loss of landscape areas and agricultural yield. In Doha it was suggested that the rise of groundwater due to recharge beneath the urban areas may have resulted in upward hydrostatic displacement of groundwater in surrounding rural areas (Eccleston, 1986).

Health hazards

In many rising-groundwater situations, potential public health risks are evident; in practice, however, it has proved difficult to attribute public health problems directly to rising groundwater. Potential public health hazards observed in the Middle East associated with rising groundwater include the following:

(a) Surface flows and stagnant areas of groundwater polluted with sewage effluent (Figure 11);
(b) Breeding of insects and other disease vectors in surface water;
(c) Polluted groundwater entering potable systems;
(d) Unauthorised use of polluted drainage water for

crop irrigation;
(e) Surface water on roads and degradation of road pavements giving rise to traffic accidents;
(f) Collapse of buildings or structures;
(g) Hydrogen sulphide generation in polluted groundwater; and
(h) The proximity of standing water and electrical equipment in basement areas.

The Qatar Study (Anon, 1983a), drew attention to possible public health risks due to contamination of groundwater by sewage effluent, and attributed the loss of life to the generation of hydrogen sulphide by anaerobic decomposition of sewage effluent in groundwater. Abu Rizaiza (1987) reported on the problem of mosquitoes breeding in surface water in Jeddah.

The author is unaware of other major studies which specifically relate public health problems to rising groundwater and considers that there is opportunity for further work in this field.

179

Water table control measures

Water table control measures on a site-specific basis are a common feature in many rising-groundwater areas of the Middle East. Site-specific water table control is often costly, not always appropriate, and often affects adjacent areas. Control on a regional scale can be more cost-effective and is now being increasingly considered, as in Qatar, Kuwait and Riyadh.

Quite commonly, significant costly damage has already been done to structures before remediation is implemented, and the remedial measures employed are not always totally effective; in some cases they have been inappropriate and have exacerbated the problems. Inappropriate remedial measures have resulted in the removal of fines or soluble salts from beneath or around the structures and have caused building settlement and damage. Inappropriate drainage systems, which rapidly clog with precipitated salts and lose efficiency, have also been constructed. Even where dewatering systems are successful at relieving hydrostatic forces on basements, grouting of sub-foundation voids and repair of chemical damage to concrete and reinforcement has been necessary.

Although other potential remediation systems have been considered for site-specific problems, on a regional basis drains or wells in one form or another are the usual options considered. In Jeddah, Kuwait and Riyadh, in areas where relatively permeable strata exist, dewatering by the use of wells has been shown to involve significantly lower capital costs than drainage systems. The principal disadvantages of well systems are the on-going running and maintenance costs, together with a greater potential to cause settlement and thus damage to adjacent structures and services. Wells may in some cases result in differential settlement due to high groundwater gradients and loss of bearing capacity due to the removal of fines or soluble material from beneath or around foundations.

In the higher transmissivity aquifers, such as the shallow limestone aquifer of Riyadh, widely spaced wells may be effective in controlling water levels. Consideration can therefore be given to the location of wells in open areas, such as parks, away from buildings and services. Where appropriate wells can also be sunk into confined aquifer zones with producing sections at depth, thus reducing the chance of significant settlement due to removal of fines or rock solution.

Good well construction practice can reduce the removal of fines from most granular deposits but, for fissured limestone, fines removal is much more difficult to control. Figure 12 illustrates the distribution of suspended solids in discharge water from wells drawing water from fractured limestone beneath Riyadh: although most of these yield water with a low total suspended solids content, four wells were found to produce 60 mg/litre or more. This is equivalent to 22 tonnes or more of suspended solids per year per 1000 m³ per day extracted. Long term abstraction at this rate could clearly result in serious settlement problems in adjacent structures.

Removal of soluble salts by flowing groundwater and drainage systems has been a problem in several areas of the Middle East. Notable examples are halite and anhydrite in Jizzan and gypsum in Kuwait. James and Lupton (1978) concluded that solution of gypsum and anhydrite can result in the formation of caverns, attack concrete grouts, increase the permeability of granular zones, enlarge fissures, and cause excessive settlements.

James and Lupton (1978), James and Kirkpatrick (1980) and James (1981a) established relationships for the dissolution of anhydrite, gypsum and carbonate rock. They described the effect of flow velocity upon the dissolution rate constants, which may enhance the rate of solution in high groundwater-velocity zones such as those adjacent to abstraction wells. The velocity rate dependence of the dissolution rate constant for carbonate rocks was given by James and Kirkpatrick (1980), who showed that the rates of solution increase rapidly at the onset of turbulent flow conditions. Increased temperature, saline solutions or those of high ionic strength may also increase the rate of solution.

James (1981b) reported work on the solution of carbonate rocks and the rate of increase of fissure size. He stressed that, before deciding how to safeguard foundations, the composition of the inflowing seepage water must be known as well as the fracture spacing and size. In urban rising groundwater situations, difficulties have been found in determining the composition of the

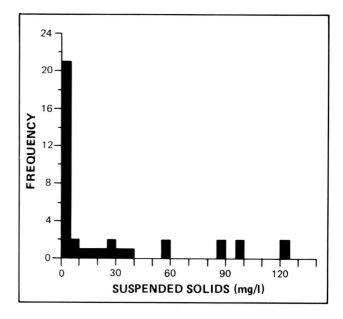

Figure 12 Distribution of concentrations of suspended solids in Riyadh well waters.

inflowing seepage water, particularly in assessing the effects of the decomposition of sewage effluent, which has a marked effect upon the carbon dioxide concentration and pH of the groundwater.

The solution and removal of limestone strata under urban rising groundwater conditions are therefore difficult to predict. The principal problems are determining the nature of the source of the water, which may be a mixture of natural groundwater, potable water, sewage effluent, irrigation returns and natural recharge; this results in a complex hydrochemical situation which is difficult to simulate and thus to predict the rate of solution of limestone.

Mitigation measures

The severity and occurrence of rising groundwater problems in some urban developments in the Middle East has shown that, in general, the potential problems of groundwater rise were not identified at an early stage, nor have effective mitigation measures been implemented to reduce the potential harmful effects in many cases.

Given the high cost and problems associated with remedial measures, it is suggested that measures taken to reduce the impact of rising groundwater prior to and during development may be extremely cost-effective. With hindsight, from the Middle East experience it has been possible to identify measures which, if implemented, could help to reduce the impact of rising groundwater. These include:

(a) Water conservation measures, including
- control of water use and storage;
- leak detection and maintenance of potable system;
- wastewater re-use;
- appropriate use of water-table control drainage water.
(b) Adoption of preventive planning strategies, including
- investigation of development sites on a regional scale to identify areas at risk;
- giving priority to the installation of foul and stormwater drainage networks in susceptible areas;
- implementation of appropriate building regulations concerning basement depths, construction materials and construction methods; and
- implementation of water-table control measures where necessary.

The water conservation measures necessary for the mitigation of rising groundwater problems can be seen as part of an overall water management plan. In some cases, water conservation and appropriate use measures may be difficult to implement, as they may involve changes in the

habits and perspective of the inhabitants of the urban areas. The use of potable water for irrigation of gardens, for example, may not be considered appropriate given the potential impact upon the water table; instead, the use of treated water from sewage treatment plants and the abstraction and use of water from strata beneath the urban area may result in a reduction in the overall water demand for the development and thus a reduction in the drainage requirements.

Brassington (1985) noted that the British Standard Code of Practice for Site Investigations (BS 5930:1981) does not contain any recommendations that site investigations should include an examination of local groundwater abstractions, which would enable possible future trends in water tables to be predicted; he made a recommendation that the Code of Practice be amended to include these omissions. As rising groundwater can be a regional as well as a local phenomenon, it is considered that, where appropriate, major urban developments should include an evaluation of its likely effect upon the water table, and strategies and regulations implemented to minimise the impact. There is evidence to suggest that uncertainty concerning future groundwater levels is now leading to designs which may be unnecessarily conservative, resulting in additional design and construction costs which could be avoided. An important element of this strategy could, therefore, be a water management plan and the setting of water table target levels to enable structures and services to be designed to appropriate standards.

The formulation and adoption of the planning strategies necessary to reduce the impact of rising groundwater would require the investigation of the geotechnical and hydrogeological properties of the strata beneath the development site. This would enable susceptible areas to be identified and appropriate development and drainage priorities to be established. As part of the above exercise it is envisaged that groundwater budgets should be established for development areas, and rates of rise predicted, allowing appropriate water-table control measures to be planned.

Based upon the above information it would be possible to implement appropriate planning and building regulations for the proposed development. The regulations in areas where high water tables are predicted may include: restricting the depths of basements in some areas of the development; specifications for the type or sources of materials to be used in subsurface structures and services; restrictions upon the size of landscape areas and the type of vegetation used; and the appropriate design of roads.

To be effective, the above measures would need to be based upon a knowledge of the particular environment of each development. Such investigations would, however, have the effect of bringing the potential problems of rising groundwater to the notice of the planners and developers, thus providing an opportunity to reduce its potential impact upon structures and services within the development areas.

References

Abu Riziaza, O.S., 1987. Lecture notes for the first workshop on *Lowering of Groundwater Table in Jeddah*. King Abdulaziz University, Jeddah.

Anon, 1983a. *ASCO Rising Water Table Project*. Ministry of Electricity and Water, Qatar.

Anon, 1983b. Stadium's void problems solved. *Civil Engineering*, October 27–29, 1983.

Anon, 1986. Bid to tackle Doha's rising water table. *Khaleej Times*, 27th May 1986.

Anon, 1987. Study of subsurface water rise in the residential areas of Kuwait. *Kuwait Institute for Scientific Research*.

Brassington, F.C., 1985. The inter-relationship between changes in groundwater conditions and engineering construction. *Groundwater in Engineering Geology*: preprints of papers for the 21st Annual Conference of the Engineering Group of the Geological Society, Sheffield, September 1985.

Eccleston, B.L., 1986. Causes of rising groundwater below the city of Doha in the State of Qatar. Paper presented to the Engineering Group of the Geological Society of London, 14th January 1986.

Fookes, P.G. and L. Collis, 1975. Problems in the Middle East. *Concrete*, vol 9 (7), p 12–17.

Fookes, P.G. and W.J. French, 1977. Soluble salt damage to surfaced roads in the Middle East. *Highway Engineering*, vol 24, p 10–20.

Fookes, P.G., W.J. French and S.M.M. Rice, 1985. The influence of ground and groundwater geochemistry on construction in the Middle East. *Quarterly Journal of Engineering Geology*, vol 18, p 101–128.

Hassan, J., 1989. Study suggests Riyadh groundwater remedies. *Arab News*, 26th August 1989.

Hurst, C.W. and W.B. Wilkinson, 1986. Rising groundwater levels in cities. *Groundwater in Engineering Geology*: Proceedings of the 21st Annual Conference of the Engineering Group of the Geological Society of London, Sheffield, September 1985. Engineering Geology Special Publication Number 3, p 75–80.

James, A.N., 1981a. Solution parameters of carbonate rocks. *Proceedings of the International Association of Engineering Geology Conference on Engineering Geological Problems of Construction in Soluble Rock*, Istanbul.

James, A.N., 1981b. Solution parameters of carbonate rocks. *Bulletin of the International Association of Engineering Geology*, vol 24, p 19–25.

James, A.N., 1989. Preliminary field studies of rates of dissolution of hydrated cement. *Magazine of Concrete Research*, vol 41, p 148.

James, A.N., and K.J. Edworthy, 1985. The effects of water interactions on engineering structures. *Hydrological Sciences Journal*, vol 30 (3), 9/1985, p 395–406.

James, A.N., and I.M. Kirkpatrick, 1980. Design of foundations of dams containing soluble rocks and soils. *Quarterly Journal of Engineering Geology*, vol 13, p 189–198.

James, A.N., and A.R.R. Lupton, 1978. Gypsum and anhydrite in foundations of hydraulic structures. *Geotechnique*, vol 28 (3), p 249–272.

Obika, B., R.J. Freer-Hewish and P.G. Fookes, 1989. Soluble salt damage to thin bituminous road and runway surfaces. *Quarterly Journal of Engineering Geology*, vol 22 (1), p 59–74.

Price, M., and D.W. Reed, 1989. The influence of mains leakage and urban drainage on groundwater levels beneath constructions in the UK. *Proceedings of the Institution of Civil Engineers*, Part 1, vol 86, p 31–39.

Salih, A., and L. Swann, 1989. Geotechnical problems associated with the rising groundwater in Arriyadh. *Second Symposium on Geotechnical Problems in Saudi Arabia*, Riyadh, p 235–238.

16 Factors affecting losses of soil and agricultural land in tropical countries

S. Nortcliff and P.J. Gregory

Abstract Commonly, constraints on agricultural development in tropical regions are frequently mainly soil or soil related. In the Amazon Basin, for example, the soils are commonly of inherently very low fertility, but many have relatively good physical properties. Inappropriate clearance of the forests may result in undesirable soil physical properties which impose an additional constraint on food production in this region. In addition, where such clearance takes place on sloping lands, soil erosion may be a major hazard.

In arid regions where arable crops can only be produced with the aid of irrigation, it is imperative that available water is used efficiently. The management of the soil-water store is a key feature in the development of sustainable agricultural systems in these areas, also the development of crops and crop-rooting systems which can maximise water use under these conditions. A further limiting factor to development is the nature of the water available for plant growth, as salinisation resulting from the application of poor quality water is a major constraint unless excess salts can be leached from the system.

Successful agricultural development in the tropics must take place with the full recognition of the relationship between the soil and the other environmental components.

Introduction

In contrast to many of the papers published in this volume, a key feature of the problems that arise when soil and land are developed is that the magnitude of the impacts due to this development depend to a large degree on the very nature of the soils which are being developed, and the environment within which the development takes place. Furthermore if, following development, the soil resource is not managed carefully, additional problems and hazards may arise.

Problems often occur because of the complex inter-relationships between the soil and its environment. These inter-relationships between, for example, soil and the climatic factors, and soil and the vegetation, are complex in the natural environment, but when man intervenes to develop the land for agricultural or other productive uses, the equilibrium may be dramatically disrupted with, in some cases, disastrous consequences.

Others in this symposium volume have used the number of deaths resulting from the hazard as an indicator of the relative magnitude of the hazard. This form of rating has

chiefly been used in the context of a catastrophic hazard, such as an earthquake, volcanic eruption or the like, where there are often a number of deaths recorded immediately. In matters relating to the devastation of the soil, however, this index is inappropriate, for the deaths which result often occur over the long term, in essence often slow death through starvation rather than rapid death through devastation. The death of a few hundreds as a result of an earthquake is "headline news", but the many thousands dying over a number of years as a result of mismanagement of the soil is hardly recorded!

To many people, the soil* is an inert layer at the earth's surface which receives little consideration as either a natural resource or a potential source of hazard following developmental actions. However, it may be argued that in many circumstances the soil is the key environmental component, in some respects the fulcrum upon which many of the other environmental components are

---

*For a succinct definition of "soil" in the agricultural (Soil Science) rather than Engineering Geology sense, see M.J. Reeve, chapter 6 in McCall and Marker, 1989.

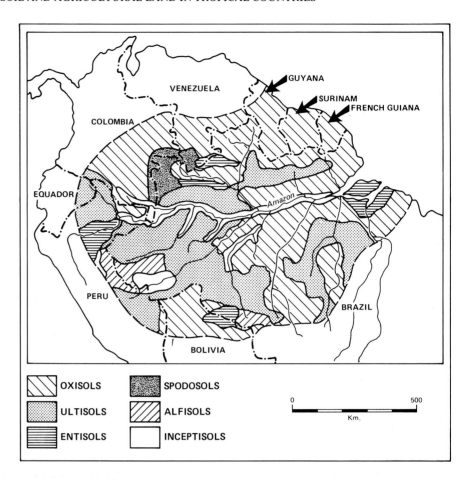

Figure 1 Distribution of soil orders in the Amazon Basin.

finely pivoted. This is because the soil, together with plants, lies at the interface of the atmospheric and terrestrial components of the environment. For example, water from the the atmosphere arrives at the soil surface, either passes over the surface as run-off, or passes through the soil as throughflow contributing to ground-water stores and river flows; alternatively it remains in the soil water store available for plant use, or to satisfy evaporative demands at the earth's surface.

In many environmental contexts, the balance between the soil and other environmental components is robust. It is possible to introduce substantial changes in these latter components, yet the soil is able to adjust to these changes without catastrophic consequences. For example, in much intensive agriculture in temperate regions there has been a tendency to manage the soil irrespective of the current consequences, in the knowledge that the resource is likely not to be damaged irreversibly, but may be renewed with ameliorative action. This action might include the need for deep ploughing in the case of compaction produced by the passage of heavy machinery when the soil is wet and vulnerable to structural damage or, at the extreme, by allowing a prolonged fallow period. In the tropics and sub-tropics, however, the soil : environmental balance is often finely tuned and disrup-

tions of this balance may have catastrophic consequences, including complete loss of the soil resource. The potential hazard of development is related to these consequences.

Two tropical systems will form the focus of this paper, the Moist Tropical Rain Forest and the Semi-Arid Tropics.

Moist tropical rainforest

The Amazon Basin of South America is one of the few remaining extensive areas of natural moist tropical rainforest, relatively little disturbed by man's activities, at least at its core; yet it is an area which is currently under considerable development pressure. Within the Amazon Basin there is a wide range of soils (Figure 1). Notwithstanding this range of soils, there are two dominant soil orders, the oxisols and ultisols (Soil Survey Staff, 1975), which impose very severe constraints on management for agriculture and related development, and furthermore their mismanagement may cause a hazard to sustained development.

The oxisols are characterised as deeply weathered soils, with very low inherent fertility, and many limitations to agricultural development (Table 1). The ultisols, whilst not offering limitations as severe as those of the oxisols, nevertheless present major limitations for development to intensive agriculture (Table 2). To the casual observer, however, the evidence on limitations to agricultural development conflicts with current relationships between the soil and vegetation, for these soils with their very low inherent fertility are in many parts of the Basin covered by luxuriant rainforest. This apparent paradox of luxuriant vegetation underlain by impoverished soil has long been recognised, and the explanation is one of an apparently almost closed nutrient cycle (Jordan, 1985), in which the bulk of the plant nutrients in the system are held in the plant biomass. Litter from the living plants is rapidly decomposed and the nutrients released are taken up by the well-developed plant root systems, with very little leakage from the system. In many tropical streams the loss of nutrients in streamflow is exceedingly small (see for example Nortcliff and Thornes, 1978; Franken and Leopoldo, 1984), confirming the small leakage and closed nature of nutrient cycling in this system. In such circumstances the natural forest system is sustainable, in spite of the very poor soils which are found below it. Remove the forest and the bulk of the nutrient store of the system is also removed; this is at the heart of the problem of developing the rainforest for agriculture.

Cultivation methods

The traditional indigenous agricultural practice of these regions was shifting cultivation. The shifting cultivator felled a small area of forest, left the debris to dry, then burnt it, thereby releasing at least a part of the nutrient store (Table 3). Agriculture was then temporarily established upon the supply of immediately available nutrients in the ash, and a slower release of nutrients from decomposing debris. The most successful shifting cultivators took further advantage of the dynamics of the system by planning their cropping programme to take advantage of the changing nutrient status of the system, planting crops with high nutrient demand (e.g. maize) in the immediate post-clearance phase, and crops with lower demands (e.g. cassava, which will give an adequate yield even under low nutrient status conditions) later in the sequence. The key to the success of this system is establishing the synchroneity between the plant demands for nutrients and the availability of those nutrients. After a small number of years (often less than five), the site of cultivation was abandoned and allowed to revert to forest, this being the fallow recovery period which, in sparsely populated areas where there was no shortage of available land, may be in excess of 50 years or more.

Whilst this form of agricultural practice has been successfully operated throughout the humid tropics for many centuries, there are pressures—for example, population pressure leading to a shortage of land—

Table 1 Oxisols: major soil constraints

1. Low plant nutrient levels
2. Low cation exchange capacity
3. Weak retention of bases
4. Strong fixation or deficiency of phosphate
5. Nitrogen losses through leaching
6. Soil acidity
7. High levels of exchangeable aluminium
5. Nitrogen losses through leaching
6. Soil acidity
7. High levels of exchangeable aluminium
8. Very low calcium content

Table 2 Ultisols: major soil constraints

1. Low plant nutrient levels
2. High levels of exchangeable aluminium
3. Nitrogen losses through leaching
4. Essential to continue to maintain soil organic-matter levels
5. Weak structure in the surface layers
6. Decrease in permeability of argillic B horizon.

Table 3 Nutrient contribution of ash after burning (in kg/ha)[a]

Element	Manaus Acrorthox	
	Forest fallow (12 years)	Virgin forest
Nitrogen	41.0	80.0
Phosphorus	8.0	6.0
Potassium	83.0	19.0
Calcium	76.0	82.0
Magnesium	26.0	22.0
Iron	22.0	58.0
Manganese	1.3	2.3
Zinc	0.3	0.2
Copper	0.1	0.2

[a] Source: Sanchez, 1981

or governmental practices (often also in response to population pressure), to discourage this form of agriculture and to shift towards a more sedentary form. Given the often inherently low nutrient status of many of these soils, this policy will give rise to major problems of sustainability without substantial inputs of imported fertilisers.

Furthermore, other problems may arise, particularly where the method of forest clearance shifts from manual felling and clearance, characteristic of the shifting culti-

Table 4 Effects of two land clearing methods on dry bulk density (g/cm³) of an oxisol measured six months after clearance, Manaus, Amazonas, Brazil[b]

Depth (cm)	Mean dry bulk density (standard error in brackets)		
	Virgin forest	Slash and burn	Bulldozed
0–5	0.79 (0.02)	0.84 (0.02)	1.31 (0.01)
5–10	1.04 (0.01)	1.06 (0.01)	1.25 (0.02)
10–15	1.12 (0.02)	1.13 (0.01)	1.17 (0.02)
15–20	1.11 (0.02)	1.12 (0.02)	1.18 (0.02)

[b] Source: Dias and Nortcliff, 1985.

Table 5 Run-off and sediment yield on experimental plots under natural forest, partially cleared forest and totally cleared forest, 27/05/87 to 22/11/87, Maraca, Roraima, Brazil[c]

Treatment	Slope position	Run-off (litres)	Sediment yield (kg/10 m²)
	Top	763.1	1.44
Forest	Middle	983.6	1.28
	Bottom	1282.5	2.42
	Top	898.0	2.72
Partially cleared	Middle	1310.5	5.17
	Bottom	650.0	2.37
	Top	802.0	3.93
Completely cleared	Middle	3611.0	52.4
	Bottom	2552.0	37.5

[c] Source: Nortcliff and others, 1990

vator, to a more mechanised clearance using bulldozers. Dias and Nortcliff (1985) investigated selected soil physical properties of an oxisol in the Amazon Basin, near Manaus, under virgin forest, traditionally cleared land, and land cleared using a bulldozer. The soil, an oxisol, was characteristic of many of the region, with extremely low plant nutrient availability, but with good stable physical properties. Selected soil physical properties under the virgin forest and two cleared plots were measured (Table 4); there were dramatic changes when mechanical clearance practices were adopted on this soil. The bulk density was increased due to the weight of the machines on the soil in the second method, resulting in reduced porosity, reduced infiltration rates and therefore more run-off erosion.

A related problem often associated with forest clearance on sloping lands is that, in many cases, the land may be subject to a high erosion hazard. Recent work by Nortcliff and others (1990) compared rates of run-off and erosion over a five-month period on three groups of plots, natural forest, partially cleared forest, and totally

cleared forest (Table 5). In this study, soil erosion from the plots where an understorey and ground litter were retained showed a rate only marginally higher than under the natural forest. Where the vegetation and litter were removed completely, the run-off and erosion were substantially increased compared to the other two treatments. These results suggest that, if the mode of management of forest clearance were modified, then the erosion hazard might be avoided.

Trees—a key component
These and other studies show that trees are a key component of the natural environmental balance in many humid tropical areas, both with respect to the cycling of nutrients through organic matter, and through the protection of the soil surface. If these areas of the globe are to be developed productively, then this inter-relationship between vegetation and soil must be recognised, and the development designed accordingly.

Currently, three types of development programme have been proposed which incorporate trees at least to some degree, and may therefore be more appropriate for this region than conventional agricultural development for arable cropping systems or grass for grazing. These may be broadly summarised as:

1. Plantation for timber and related products
2. Plantation for extractive products such as rubber, cocoa or oil palm
3. Agroforestry

The first two closely mimic the natural forest system, the major difference being the greater species diversity of the natural forest. If established with great care, using slash and burn practices to minimise the loss of nutrients from the forest biomass on clearance, all but the most extractive of systems seem to be sustainable with only low levels of additional inputs of soil nutrients. The third option, agroforestry, is widely promoted in the African and Asian Tropics, but to date has been relatively little used in the American Tropics. The method is intermediate between the natural forest and the arable or grazing option, and has a spatial mix of trees and crops/grass. Figure 2 diagrammatically contrasts the functioning of natural forest, monoculture agriculture and agroforestry systems.

The rationale of the agroforestry system is that trees and crops are grown continuously in close association, and that their interactions are beneficial. The system builds on the apparent success of the shifting cultivator in maintaining production at a site by having a short cropping period followed by a long fallow period; in essence, it attempts to mix continuous cropping and continuous fallow in a spatial association. Whilst there are relatively few detailed experimental data to support

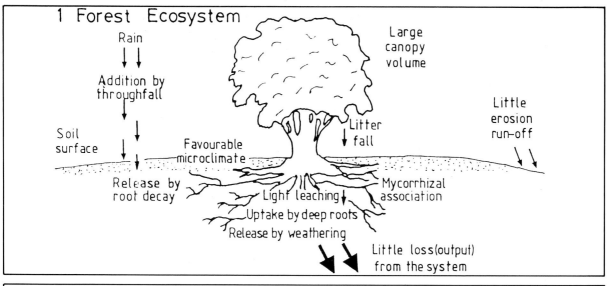

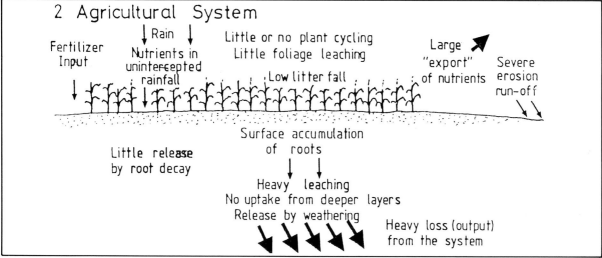

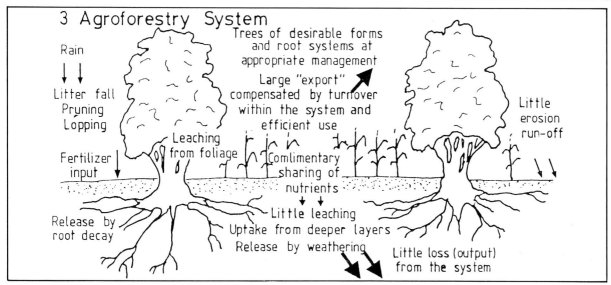

Figure 2 Schematic representation of nutrient flows and erosion losses in (1) Forest ecosystem, (2) Agricultural system, and (3) Agroforestry system (adapted from Nair, 1983).

the many assertions of the long-term sustainability of this system, Young (1989) has recently provided a comprehensive summary of the available data which suggests that, in conjunction with other traditional agronomic practices, moderately high levels of agricultural output may be sustained over a number of years with some form of agroforestry system.

Arid and semi-arid regions

The humid and semi-arid tropics provide many contrasts, not least in the potential problems facing agricultural development. In the humid tropics many of the constraints on agricultural development are associated with an excess of rainfall or an excessive intensity of rainfall, whereas in the semi-arid tropics the limitations to plant growth and land use are principally due to moisture deficiency. As in the humid tropics, the agricultural problems have traditionally been accommodated by a nomadic lifestyle. Livestock (cattle in Africa and sheep and goats in the Near East), with some cropping, have formed the mainstay of agricultural production, with animals and people moving seasonally to areas where water and pasture were available. This stable system has come under increasing pressure as the increasing population has resulted in the intensification of what was traditionally an extensive system of agriculture.

There is a close relationship between plant growth and water use which has been demonstrated by many researchers (e.g. Day and others, 1978). The importance of the soil as a store of water for crops can be simply demonstrated: for most temperate and tropical crops, the rainfall received during the growing season is less than the evaporative demand of the atmosphere (the upper limit of water demand) and the actual crop water use. To sustain growth, therefore, the crop must rely on water stored within the soil. Table 6 illustrates the importance of stored water for a range of crops grown in several environments. In all cases the demand for water by the crop exceeds the input by rainfall; in the example of groundnut production in India, the water requirements of the crop are met completely from stored water.

These measurements indicate that any development that results in a loss of soil may also adversely affect the supply of water to vegetation.

Soil water storage and irrigation
The supply of water in the soil available to satisfy the plant evaporative demands is dependent upon the texture (the relative proportions of sand, silt, clay and organic matter content) and, in many cases to a lesser degree, the structural arrangement of the soil. Generally, as the proportion of fine particles increases, so also does the amount of water stored in the soil. However, the

Table 6 Crop water use, evaporative demand and rainfall for crops in a range of environments.

Crop	Location	Seasonal water use (mm)	Potential evaporation (mm)	Rainfall (mm)
Winter wheat	UK	188	270	68
Barley	UK	220	270	125
Barley	Syria	154	300	93
Millet	India	87	300	30
Groundnut	India	102	554	0

Table 7 Values of D_i for two soils and five crops

	Crop Values of D_i	
	Rothamsted soil (clay loam)	Woburn soil (sandy loam)
Barley	100	40
Beans	80	30
Potatoes	80	35
Sugarbeet	—	100
Wheat	140	30

water stored in very fine pores (between clay particles) is not available to plants, so the soils that hold the most plant available water are the silty loams and organic-rich soils. The efficiency with which the water may be extracted from the soil is also dependent upon the size and shape of the crop root system (Gregory and Brown, 1989) and in many semi-arid regions is crucially dependent on the depth of rooting (Monteith, 1986). Many native species growing in semi-arid regions have evolved deep taproots to allow the extraction of water from considerable depth (some greater than 10 m).

Where water is available in rivers or aquifers, the soil storage may be supplemented by irrigation. Again, though, a knowledge of the ability of soils to store water is essential for the efficient management of irrigation water. Penman (1970) postulated that crop growth will occur until the water content in the soil profile falls to a certain value, the limiting deficit D_i when growth will cease. This limiting deficit can be used in practice as the basis for irrigation scheduling, where additions of water from outside sources are made to satisfy the demands for continued crop growth. Table 7 shows the values of D_i for two soils in the south of England for a range of crops; it illustrates the complex interaction between the texture of the soil and the nature of the crop and its rooting system under broadly similar climatic conditions.

Efficient irrigation is also limited in practice because of the long-term accumulation of salts. Salinisation is an inevitable consequence of irrigation unless excess salts are removed from the system, usually in a period when excess throughflow (drainage) is deliberately encouraged.

Table 8 Water use efficiency in contrasting cropping systems

Crop	Location	Season	Evaporation E (mm)	Transpiration T (mm)	E/(E+T) (%)
Maize + pigeonpea	India	Monsoon	172	212	45
Pigeonpea	India	Dry	28	220	11
Barley	Syria	Winter/spring	162	108	60

In many parts of the world, irrigation is not possible because of the lack of suitable supplies of water; in areas where water is scarce, it is important that what is available is used efficiently. Growth can only occur if water is transpired from the plant, and water lost by other routes (e.g. run-off or evaporation directly from the soil surface) reduces the amount of biomass that can be produced. In some areas, in particular in areas subject to intense rainfall, much of the rainfall may not penetrate the soil, but instead may run off. The water which runs away will not only not be available to recharge the soil water store (thereby reducing crop growth) but may also give rise to soil erosion and further reduction in the potential soil water store. Even where water does penetrate the soil, evaporation from the soil surface may be a substantial cause of losses. For example, in Mediterranean regions where the soil surface is kept moist by winter rain, up to 70% of the rainfall may be lost by evaporation and thus contribute little to crop growth (Cooper and others, 1987). The contrasting magnitudes of evaporative and transpirational losses are shown in Table 8 for three contrasting sites.

In practice most farmers adopt systems which attempt to increase transpiration relative to other pathways of loss by using a variety of agronomic practices such as mulching, early sowing, modifying plant populations and spacing, and by applying manures and fertilisers. Work in Syria at the International Centre for Agricultural Research in the Dry Areas (ICARDA) has shown that applying fertilisers can substantially increase crop growth even though the water supply is often considered to be the limiting resource (see Table 9). Gregory (1990) and Cooper and others (1987) have consistently shown that applications of phosphate fertilisers to soils deficient in

phosphate can increase the efficiency of water use by reducing evaporation.

Crops grown in many arid and semi-arid areas are frequently used to support livestock; sometimes the whole crop may be grazed, but frequently it is the stubble that provides a reserve of food during the dry season when native pasture is exhausted. This integrated system has frequently been shown to be sustainable, but degradation of the land will occur when the number of animals is out of balance with the production of crops or natural vegetation. Sustainability in these mixed systems therefore depends on both crop *and* animal husbandry.

Conclusions

In conclusion, it is apparent that development of land in the humid and semi-arid tropics will continue as land pressures grow as a result of the increasing population and the wastage of land from previous cases of mismanagement. If the rate of wastage of one of our most important non-renewable natural resources is to be reduced, it is imperative that land development be sustainable and that the minimum of environmental disruption be caused by it.

Land development must also involve a conversion process from natural vegetation to managed vegetation which does not irreversibly damage the system, thereby making any subsequent agricultural practice impossible. The most sustainable form of lower-input agricultural development in both arid and humid environments is most probably an integrated system with trees, crops and possibly livestock. No single system of development can be recommended for all locations, however: to be successful the system chosen must take account of the inter-relationships between the soil, its natural environment and the adjustments brought about by the intervention of man.

Table 9 Crop yield and water use with fertiliser additions (N at 20 kg/ha and P at 13kg/ha) for barley in Northern Syria (1982–3)

Fertiliser application	Grain yield (tonnes/ha)	Water use (mm)	Yield/water use (kg/ha/mm)
With	1.74	236	7.37
Without	1.04	228	4.56

References

Cooper, P.J.M., P.J. Gregory, D. Tully and H.C. Harris, 1987. Improving water use efficiency of annual crops in the rainfed farming systems of West Asia and North Africa. *Experimental Agriculture*, vol 23, p 113–158.

Day, W., B.J. Legg, B.K. French, A.E. Johnstone, D.V. Lawlor and W. de Jeffers, 1978. A drought experiment using mobile shelters: the effect of drought on barley yield, water use and nutrient uptake. *Journal of Agricultural Science*, Cambridge, vol 91, p 599–623.

Dias, A.C.P., and S. Nortcliff, 1985. Effects of two land clearing methods on the physical properties of an Oxisol in the Brazilian Amazon. *Tropical Agriculture*, vol 62, p 207–212.

Franken, W., and P.R. Leopoldo, 1984. Hydrology of a catchment area of Central-Amazonian forest streams. In: H. Sioli (ed.), *The Amazon: Limnology and Landscape Ecology of a Mighty Tropical River and its Basin*, W. Junk, Dordrecht.

Gregory, P.J., 1990. Plant and management factors affecting the water use efficiency of dryland crops. In: P.W. Unger, W.R. Jordan and T.V. Sneed (eds), *Proceedings of the International Conference on Dryland Agriculture* (in press).

Gregory, P.J., and S.C. Brown, 1989. Root growth, water use and yield of crops in dry environments: what characteristics are desirable? *Aspects of Applied Biology*, vol 22, p 235–244.

Jordan, C.F., 1985. *Nutrient Cycling in Tropical Forest Ecosystems*, John Wiley and Sons, Chichester.

Monteith, J.L., 1986. How do crops manipulate water supply and demand? *Philosophical Transactions of the Royal Society*, London, vol A316, p 245–259.

Nair, P.K.R., 1983. Tree integration on farmlands for sustained productivity of small holdings. In: W. Lockeretz (ed.), *Environmentally Sound Agriculture*, p 333–350, Praeger, Oxford.

Nortcliff, S., and J.B. Thornes, 1985. Water and cation movement in a tropical rainforest environment. I. Objectives, experimental design and preliminary results. *Acta Amazonica*, vol 8, p 245–258.

Nortcliff, S., S. Ross and J.B. Thornes, 1990. Soil moisture, runoff and sediment yield from differentially cleared tropical rainforest plots. In: J.B. Thornes (ed.), *Vegetation and Geomorphology* (in press), John Wiley and Sons, Chichester.

Penman, H.L., 1970. Woburn irrigation, 1960–8. IV. Design and interpretation. *Journal of Agricultural Science*, Cambridge, vol 75, p 69–73.

Reeve, M.J., 1989. Soils. In: G.J.H. McCall and B.R. Marker (eds), *Earth Science Mapping for Planning and Development*, Graham and Trotman, London.

Sanchez P.A., 1981. Soils of the Humid Tropics. In: *Studies in Third World Societies: 4 (Blowing in the wind; Deforestation and long range implications)*, p 347–410.

Soil Survey Staff, 1975. *Soil Taxonomy: a basic system of soil classification for making and interpreting soil surveys*. US Department of Agriculture Handbook 436, Washington.

Young, A., 1989. *Agroforestry for Soil Conservation*, CAB International, Wallingford.

17 Reduction of biodiversity— the ultimate disaster?

W.S. Fyfe

Abstract Complex biological processes maintain the stability of the global climate and bioproductivity, which affect the atmospheric gases; anything which adversely affects this balance can reduce the earth's biodiversity and disturb the functioning of its thermostat. The explosive growth of human population lies at the root of the "quiet hazards", those that threaten huge areas of the globe with disasters such as soil erosion and ecological destruction.

Accelerating changes in the life support systems have impacts on human society that are difficult to predict. Topsoil erosion at 0.7% per year is a serious problem for sustainable food production. Expanding population, the need for increased production and monocultures are major causes of soil erosion with forests destroyed, poor soils invaded to expand production, and irrigation leading to soil salinization. Climatic fluctuations coupled to soil erosion are a major cause of falling crop yields. Population growth has been assisted by the use of energy in food production, but this has resulted in increased carbon dioxide in the atmosphere and consequent temperature rises. Solutions to the problem lie in the use of cleaner energy sources, including nuclear, solar and geothermal. With human activity being the greatest factor for change on the earth, earth scientists have an expanding role to play in understanding the limits to resources and the global thermal balance. An ecocentric philosophy of development requires such an understanding, and the necessity of facing realistic population control.

Introduction

I am writing at the beginning of the final decade of the 20th century. I listened to the Christmas broadcasts from the Queen, the Pope and our local Canadian leaders. All mentioned the need to protect the planet Earth and our local environments. Over the past decade, the environment has moved to centre stage of the concerns of humankind. Slowly (yes, slowly!), scientists are becoming involved in the enormously complex task of understanding the interactions of the systems which have allowed life to develop on our very special planet and how the systems are changing. A new question has appeared in all aspects of our human development—*is the development sustainable?*

I think we are truly at a turning point in our views on our relationship with our surroundings. Perhaps it was the invention of nuclear weapons that first brought home to us the fact that we can destroy ourselves totally. And the exploration of the solar system provided us with a new vision of the earth and a new vision of its most unique feature: a hydrosphere which has not boiled or totally frozen for at least four billion years. This recognition has made us consider how our global thermostat works, the thermostat of the system

$$Sun \rightarrow Living\ Cell \rightarrow Earth$$

We know that to maintain our shell of liquid water, even while the sun changes, and deep earth changes, requires other complex processes which can only be biological and which may require biological diversity. As earth scientists we have been slow to appreciate the simple fact that our atmosphere in many critical ways (carbon dioxide, methane, oxygen, ozone . . .) is a result of processes in living cells. The recent spectacular results from the study of ice cores and the like have shown that the global climate and global bioproductivity are coupled in remarkable detail.

I think there is no doubt that the recent growth of scientific endeavours like the International Geosphere/ Biosphere Project (IGBP) and the International Decade of Natural Disaster Reduction (INDNR) are all linked to the driving force of human population growth. A figure that always impresses me is that, when Christ was born, world population was at about 200 million and, by 1500 AD, this population had doubled to 400 million. The last doubling which took us from 2500 million to above 5 billion took 37 years (and the last doubling of the population of the Arabian Peninsula took 12 years!).

We are worried that this situation is not sustainable. We are worried that over 40,000 children under five die each day. We are worried that 75% of the population of Egypt, and 68% of the population of India, are anaemic (Sadik, 1989). We are worried that we could so reduce the earth's biodiversity that we could disturb the functioning of the earth's thermostat. We need to know, and to know fast, how this system functions.

Observation

The wonderful situation today is that we have the tools to study this complex question with some hope of success. From satellites we can observe and monitor the earth's atmosphere, oceans, and bioproductivity in ever increasing detail, with ever increasing precision. There is no longer any national privacy, we can watch how humankind is changing the system. With our modern tools for the analysis and description of the chemical elements and their isotopes at the atomic scale we can produce detailed records of past changes in the environment. With the expanding power of computers, we can manage the necessary vast data-sets and make them available to all scientists. And, of great significance, we now have more than one planet to study and on which to test our models of environmental change and forcing parameters.

Finally, as our knowledge improves, we have the communication systems to make this knowledge available to all. Political and social response to this growing knowledge will shape the future of our species.

The 'Quiet Hazards'

A large number of changes in our fundamental life-support systems are occurring and for most of them the rate of change is accelerating. It is difficult to rank such changes in order of importance, for their impacts on human society are difficult to predict as are society's reaction to change. It is also difficult to predict how new technologies may be rapidly developed and deployed to reduce impacts of change but, as population continues to grow, the resources needed to apply a new technology become huge. We have recently seen good examples of how difficult it is to control "simple" problems like acid

rain or ozone depletion, and these difficulties will surely increase as the number of humans living in grinding squalor increases. For these unfortunates, survival for days is the issue, not long visions of decades or centuries.

Food and fibre: soil

For most humans who live on the surface of Earth, our most fundamental requirements for food and shelter, and opportunity for creative life, are derived from life that lives on soil. Climate, water supply, and soil are coupled to produce food and fibre (and often the local fuel system).

The Worldwatch Institute of Washington recently reported that topsoil erosion is occurring globally at a rate of 0.7% per year. Topsoil is that complex, porous, biologically-alive material that serves as the nutrient bank and physical support system for most of the food we eat. To support crops for sustainable food production, soils must have diverse chemical and biological components and thickness (see Fyfe, 1990). In the past decades loss of nutrients, or additions for greater productivity, have involved many types of fertilizers, herbicides, pesticides, etc.

If the Worldwatch figure of 0.7% soil loss per year is accurate (or even near accurate) then the future of world food production must be of grave concern and, already, in many regions (e.g. most of Africa for decades, now Brazil) the signs of stress are apparent as food production per capita declines. There are signs that longevity is also decreasing.

Soil thickness and bioproductivity provides a perfect example of non-linear response to change. I recently heard a lecture by H.E. Dragne of Texas Technical University on soil erosion and crop productivity in a part of Texas. For years soil erosion was rapid and crop yields had fallen moderately. Suddenly, the yields dropped to near zero—the soil was too thin to allow the plants to stand up in the winds of the region.

The causes of accelerating soil erosion are many but ultimately most are related to population and the forcing of increased production of monocultures by various technologies. Man is invading every part of the surface which can produce food. In this process protective forests are destroyed, steep slopes and regions of poor soils and poor water-supply are invaded. Enhanced irrigation practice has led to salinization. Over-use of fertilizers, pesticides, and herbicides has produced toxic soil conditions and reduced biodiversity.

Climate undergoes fluctuations on many scales and for many reasons. Recently Grove (1988) summarized the climate history of Europe and elsewhere over the past 500 years or so. Bad summers occur at a rate of about five per century. Famines in Europe were common in the period 1500–1800. In 1696–97 a famine in Finland killed up to one third of the population (Warnock, 1987). As Grove showed, many of Europe's frost events were related to major volcanic events.

In 1988 the USA had a grain deficit of 10%, something one can hardly imagine. Corn yields fell to the values of the mid 1960s. Today the world grain surplus stands at a mere 54-day supply, the lowest for decades. Climate fluctuations coupled to soil erosion are seen as the major causes of the present situation; and here it is that it becomes absolutely necessary for geologists to improve the prediction of major volcanic events, as in the period 1817–1837 (see Grove, 1988).

I often wonder what would happen if the world had a year without summer next century—a possibility foreshadowed by Rothery (this volume). How would we respond to the social chaos caused by 1–2 billion deaths? For sure, history shows the necessity for surplus which provides security. It is here that I reflect on the words of David Pimentel of Cornell (1987): "World resources and technology can support an abundance of humans, e.g. 10–15 billions living at or near poverty, or support approximately one billion humans with a relatively high standard of living."

Soil erosion, overpopulation and reduction in biodiversity go hand in hand. If the present deterioration accelerates, the reduction in the capacity of the terrestrial biomass to fix carbon can only lead to an even more rapid build-up of carbon dioxide in the atmosphere and the potential for rapid global warming.

Energy and climate change

There is no doubt that one of the factors that has allowed the phenomenal growth of population is the use of fossil fuels for energy production (coal-oil-gas). World population could not have grown without the use of energy in food production and the related food-distribution systems. But there has been a cost which only now are we beginning to grasp. These technologies have increased carbon dioxide, carbon monoxide, and methane in the atmosphere, and the rate of increase is accelerating. The "Greenhouse Effect" (I prefer the name "Venus effect"!) is on the minds of all climate experts. All predict a global temperature rise and changes in rainfall distribution and soil moisture. As Brown of Worldwatch stressed (1989), it is not just averages that count, it is the peaks that kill. While there is great debate about how much warming will occur, it is a fact that the five warmest years of the past century have all fallen in the 1980s and most in a period of slightly declining solar energy input.

Soil degradation, drought, etc., have led to a new class of refugees, the "environmental refugees". And perhaps the flood is only starting. When we use—and often waste—energy, when we contribute to atmospheric change, are we prepared for the potential costs? For example, who will accept the refugees from the low-lying regions of the planet (e.g. Bangladesh) which may be flooded next century?

Here I can never forget the prophetic words of Aldous Huxley, written in 1948 (in *Themes and Variations*):

"Industrialism is the systematic exploitation of wasting assets. In all too many cases, the thing we call progress is merely an acceleration in the rate of that exploitation. Such prosperity as we have known up to the present is the consequence of rapidly spending the planet's irreplaceable capital.

"Sooner or later mankind will be forced by the pressure of circumstances to take concerted action against its own destructive and suicidal tendencies. The longer such action is postponed, the worse it will be for all concerned. . . Overpopulation and erosion constitute a Martian invasion of the planet. . .

". . . Treat Nature aggressively, with greed and violence and incomprehension: wounded Nature will turn and destroy you. . . if, presumptuously imagining that we can 'conquer' Nature, we continue to live on our planet like a swarm of destructive parasites—we condemn ourselves and our children to misery and deepening squalor and the despair that finds expression in the frenzies of collective violence."

Huxley chose his words carefully. So much of our schemes of development have been based on the three words: greed, violence and *incomprehension*. I stress incomprehension—this is what the IGBP is all about.

One thing is certain, if the whole World used energy technologies the way the rich world does, present knowledge shows that all present predictions on climate change would be far too conservative. Venus is an intriguing planet and we do not understand its history. Could we produce a Venus here by burning all the fossil carbon on Earth in a few hundred years?

But there is hope. We only solve problems when we are aware of their magnitude. There are cleaner energy sources available. Much more secure nuclear energy seems possible now; solar-electric devices are now being developed with over 30% efficiency; geothermal energy is almost infinite. In all these developments the earth scientist will play a major role.

Garbage

On average, we produce about 1 kg of refuse per person per day. For 10 billion people, this amounts to over 3 km^3 per year, about the mass of material produced annually by continental volcanism. We do not know how to deal with this garbage. The problem becomes increasingly complex as over 50,000 strange, xenobiotic compounds are included in the mass.

Some of the present technologies are obviously dangerous or even stupid! For example, we cut a tree which fixed carbon dioxide, we change it into newspaper or packaging material, we bury the waste in a landfill where anaerobic bacteria convert the carbon to methane, a much more potent greenhouse gas than the original carbon dioxide. One can add a host of such examples.

It is always easier to clean the front end of a system rather than clean up the dispersed debris. For all materials we need greater front-end quality control whether it be coal, cadmium-enriched natural phosphate, or paper. The geochemist with his great analytical skills has a key role to play. The future must include such quality control, sorting of the refuse and recycling. The cost of care at the beginning, versus the cost of cleaning up dispersed pollution, are surely not comparable.

Water

Clean water is one of the most vital resources for human development and it is becoming a rare commodity. At the root of all strategies to use and conserve water must be our knowledge of the global water cycle. The mass of water used to sustain life rises far faster than population as industrial and agricultural uses expands.

I am always amazed by the present data. The Mississippi River flux to the Atlantic Ocean is 580 km^3 of water per year, about 1.5% of the total discharge into the oceans from all the world's rivers. Each year 3300 km^3 of water is removed from rivers and aquifers for agriculture alone, about 9% of the total flux! The global water cycle is being seriously perturbed by man.

I am very impressed by the recent work of Wallace Broecker and his colleagues on the impact of a river diversion on climate, the Younger Dryas Cold Event (Broecker and others, 1989). They have shown that fresh light water input plays a key role in ocean circulation patterns. We are changing this situation and will change it even more in the future.

Perhaps the greatest future impact will come from the giant dam systems planned for electricity production in the Amazon River which supplies 17% of total discharge to the oceans. Do we have the knowledge to predict the impact of such developments?

Concluding statement

Human activity has become perhaps the greatest factor forcing change on our planet. The scale of such forcing will increase fast as human population continues to increase into the next century (at present 90 million more people per year). As humankind increases, other species are destroyed or decline, a fact recognized by Charles Lyell more than a century ago.

The earth and planetary scientists have an expanding role to play in the future developments of our major technologies and the education of our societies. We know a little of the history of our planet. We understand the limits to resources perhaps better than any other scientific discipline.

Man faces a choice. We can continue to rush forward with a purely "anthropocentric" philosophy of development or we can move to an "ecocentric" philosophy of development. At present, we do not have a real understanding of the consequences of the "anthropocentric" approach, the consequences of decreased biodiversity on the stability of the planet. Without such knowledge, surely the "ecocentric" approach is that of maximum security but with this goes the necessity of facing realistic population control and even population reduction.

At Christmas, 1989, we heard messages from many world leaders with focus on saving the children. But surely we now know that to save our *future* children and provide a life of security and opportunity and creativity, requires that we produce *fewer* children. As recently stated by Bill McKibben in a most controversial book "The End of Nature" (1989): "As birds have flight, our special gift is reason. . . Should we so choose, we could exercise our reason to do what no other animal can do: we could limit ourselves voluntarily. . .".

References

Brown, L.R., 1989. *The State of the World, 1989*. W.W. Norton and Company, New York, 256 p.

Broecker, W.S., J.P. Kennett, B.J. Flower, J.T. Teller, S. Trumbose, G. Bonai and W. Wolfli, 1989. Routing of meltwater from the Laurentide ice sheet during the Younger Dryas cold episode. *Nature*, vol 341, p 318–321.

Fyfe, W.S., 1990. The International Geosphere/Biosphere Programme and global change: an anthropocentric or an ecocentric future? A personal view. *Episodes*, vol 13 (2), p 100–102.

Grove, J.M., 1988. *The Little Ice Age*. Methuen, London, 498 p.

Huxley, A., 1948. *Themes and Variations*. Chatto and Windus, London.

McKibben, W., 1989. *The End of Nature*. Random House, New York, 226 p.

Pimentel, D., 1987. Technology of natural resources. In: D.J. McLaren and B.J. Skinner (eds), *Resources and World Development*, Wiley Interscience, New York, p 679–710.

Sadik, N., 1989. *The State of World Population*. UNFPA, New York, 34 p.

Warnock, J.W., 1987. *The Politics of Hunger*. Methuen, Toronto, 334 p.

Part Five

What Can Be Done?

18 Policy in response to geohazards: lessons from the developed world?

D. Brook

Abstract The developed world has long experience of responding to hazards within a policy and legislative framework, and these are examined to see whether this experience can be applied to the developing world. It is important, however, that it should be applied within the cultural, legal and political framework of the countries at risk.

The typical response to a geohazard is reaction after an event, and the opportunity should not be lost at this stage to introduce measures to reduce vulnerability to future events. For some hazards, such as mining subsidence, the hazard can be identified before it causes damage, and the response is to limit activities which might trigger the hazard. The effects of a hazard may be mitigated by engineering controls or by planning responses, such as foundation design regulations in earthquake-prone areas, and diverting development from high-risk areas.

The International Decade for Natural Disaster Reduction is expected to produce significant strides in the application of lessons already learned on a much wider scale. These must be applied as wisely as possible, particularly in the light of the growing world population and its increasing concentration in areas which are prone to geohazards.

Introduction

With the growth of an increasingly urbanised world population, the potential for natural disasters resulting from geohazards is growing daily. This potential has been recognised by the United Nations declaration of the 1990s to be the International Decade for Natural Disaster Reduction (General Assembly resolution 42/169 of 11 December 1987: reproduced in Scott, this volume).

This International Decade aims to assemble, disseminate and apply scientific and technical knowledge of natural hazards in order to reduce catastrophic loss of life, damage to property, and social and economic disruption.

It is timely, therefore, to examine briefly the nature of responses to geohazards by central and local governments, non-governmental voluntary organisations and private- and public-sector landowners and developers, and to begin to assess whether they can be improved. Examples from the developed world, which has long experience of responding to hazards within a policy and legislative framework, serve to illustrate some of those varying responses. In particular, the development of responses to geohazards within the United Kingdom will be examined briefly.

Finally, the aims and objectives of the International Decade will be examined, and the possible application of lessons learned in the developed world assessed. Nevertheless, it is important that at all times *these lessons should be applied within the cultural, legal and political framework of the region or country at risk*.

The nature of responses to geohazards

The range of responses to geohazards varies with the nature and severity of the hazard, and with the public tolerance of that hazard within the affected society. Table 1 shows a simple classification of responses, modified after Cooke (1984). In order that the appropriate responses can be determined, it is essential to improve understanding of the hazard, both scientifically and technically, and in its perception by the general public.

Table 1 Classification of responses to slope failure (modified after Cooke, 1984)

Class of Response	Examples
Emergency response and crisis management	Evacuation Emergency procedures Public relief and rehabilitation
Plan for losses	Insurance Legal responses Tax adjustments
Modify the hazard	Prevention Correction
Control the effects	Engineering controls Planning responses
Improve understanding	Hazard evaluation Monitoring Prediction Forecasting

Emergency responses

Perhaps the typical response to a geohazard is that of reacting after an event. Immediate mobilisation of rescue equipment and personnel and provision of relief aid has resulted in many lives being saved, from a few days to a few weeks after the event, and in harrowing stories of those that could not be saved. Few who saw them will forget the television pictures of 12-year old Omayra Sanchez, trapped in mud up to her neck beside the submerged body of her aunt amid a sea of coffee beans spilt from a sack, following the eruption of the Nevado del Ruiz volcano in Colombia on 13 November 1985 (Allen-Mills, 1985; Matthews, 1985). Unfortunately she died after three days of rescue attempts. In contrast, Senora Maria Rosa Elvira Eschevarria de Cardona was rescued from her shack in the centre of Armero after 24 days (Anon, 1985a).

Rescue of victims, however, is not the end of the matter; victims may well be traumatised by the events which they have suffered. After the Mexico City earthquake of 1985, for example, there were a number of people rescued from the rubble of collapsed buildings. The forty or so "miracle babies" rescued from the hospital did not appear to suffer any long-term effects, but this was not the case for some other survivors. A seamstress in her mid-twenties was buried for eight days under the rubble of a factory with the decomposing body of her boyfriend beside her. No adult was buried longer and survived. Physically she lost an eyelid and her eye is permanently bandaged, but otherwise all is in working order; mentally, however, it is a different story—she has undergone infantile regression and does not talk, but whimpers and spends her days and nights in bed only half alive (Carlin, 1986).

Relief aid provision of the basic necessities of food, medicines, clothing and shelter changes with time to replacement and reconstruction activities. Too often, however, the opportunity is lost at this stage to introduce measures designed to reduce the future vulnerability of communities to similar events. Hazard perception is at its highest immediately following a major event and certainly this would be the ideal time to introduce ideas and measures related to hazard control.

Effective provision of relief aid requires the establishment of emergency procedures on both national and international scales. Coordination is required of activities within the recipient area as well as of the equipment and personnel supplied by donor countries and charitable relief organisations. It is often counter-productive for concerned people, however rightly motivated, to rush off immediately to a disaster area.

Key problems are often those of logistics and communications. In remote areas, it may be several days before the true geographical extent of a disaster becomes evident. For example, in Armero, Colombia, bridges were washed out by the landslides and the weather was not conducive to flying operations, so the town was cut off for several hours before relief operations could begin (Linton, 1985; Anon, 1985b).

The provision of warning systems and arrangements for evacuation of the population are an important element of crisis management which may assist considerably in saving lives. Certainly, the Pacific-wide tsunami warning system seems to work quite well, though there will always be the odd individual who goes down to the beach to watch the wave coming in. Similarly, at Mount St Helens in the United States, the loss of life was reduced by evacuation though, again, one or two geologists and others who stopped to observe or refused to move were killed.

Armero, Colombia The major difficulty with such systems is the identification of the correct trigger for emergency procedures, and the reaction to it by the local population. At Armero, for example, geologists had predicted the outcome of an eruption of the Nevado del Ruiz Volcano (Anon, 1985c; McDowell, 1986; Hall, this volume). The Colombian Institute of Geological and Mining Research (INGEOMINAS) had prepared a hazard assessment one month before the eruption; mudflow routes had been mapped and emergency procedures for residents of nearby towns and villages were being drawn up. Some weeks before the eruption, the Civil Defense Department had tried to evacuate the region around the volcano after the crater began to throw out clouds of ash. However, the local people, who were mostly coffee growers, refused to abandon their crops (Linton, 1985). Significantly, an international committee which was called together to study the volcano at the time said in their report that an eruption was unlikely (or a least so it was quoted in the international press, though subsequent reports of a 67% probability may not be interpreted that way by some; Anon, 1985d).

The enhanced perception of hazard immediately after an event is also exemplified by Armero in that, in early January 1986, 5000 people were evacuated from low-lying ground after fresh emissions of ash and smoke from Nevado del Ruiz (Anon, 1986). However, only seven months later than that, the volcano was once again becoming active; the government ordered permanent evacuation of areas within 6 miles of the volcano and of the lowest parts of the valleys many more miles away but press reports indicated that "no-one is moving" (Taylor, 1986).

New Britain Bunker (1985) reported on volcanic prediction at Rabaul, capital of West New Britain with a population of 60,000, and the only town in the world built within the crater of an active caldera. Because of the absence of suitable port facilities elsewhere, it has been accepted that the town will remain in its present position and that plans for evacuation in the event of eruption will have to be made. The Vulcanological Observatory at Rabaul issues public warnings of impending eruptions in relation to notional stages:

1. An eruption is expected in the next few years;
2. An eruption is expected in the next few months or years;
3. An eruption is expected in the next few days or weeks;
4. An eruption is imminent.

Regular briefings are held with a "disaster committee" consisting of those responsible for civil defence measures and including representatives from medical, communications, security and other interested groups. Using the information from the vulcanologists, however, is a matter for the decision-makers of society, who have to balance costs against risks, and is not a job for scientists.

Bunker considered that the situation had already been reached where it is unlikely that the volcano will erupt without there being time to evacuate the town. He therefore considered it sensible for those who have to work in the town to stay for the present. He identified the great challenge for those who have to issue the final alerts (and stressed that this is a political decision, not a scientific one, though the politicians can be given the best possible advice) as *avoiding giving the final evacuation warning too soon*: if given too soon, people will drift back into town on the assumption that the warning was a false alarm, with the ensuing possibility of great loss of life. He also identified the potential problems from prevention of looting during a prolonged evacuation. Bunker concluded that, in these circumstances, the vulcanological observatory has to be limited in its role to advising disaster committess and politicians, rather than having to act in the political arena itself.

Planning for losses

In some cases, the risk to life may be reduced, either because of the nature of the hazard or because emergency procedures are sufficiently well established and have been tested. In such cases, losses to property and social and economic disruption may have to be accepted because communities have been long established in hazardous areas; the losses to individuals and organisations may, however, be spread more widely through society as a whole by some form of *insurance*. Such a response is designed largely to mitigate the effects of a hazard rather than to reduce the vulnerability.

Difficulties may be experienced in selling such insurance to the general public and it is likely that only those at greatest risk are likely to apply for it. It may also be difficult to persuade private insurance companies to market suitable policies. Private companies are generally unwilling to insure a risk which is not predictable, measurable, geographically spread and causing limited damage. This has often led governments to take on schemes of hazard insurance or to act as re-insurers.

In the *United States*, mine subsidence insurance programmes are in operation in seven states (Murphy and Yarbrough, 1988). They differ in their organisation and management procedures, but all have the common objective of providing economic protection to property owners for damage incurred to their structures over mine workings.

For example, the Illinois Mine Subsidence Insurance Fund was brought into being in 1979 by the Illinois General Assembly. In a situation where conventional property insurance does not provide for any coverage that might be categorised as "ground movement", the Fund provides reinsurance for loss caused by mine subsidence for all insurers writing mine-subsidence cover within the state of Illinois. Insurance companies are required to offer residents cover for mine subsidence as part of their property insurance; claims are adjusted through the industry's conventional systems, and losses are paid by insurance companies. The Fund then re-imburses the companies.

The Fund's statutory responsibilities include providing reinsurance for losses from mine subsidence to insurers, establishing rates and rating schedules, providing under-writing guidance to insurers, and investigating reported mine subsidence claims. All its activities are funded by the premiums ceded by the insurance companies and no financial subsidy is provided by the insurance industry or the state of Illinois.

In its first eight years of operation, the Fund has averaged 380 claims per year, of which 16% have been shown to be mine subsidence and US$7.25 million has been paid in losses and expenses related to those claims.

In *New Zealand*, the Earthquake and War Damage Commission is the government body responsible for the provision of insurance against most geohazards. Under the terms of the Earthquake and War Damage Act, 1944, all property insured against fire damage is also insured at the indemnity value or the sum insured (whichever is the least) for damage resulting from earthquake or war damage. The Commission is statutorily responsible for the administration of the Earthquake and War Damage Fund, financed by an insurance levy which is additional to the fire premium collected by insurance companies.

The original scheme has undergone several changes. In 1948, cover was extended to include extraordinary disaster storm and flood damage. In 1956, volcanic eruption damage was included in the scheme and, in 1970, cover was further extended to include landslip damage. In 1984, cover was again extended to include hydrothermal activity and a restricted form of cover for land itself was introduced. At the same time as the Commission extended its activities to include land insurance, its previous role in covering insured structures from storm and flood under its Extraordinary Disaster Damage cover was relinquished. All flood and storm damage is now handled by the private insurance industry.

As well as damage to the insured building, the Earthquake and War Damage (Land Cover) Regulations, 1984, provide for the insurance of land against earthquake damage, disaster damage (damage caused by storm, flood, volcanic eruptions or hydrothermal activity) and landslip damage. The cover includes:

(a) the land on which the insured building is situated;
(b) all land within 8 m of the insured building:
(c) the main access way and all services to the insured building within the land holding on which the insured building is situated, including any bridges and culverts; and
(d) all retaining walls and their support systems within 60 m of the insured building.

For the year ended 31 March 1984, the Commission's annual income was NZ$158.7 million (US$95 million), of which NZ$70.4 million represented nett premiums and the remainder was largely interest on investments. Claims totalled 4363 and total payments were NZ$8.5 million. Of these claims, 1067 were for earthquake damage, 105 for landslip damage and 3191 for extraordinary disaster damage, excluding landslip.

Modifying the hazard

For some hazards, such as landslip or mining subsidence, it may be possible to identify the hazard before it causes damage. If the risk of losses is significant and the benefit-cost analysis is favourable, it may be appropriate to modify the cause of the hazard. Such responses could include limitation of activities which might trigger or exacerbate the hazard, and preventive measures or remedial works to stop its further development.

Examples of such responses include restrictions on the cutting and engineering of slopes during the rainy season (e.g. Los Angeles; Cooke, 1984); restraints on mining activities such that subsidence is not caused or is limited in its effects; and corrective measures on landslides, backfilling of abandoned mine voids and attempts to divert lava flows, as at Mount Etna in Sicily and Hecla in Iceland (Matthews, 1984; Grove, 1973). Lava-flow diversion often makes good television, and increases the public awareness of geohazards.

Of necessity, this response tends to act on an individual-site basis where the consequences of an event and the costs of preventing or correcting it can be readily quantified. It is also of limited applicability: some geohazards, such as earthquakes, are neither sufficiently predictable in space and time, nor amenable to modification of the hazard. Such hazards give rise to events which are unavoidable, but the consequences can be modified to reduce the hazard to people and property.

Controlling the effects

An alternative means of reducing losses to an acceptable level is to control the effects of a hazard. This may be achieved by engineering controls or by planning responses.

Engineering controls These tend to be specific to particular structures and are designed to make them better able to withstand the effects of a hazard. Foundation design regulations are typical of such responses in earthquake-prone areas.

In Italy, for example, Law No 64 of 1974 defined the measures to be taken in those areas prone to seismic activity or where settlements have been identified as being threatened by slope instability (Fulton and others, 1985). As modified after the 1980 earthquake, this law controls development by:

(a) defining the investigations and documentation to be produced before construction in seismic hazard areas;
(b) by stressing the importance of an overall view of the context of any development; and
(c) by controlling the types of retaining structures employed.

Foundation design regulations have, of course, been around for a long time. For example, Vitruvius in his *De Architectura* (the Ten Books of Architecture) in the 1st century BC commented that "the foundations of these works should be dug out of the solid ground, if it can be found, and carried down into solid ground as far as the magnitude of the work shall seem to require". This work was rediscovered in the 15th century, after which the author's precepts on design and proportion were treated almost as law until at least the 17th century.

Most countries have anti-seismic building codes (e.g. USA, Japan, Turkey, Mexico, Iran, Algeria, China) and the development of international codes (e.g. Eurocode 8) is under discussion.

Planning responses These responses are those which seek to divert development from high-risk areas in order to minimise the effects of geohazards. They are concerned with general land use, with detailed developments within specific land-use zones, and with the nature of operations and buildings within developments. They may operate at all levels from national government down to the local area.

In the United States, such responses are incorporated within zoning regulations, subdivision regulations or building regulations. Zoning ordinances seek to direct development away from hazardous areas by the mechanism of allocating certain land uses to pre-determined zones, where they are further controlled by subdivision ordinances which specify the usage of the lots within zones. If development is allowed within a potentially hazardous zone, precautions are usually pre-scribed, and building codes seek to control the manner in which hazardous zones are affected by development and construction. They cover such matters as the develop-ment, the timing of works on slopes, the issuing of grading permits, building permits, etc.

Cooke (1984) noted that the County Zoning Code for the County of Los Angeles demonstrated a clear response to flood hazard by prohibiting most public building in flood zones, but had little relevance to sloping terrain. This is, however, dealt with in the County Sub-division Ordinance which requires the submission of a geological report with a development application where a geological hazard exists or may potentially exist.

In New Zealand, the Local Government Act 1974 specified that subdivision shall not be permitted in certain circumstances, in particular (Section 274 (1) (f)) where:

(i) the land or any part of the land in the subdivision is subject to erosion or subsidence or slippage or inundation by the sea or by a river, stream or lake or by any other source, or

(ii) the subdividing of the land is likely to accelerate, worsen or result in erosion or subsidence or slippage or inundation by the sea or by river, stream or lake or by any other source, of land not forming part of the subdivision;

provided that this paragraph shall not apply if

provision to the satisfaction of the council has been made or is to be made for the protection of the land (whether part of the subdivision or not) from erosion or subsidence or slippage or inundation.

This Act further specifies, in Section 641, that the council shall refuse to grant a building permit where:

(b) the proposed building or alteration is, or within the useful life of the building or alteration is likely to be, subject to damage arising directly or indirectly from:

(i) erosion subsidence or slippage of the land on which the building or alteration is proposed to be erected or any other land, or

(ii) inundation arising from such erosion, sub-sidence or slippage, unless the council is satis-fied that adequate provision has been made or is to be made for the prevention of that damage.

Section 641A granted powers to issue a building permit on land subject to erosion subsidence, slippage or inundation under certain conditions. It is perhaps signifi-cant that Section 641A (3)(d) specifies that, where a building permit has been issued in such circumstances and the building or alteration to which the building permit relates later suffers damage arising directly or indirectly from erosion, subsidence or slippage or inundation arising from such erosion subsidence or slippage—

"The council and every member, employee or agent of the council shall not be under any civil liability to any person having an interest in that building on the grounds that it issued a building permit for the build-ing in the knowledge that the building or alteration for which the permit was issued or the land on which the building or alteration was situated, was, or was likely to be, subject to damage arising directly or indirectly from erosion subsidence or slippage or inundation arising from such erosion subsidence or slippage."

Improving the understanding of hazards

Whatever response to geohazards decided upon in the context of the political, economic and administrative environments in which they occur, an essential element in helping to reduce losses must be improved understanding of the hazard. Clearly, except for the purely reactive relief aid to an emergency, it is not possible to respond to the likelihood of an event occurring unless one knows *where* it is likely to occur. This requires some form of hazard evaluation of the sort described by other authors in this volume. In general terms, this needs an ability to monitor, predict or forecast the timing, location and nature of a particular hazard. The purpose of such hazard evaluation should be to allow the decision-makers in a society to have the answers to questions such as those below:

1. What is the hazard?
2. Where is it going to occur?
3. What will it affect?
4. How widespread is it, or the potential for it?
5. Can it be avoided?
6. If not, what can be done to reduce the hazard?

In addition, two further questions need to be answered:

1. Who is responsible?
2. Who should pay?

It is undoubtedly true that major advances have been made in hazard evaluation in recent years. However, it is nowhere near perfect. It is of interest to note comments (Geomorphological Services Limited, 1987) in relation to landslide evaluation that *monitoring, prediction and forecasting of slope failure are rudimentary* and of limited value in management. This is chiefly because of the complexity of the factors affecting slope failure, the fact that understanding of slope failure is often quite primitive, the methods are often not standard and there is commonly too slender a historical data base upon which analysis can be developed.

Planning responses to geohazards in Britain: the mining industry

Having now established the types of responses that can be made to geohazards in general terms, it is of interest to look at one particular example to see how these are operated in practice. For this purpose the example of Britain is chosen, where the development of responses to geohazards has developed over a long period of time. Much has been influenced by developments in the mining industry—particularly for coal—and hazards associated with it. Early responses related largely to underground hazards to those employed in mining, and tended to follow after disasters of one kind or another. It is from this type of development that virtually the whole of our Health and Safety legislation is based. More recently, there have been developments in both the legislative and policy fields in respect of both natural and man-made geohazards through the operation of the Town and Country Planning system.

Responses to mining hazards   Britain has a long history of mining on which the industrial revolution was based, and an equally long history of mining disasters due to different types of geohazard (Duckham and Duckham, 1973).

A common hazard in mining is that of breaking into old flooded waste left by miners of a previous generation, resulting in inundation of the working mine and presenting serious hazard to the miners. Such inrushes have been recorded many times in the history of

the British coal mining industry, one of the earliest being at Gallow Flat, Northumberland, in 1658. In 1938, the Royal Commission on Safety in Coal Mines estimated that fatalities due to inrushes had averaged some 8 per year between 1850 and 1938. Most such accidents were, however, very much in the "avoidable" category, dependent on the availability (and consultation) of adequate records and testing by boreholes when the presence of underground reservoirs is suspected.

As early as 1740, the Scottish coal master, Sir John Clerk of Penicuik, emphasised the need to "keep exact plans . . . of all the work that has been done in a coaliery". Efforts to encourage voluntary registration of mine plans date from the end of the 18th century but little or no action resulted. Evidence to the 1835 Select Committee on Accidents in Mines stressed the importance of reliable plans and, in 1849, the House of Lords Select Committee on Dangerous Accidents also paid attention to inrushes. The Mines Inspectorate was established by Act of Parliament in 1850 and this Act also made obligatory the keeping of an accurate plan at the mine owner's expense. Only by Acts of 1872 and 1887, however, was the depositing of mine abandonment plans with the Secretary of State made compulsory.

Whilst this was a major development which was to have its effect both on safety within the mine and on surface development, there were still problems. Poorly kept plans, imperfectly understood geology and human error still led to disasters due to inrushes of water from old mine workings or of near-surface peat. Examples include Audley, North Staffordshire (1895), Reddings Colliery, Stirlingshire (1923), Knockshinnock Castle, Ayrshire (1950) and the Lofthouse Colliery inrush of 1973.

These and other disasters due to different causes (e.g. explosion, fire, entombment) have been followed by inquiries of one sort or another and often subsequently by legislation, or at least policy changes. Perhaps the most memorable example of a disaster leading directly to legislation was the Aberfan tip slide of 21 October 1966 in South Wales, when 144 people died including the majority of the children gathered at the local primary school for the morning assembly. This was investigated by the Franks Tribunal in its report of 3 August 1967 and followed by the enactment of the Mines and Quarries (Tips) Act 1969. This Act was perhaps unusual in that it was inspired by the threat to people outside the workplace rather than to the health and safety of those at work.

There has been considerable progress in the last century in reducing the hazards to underground miners and to others affected by mining activities. This progress has not been simply a matter of legislation but has depended also on stricter enforcement of regulations, better and more uniform training, and more effective safety propaganda, as well as on technical progress within the mining industry. Between 1939 and 1968, the underground fatality rate in coal mines in Britain was more than halved due to these and other responses to the hazards presented in mining.

Responses to hazards to surface development A number of hazards threaten surface development in Britain. Mine shafts, mining subsidence, rockfalls, landslides, natural underground cavities, made ground, compressible ground, landfill gas and earth tremors all pose a potential threat of greater or lesser magnitude. Consideration of these hazards has always been implicit in the operation of the various Town and Country Planning Acts since the planning system was established in 1947. This system, which has not changed in its essentials, is now contained in the Town and Country Planning Act, 1990. The fundamental requirement is that development —including building, engineering or mining operations at the surface or underground—as defined by the Act, may not be undertaken without planning permission (Department of the Environment, 1988).

From the earliest days of the planning system, government policy statements have attempted to clarify this situation in recognition of the different interpretations that have operated in practice as to whether such hazards are indeed covered by the planning acts or by other legislation.

For example, Ministry of Town and Country Planning Circular No 65 (1949) related to the transfer from central to local government of consultation on proposals for development by local authorities and statutory undertakers (suppliers of water, gas, electricity, etc). Where such matters as subsidence had previously been considered by the Ministry's regional offices, this circular made known to local planning authorities that advice was available in cases involving subsidence from the Mineral Valuers of the Valuation Office, Board of Inland Revenue. This advice is still available to local authorities today.

In the UK, subsidence is due largely to underground coal mining but, historically, subsidence due to brine extraction has been significant in some areas. Both the Cheshire (Brine Subsidence) Compensation Act 1954 and the Coal Mining (Compensation for Subsidence) Act 1957 embody a response which entails planning for losses and making recompense to those who suffer the losses. Both Acts do, however, include provision for consultation on planned surface development above the area affected by the particular form of mining subsidence. The 1957 Act and the Coal Industry Act 1975 also include provisions for the modification of the effects of subsidence through preventive work on existing buildings and precautionary measures for new development.

In view of the different considerations which were being given by different planning authorities to the question of subsidence, further advice was issued in Ministry of Housing and Local Government Circular 44/61 (1961). This stated the Minister's opinion that decisions should not be taken to permit surface development without taking cognisance of what is known or can be conjectured about the stability of the site. Whilst this circular was addressed to authorities in coal mining areas, this statement is regarded as one of principle which is more generally applicable, both to other areas and to other hazards.

Under the planning acts, local authorities are empowered to prepare development plans. As part of plan preparation, they will normally carry out a survey of the principal physical and economic characteristics of the area of a plan. This provides an opportunity for local authorities to take due account of geohazards in their forward planning as well as in considering individual applications for planning permission.

Despite this long history of legislation and policy covering a planning response to both natural and man-made geohazards, there is still some confusion as to whether such hazards are a planning matter. For this reason, the Department of the Environment (1990) has recently issued guidance which makes it clear to local planning authorities and developers that, insofar as it affects land use, the stability of the ground is a material planning consideration of which they should take due account in both forward planning and in individual decisions.

As a basis for this policy, the Department carries out research designed to review the scale, nature and extent of the principal physical constraints on the development of land and what can be done about them. Particular effort is placed on the development of mapping and presentational techniques which will enable local planning authorities and others, who may not have the relevant earth science or engineering expertise, to take account of such constraints in the course of their day to day work.

The International Decade for Natural Disaster Reduction

We have seen, above, the types of responses which can be made to geohazards and of how these responses apply in the context of health and safety and planning legislation and policy in the United Kingdom. However, it is not yet clear that any lessons learnt from experience in Britain and elsewhere are being applied in the areas where such hazards are both more common and more hazardous. To do so we need to examine briefly the international response to geohazards.

Considering that natural disasters have claimed about 3 million lives in the past two decades, adversely affected

800 million more people and resulted in immediate damages in excess of US$23 billion, and recognising the potential positive effects of scientific and technical advances in the field of natural disasters, the United Nations General Assembly decided in resolution 42/169 of 11 December 1987:

"to designate the 1990s as a decade in which the international community, under the auspices of the United Nations, will pay special attention to fostering international cooperation in the field of natural disaster reduction."

As modified by resolution 43/202 (United Nations, 1988) and 44/236 (United Nations, 1989), the objective of this International Decade for Natural Disaster Reduction is:

"to reduce through concerted international actions, especially in developing countries, loss of life, property damage and social and economic disruption caused by natural disasters such as earthquakes, windstorms (cyclones, hurricanes, tornadoes, typhoons), tsunamis, floods, landslides, volcanic eruptions, wildfires, grasshopper and locust infestations drought and desertification and other calamities of natural origin."

The General Assembly also established five goals for the Decade:

(a) "to improve the capacity of each country to mitigate the effects of natural disasters expeditiously and effectively, paying special attention to assisting developing countries in the assessment of disaster damage potential and in the establishment of early warning systems and disaster-resistant structures when and where needed;

(b) "to devise appropriate guidelines and strategies for applying existing scientific and technical knowledge, taking into account the cultural and economic diversity among nations;

(c) "to foster scientific and engineering endeavours aimed at closing critical gaps in knowledge in order to reduce loss of life and property;

(d) "to disseminate existing and new technical information related to measures for the assessment, prediction and mitigation of natural disasters; and

(e) "to develop measures for the assessment, prediction, prevention and mitigation of natural disasters through programmes of technical assistance and technology transfer, demonstration projects, and education and training tailored to specific disasters and locations and to evaluate the effectiveness of those programmes."

Day-to-day activities of the Decade are to be co-ordinated by a small secretariat in association with the UN Disaster Relief Organisation (UNDRO) in Geneva. Preparations for the Decade began with the establishment in 1988 of:

(a) a UN Steering Committee, composed of all the relevant entities of the United Nations system; and

(b) a Scientific and Technical Committee on the International Decade for Natural Disaster Reduction of 20-25 scientific and technical experts to develop overall programmes to be taken into account in bilateral and multilateral co-operation for the Decade, paying attention to priorities and gaps in technical knowledge identified at the national level, to assess and evaluate the activities carried out in the course of the Decade and to make recommendations on the overall programmes in an annual report to the Secretary-General.

In addition to these organisational arrangements, Resolution 44/236 also established the policy measures to be taken at national level. These include:

(a) formulation of natural disaster mitigation programmes and economic, land-use and insurance policies for disaster prevention, all fully integrated into national development programmes;

(b) participation in concerted international action for the reduction of natural disasters, with the establishment, where appropriate, of national committees in co-operation with the relevant scientific and technological communities;

(c) encouragement of support from both the public and private sectors in contributing towards achieving the aims of the Decade;

(d) keeping the Secretary-General informed so that the United Nations may become an international centre for the exchange of information and co-ordination of international efforts;

(e) taking the necessary measures to increase public awareness of damage risk probabilities and of the significance of preparedness, prevention, relief and short-term recovery activities with respect to natural disasters;

(f) paying due attention to the impact of natural disasters on health care and to activities to mitigate the vulnerability of critical social and economic infrastructure;

(g) improving the early international availability of emergency supplies through the identification and storage of such supplies in disaster-prone areas.

Through the activities of the International Decade for Natural Disaster Reduction, it is expected that significant strides can be made in the application of some of the lessons already learnt in different countries to be applied on a much wider scale in order to reduce the impact of geohazards. Whilst it is not expected in the majority of cases that the events which lead to hazards can be avoided, it is expected that the hazard involved can be reduced. It is perhaps of interest in this context to examine briefly the reaction in Britain to the International Decade and how that could assist the achievement of its objective and goals.

Britain and the International Decade for Natural Disaster Reduction

The British Government is strongly supportive of the aims and objectives of the International Decade for Natural Disaster Reduction (Burton, 1990). It will work with other nations and with the international organisations to assist in meeting, as far as is reasonably possible, the goals established by the General Assembly.

There is a strong British tendency towards concentration on the donor aspects of natural disasters, not just in terms of the relief provision for overseas disasters but also the provision within overseas aid for development of disaster prevention and mitigation actions, e.g. the International Conference in London in December 1989 on combatting the annual flood cycle in Bangladesh.

This concentration on donor aspects is perhaps understandable. Under present legislation and policies regarding its enforcement, the potential for disasters due to geohazards in Britain is relatively low, depending (of course) on how one defines disaster. It is important to remember that, to those affected, the loss of a single loved one's life or irremediable damage to a single property constitutes a disaster. This differs from the type of disaster intended to be covered by the International Decade, which involves the disruption of the social and economic life of a large part of a community or nation.

Civil emergencies in the United Kingdom, such as those that might arise from geohazards, are the responsibility of the Home Office. During 1989, a major Home Office review of the machinery and procedures for dealing with national disasters of all kinds culminated in a number of important steps, including the appointment of a Civil Emergencies Advisor.

Provision of overseas relief and development aid is the responsibility of the Overseas Development Administration. The ODA's Disaster Unit was set up in 1974 to administer official relief aid to natural and man-made disasters in the developing world and to natural disasters in the developed world. In the last 15 years it has responded over 650 times and spent over £123 million.

The ODA's Disaster Unit has been nominated as the UK focal point in respect of the activities of the International Decade. This unit, with the collaboration of the Home Office's Civil Emergencies Advisor, and liaison when necessary with other Government departments, will provide the mechanism to co-ordinate the UK's strategy and policy towards the Decade. It will thus be the focal point for harnessing and monitoring UK private and voluntary sector expertise and involvement.

It will also maintain liaison with any committees established by the scientific and technological community. Agreement has been reached, for example, with the Royal Society which, in collaboration with the Fellowship of Engineering, will set up a science, technology and engineering committee to cover the interests of the private and voluntary-agency sectors in those subjects. Further arrangements will be adopted as necessary to ensure that all sectors with a part to play in disaster preparedness, prevention and mitigation can effectively operate and make Britain's influence fully felt to assist in achieving the Decade's objectives.

In March, 1989, proposals to improve the decision-making and co-ordination structure in Britain with respect to disaster relief and mitigation were set out in a major speech by the then Minister for Overseas Development (Patten, 1989). These included:

(a) Home Office review of responses to disasters in the UK, followed by the appointment of a Civil Emergencies Advisor;

(b) ODA review of disaster relief operations (completed: a number of changes are now in train);

(c) seeking ways of improving the effectiveness of UNDRO in speed of response, management on the ground and co-ordination with donors and other UN bodies;

(d) establishment of a register of donor expertise (this UK initiative has already been taken up by UNDRO);

(e) maintenance of links with non-governmental organisations;

(f) examination of the potential role of the military in disaster relief;

(g) enhancement of disaster prevention and preparedness through UNDRO and under the British aid programme;

(h) encouragement to private sector organisations to sponsor and finance activities as their contribution to the Decade.

Examples of UK activities overseas in disaster prevention and preparedness are described elsewhere in this volume (Reynolds, Scott). The UK has provided substantial financial assistance to such projects from the aid programme, both bilaterally (country-to-country) and through UNDRO. This assistance covers a wide range of activity including the establishment of early-warning systems, food security studies, prevention and preparedness strategies, and disaster management, research and training. Particular mention should be made here of the co-operation with the Asian Disaster Preparedness Centre, the Bangladesh Flood Alleviation Project and the Pan-Caribbean Disaster Preparedness and Prevention Project. The UK aid programme has also co-funded the production of a UN documentary film aimed at increasing the awareness of policy makers and aid donors of the problems and consequences of natural disasters.

The ODA will continue to consider requests from developing countries for assistance under the aid programme for disaster prevention and preparedness projects and will also increase support for UNDRO projects in this area.

Finally, the ODA and other Departments will be publicising their "Decade" activities through a range of publicity material and will be encouraging UNDRO to increase its activities and profile in this area. Whilst this will not eliminate the disaster potential of geohazards it could assist greatly in improving awareness of those who may have a contribution to make to the achievement of the objectives of the International Decade.

Conclusions

It has clearly not been possible to review the whole field of policies in response to geohazards. All that has been attempted is an explanation of the types of response which are operated, with some examples from the developed world of the legislation which has been enacted to implement those responses. In particular, brief examples have been quoted from Britain of the development of health and safety legislation with respect to underground mining and of the planning and related legislation and policy statements in amplification of that legislation.

During the 1990s, there is considerable potential for advances to be made on a world scale through the activities of the International Decade for Natural Disaster Reduction. Whilst there may be lessons to be learnt from those countries in the developed world which have a well developed legislative and policy framework for dealing with natural disasters, whether due to geohazards or not, it will be important not to try to impose solutions on countries which do not yet have this framework. To attempt to do so will undoubtedly, and justifiably, meet with political resistance and will be unlikely to be commendable to the population of the country affected.

It is important, however, that the lessons that can be learnt are applied as wisely as possible, particularly in the light of the growing world population and its increasing concentration in both urban areas and in areas which are prone to geohazards.

In South America, for example, the population has increased by a factor of three in the last four decades. Hermelin (1989) illustrated these factors in Colombia, where, between 1960 and 1983, population increased from 15.5 million to 27.5 million whilst experiencing an enormous increase of urban population. At the end of the 1950s, rural population in Colombia was 73%; it is now 34%, and the country has now more that 30 cities with more than 100,000 inhabitants. Medellin, the second largest city, has grown from 500,000 in 1951 to 2.3 million in 1988. Needless to say, much of this urban increase has occurred through the occupation of areas which are not necessarily suitable for such uses, e.g. because they are too steep, unstable, subject to flooding, etc.

It is also evident that some of these lessons are being learnt and Hermelin (1989) gave a useful resume of the development of environmental management in Colombia in the last 20 years or so from the adoption of the National Code on Natural Resources and Environment in 1974 to the adoption of the Urban Reform Law in 1989, under which cities and villages must prepare a Development Plan which necessarily has to include a set of regulators on urban land uses and the identification of natural risks. However, he also stated that, at the time of writing, only one province in Colombia had taken action to enforce the Urban Reform Law.

Thus, although some significant steps have been taken, there remain problems in the implementation and enforcement of the necessary policies and legislation which will bring about a much-needed reduction in the exposure of human society to the environmental risks arising from geohazards. These problems arise both from the technical nature of the hazards and from the social and economic structure of the societies affected.

For example, the trigger level for any evacuation procedures for an impending disaster will be difficult to establish, and it will be difficult to persuade people whose immediate concern is merely staying alive that they should not develop (often illegally) shanty towns in the most hazardous areas, which is where they tend to be located. The Superbarrio (Superman) figure who appeared in Mexico City after the 1985 earthquake may be an effective means of protest after the event to ensure that the needs of the people are given more consideration in reconstruction, but to a large extent that is too late.

Much more effort needs to be concentrated on disaster prevention and preparedness, and this has been rightly stressed by the United Nations with respect to the International Decade.

In many countries, as in the UK, the targeting of new development away from hazardous zones may be relatively straightforward, provided that such zones have been identified and the procedures are in place to consider them in the planning of new development. However, it is certain that, in many places where development already exists, there are geohazards which have either only been recognised subsequent to the development or were not considered to be hazardous at the time of the development. In such situations there may be serious difficulties in avoiding serious personal disruption in developing a hazard management programme. It is likely that, in many such cases, crisis management and planning for losses may be the only viable option.

Two particular elements of policy are seen as essential in dealing with geohazards and these have been recognised in relation to the International Decade. They are:

1. Enhancement of scientific advances to improve hazard management procedures; and
2. Dissemination of both scientific knowledge and policy mechanisms.

It is equally clear that public awareness is all important. If the general public are aware and accept what is needed in particular hazard management situations, then it is more likely that the politicians and decision-makers will also be aware, and a response will be forthcoming. It is perhaps important, therefore, that the level of publicity and education in respect of the various scientific and policy aspects of geohazards will need to be not at the level of the scientific or technical journal or conference (though these are obviously of value), but at the level of the popular press.

At the beginning of the International Decade for Natural Disaster Reduction, it is important that we look forward with optimism to the time when the press will be reporting geo-events as *hazards which have been overcome* or *avoided* rather than as the natural disasters which are reflected in the references at the end of this paper. We have a much greater understanding today of the causes of such hazards than ever we had in the past and we are much nearer to developing the necessary policy responses to deal with them. The opportunity is there for all to share in this understanding and to learn from both the successes of others and the mistakes that have been made.

Acknowledgements

The preparation of this paper has been greatly assisted by the work carried out by John Cowley, Ron Cooke, David Jones and Alastair Fulton under a research contract for the Department of the Environment (Geomorphological Services Limited, 1987). Particular thanks are due to my colleague, Dr Brian Marker, for a number of useful discussions during the preparation of this paper. I am also grateful for the help and advice received from Mr Peter Burton, head of the Overseas Development Administration's Disaster Unit. However, any views expressed are those of the author alone and do not necessarily represent the views of the Department of the Environment or any other organisation.

References

Allen-Mills, T., 1985. Tragedy and triumph in tomb of slime. *Daily Telegraph*, 18 November 1985.

Anon., 1985a. Armero widow rescued after 24 days alone. *Daily Telegraph*, 10 December 1985.

Anon., 1985b. Volcano buries 15,000. *Birmingham Evening Mail*, 14 November 1985.

Anon., 1985c. A trigger for disaster. *Construction News*, 21 November 1985.

Anon., 1985d. Town ignored a warning over volcano. *Daily Telegraph*, 15 November 1985.

Anon., 1986. Thousands flee from volcano. *Daily Telegraph*, 6 January 1986.

Bunker, C.A., 1985. The prediction of volcanic eruptions: some applied physics. *Geology Teaching*, vol 10(3), p 78–86.

Burton, P., 1990. The UK response to the decade and to international disasters. British Consultants Bureau Conference—"Disaster relief and mitigation", London, 7 February 1990. Preprint, 4 p.

Carlin, J., 1986. Fears still haunt the rescued victims. Mexico after the quake, Part 2., *The Times*, 29 April 1986.

Cooke, R.U., 1984. Geomorphological hazards in Los Angeles. *The London Research Series in Geography*, vol 7, George Allen and Unwin, London.

Department of the Environment, 1988. Planning Policy Guidance: General policy and principles. *Department of the Environment/Welsh Office Planning Policy Guidance Note* (PPG) 1, HMSO, London, 5 p.

Department of the Environment, 1990. Planning Policy Guidance: Development on unstable land. *Department of the Environment/Welsh Office Planning Policy Guidance Note* (PPG) 14, HMSO, London, 25pp.

Duckham, H., and B. Duckham, 1973. *Great pit disasters —Great Britain, 1700 to the present day*. David and Charles, Newton Abbot, 227 p.

Fulton, A.R.G., D.K.C. Jones and S. Lazzari, 1985. The role of geomorphology in post-disaster reconstruction: the case of Basilicata, southern Italy. In: *Proceedings of the 1st International Conference on Geomorphology, Manchester*: John Wiley and Sons, Chichester.

Geomorphological Services Limited, 1987. Review of research into landsliding in Great Britain, Series D, Landslides and policy, Vol. III, Legislative and administrative provisions, practice in England and Wales and a review of overseas practice. *Report to the Department of the Environment (Contract No PECD7/1/168)*, April 1987, 373 p + 6 appendices.

Grove, N., 1973. Volcano overwhelms an Icelandic village. *National Geographic*, vol 144 (1), p 40–67.

Hermelin, M., 1989. Problems and approaches in environmental geology in Latin America: the Colombian experience. International Meeting on Earth Sciences for Environmental Planning, Santander, Spain, October 1989. Preprint, 21 p.

Linton, M., 1985. Volcano 'kills 20,000'. Towns buried under mud. *Daily Telegraph*, 15 November 1985.

McDowell, B., 1986. Eruption in Colombia. *National Geographic*, vol 169 (5), p 640 53.

Matthews, C., 1984. Duel with Mount Etna. *Readers' Digest*, April 1984, p 114–20.

Matthews, G., 1985. Killer volcano erupts again. *The Observer*, 17 November 1985.

Ministry of Housing and Local Government, 1961. Town and Country Planning Acts, 1947–59. Surface development in coal mining areas. *MHLG Circular No 44/61*, HMSO, London, 2 p.

Ministry of Town and Country Planning, 1949. Consultation on proposals for development by local authorities and statutory undertakers. *MTCP Circular No 65*, HMSO, London, 2 p.

Murphy, E.W., and R.E. Yarbrough, 1988. Reconstruction of homes damaged by coal mine subsidence—progress report. In: M.C. Forde (editor), *Mineworkings '88, (Proceeding of the 2nd International Conference on Construction in Areas of Abandoned Mine Workings, Edinburgh, 28–30 June 1988)*: Engineering Technics Press, Edinburgh, p 185–89.

Patten, C., 1989. Disaster relief—Where do we go from here? *Address by Christopher Patten MP, Minister for Overseas Development, to the Fontmell Group on Disaster Relief conference on Disaster Relief and the Military, at the English-Speaking Union, 21 March 1989*, Overseas Development Administration/Central Office of Information, London, 16 p.

Taylor, F., 1986. Black cloak of death over volcano's 23,000 victims. *Daily Telegraph*, 7 July 1986.

United Nations, 1987. International Decade for Natural Disaster Reduction. *Resolution 42/169 Adopted at the Forty-second Session of the United Nations General Assembly, 11 December 1987*, United Nations, New York, 2pp.

United Nations, 1988. International Decade for Natural Disaster Reduction. *Resolution 43/202 Adopted at the Forty-third Session of the United Nations General Assembly, 20 December 1988*. United Nations, New York.

United Nations, 1989. International Decade for Natural Disaster Reduction. *Resolution 44/236 Adopted at the Forty-fourth Session of the United Nations General Assembly, 22 December 1989*, United Nations, New York, 3 p.

Vitruvius, 1st Century BC. *De Architectura.*

19 Some perspectives on geological hazards

R.J. Blong

Abstract The importance of geological disasters ranks with with meteorological disasters on a global scale. Although available data are of limited quality it appears that geological disasters are far less frequent, kill twice as many people per event, produce more damage, but are less widely covered by insurance—suggesting that the victims are affected more severely.

Hazard pairing is too often ignored but mapping programs need to recognise that geological hazards frequently occur together, that a wide variety of types of hazards and consequences may stem from one event, and that equivalencies of risk must be established between hazards before hazard or consequence maps will achieve their full potential for planners and administrators.

Extreme events, with frequencies of about 1 in 10,000 years, are rarely considered by those interested in geological hazards though action has already been taken in some parts of the world to reduce the effects of floods with similar return periods. Examples of eruptions and tsunamis suggest that such infrequent geological events need to be at least included in worst-case scenarios. Several lines of evidence suggest that future disasters resulting from geological hazards are likely to be worse than those experienced so far this century.

Introduction

This paper stems from two earlier papers which were produced as a result of an AGID-ILP workshop in 1984 (Johnson and Blong, 1984; Blong and Johnson, 1986). The second of these papers examined the identification, assessment and impact of geological hazards in the Southwest Pacific and Southeast Asian regions; several problems associated with definitions of geological hazards were examined and a tectonic framework for geological hazards within the region was presented. Seventeen examples of hazards in the region were presented as brief case studies in order to develop particular points that were believed to be important for an understanding of geological hazards and the role of geologists in the reduction of hazard impacts. Subsequent sections of the paper dealt with strategies for coping with disaster, hazard perceptions, and the importance of regional databases on hazards and their impacts.

The present paper summarises many of the points made in the earlier presentations and provides a more recent perspective on geological hazards, their impact and their mitigation. Although the perspectives in this paper are global, most of the examples chosen are from the Southwest Pacific and Southeast Asian regions.

Definitions of geological hazards

Natural hazards can be divided into two broad groups —geological hazards and meteorological hazards—but there is some overlap between the two. Landslides, for example, are often triggered by heavy rain associated with tropical cyclones, and the rates of coastal erosion are markedly influenced by the magnitude and frequency of coastal storms. Is a snow avalanche a geological or a meteorological hazard?

In this paper earthquakes, volcanic eruptions, tsunamis, landslides, subsidence, coastal erosion, coastal progradation, and soil erosion will be regarded as the suite of geological hazards, but most attention is focussed on the first four hazards in this list. Some of these hazards, like earthquakes, are intensive, offering no warning and lasting only a few seconds or minutes, while others, such as coastal progradation and subsidence, are slow onset (or pervasive) hazards, continuing for decades and creating a slow, progressive alteration of the physical environment.

Natural hazards have been defined as:

"extreme geophysical events greatly exceeding normal
human expectations in terms of their magnitude or

Table 1 Criteria for "selecting" hazard events (after Hewitt and Burton, 1975).

1. Property damage extending to more than 20 families, or economic losses (including loss of income, a halt to production, and costs of emergency actions) in excess of US$50,000.

2. Major disruption of social services, including communications failure and closure of essential facilities or economically important establishments.

3. A sudden, unexpected ("unscheduled") event or series of events which puts excessive strain on essential services (police, fire-service, hospitals, and public utilities) and/or requires the calling in of men, equipment, and funds from other administrative jurisdictions.

4. An event in which 10 or more persons are killed or 50 or more persons are injured.

Table 2 Global frequency of natural hazards and deaths per event (based on data in Thompson, 1982).

	Hazard frequency	% Frequency	Deaths per event
Geological hazards			
Landslides	29	2.7	190
Tsunamis	10	1.0	856
Volcanoes	18	1.7	525
Earthquakes	161	15.2	2,652
Meteorological hazards			
Cyclones	211	19.9	2,373
Tornadoes	127	12.0	65
Floods	343	32.3	571
Heatwaves	22	2.1	315
Thunderstorms and gales	36	3.4	587
Total/average	1062	100.0	1140

frequency, and causing significant material damage to man and his works with possible loss of life. As such they are recognised to result from an interaction between systems of human resource management and systems of geophysical events. Particularly serious manifestations of such natural hazards at specific times and places have been identified as natural disasters" (Heathcote, 1979).

Several points arise from this definition. While it is not clear just what "exceeding normal human expectations" means, it is certain that human societies differ in their expectations of natural hazards, and resilience in the face of them. If we regard minerals, landforms, soils and so on as resources which human populations utilise, it may be useful to regard natural hazards as *malevolent resources* (Britton, 1987).

What is meant by "causing significant material damage"? Table 1 lists four criteria for "selecting" hazard events. While the criteria listed in the table seem appropriate enough for a western community and can be readily adjusted for inflation where necessary, they seem quite inappropriate for the reality of hazard impacts on an atoll outlier in, say, Tonga or the Solomon Islands. For example, a tsunami which kills five of the ten able-bodied men, salinises much of the garden area and destroys both of the large fishing canoes of a small atoll community will not rate in terms of any of the criteria in Table 1, but the effects on both the short and long-term viability of the community are, to state the obvious, profound.

Clearly, the magnitude of the impact of a geological hazard needs to be measured in terms of the impact on the afflicted community rather than in terms of monetary loss, the body count, or disruption to (non-existent?) services. As Westgate and O'Keefe (1976) noted: "vulner-

ability embraces not merely the risk from extreme phenomena but the endemic conditions inherent in a particular society which may exacerbate the risk".

In the following discussion the term *hazard* is used to refer to the physical (geological or meteorological) event and *disaster* to refer (loosely) to the consequences of hazards. The term *risk* is used to refer to the product of hazard and vulnerability.

How important are geological hazards?

In providing some perspectives on the question "How important are geological hazards?", several viewpoints and scales need to be examined. These include global and regional scales, and a consideration of the numbers of deaths and monetary losses produced by geological hazards. In all cases, there are major problems with the adequacy of the data available, but some general conclusions can be drawn.

Global and regional scales

One data set that can be used to consider the global importance of geological hazards stems from a survey of reports of natural disasters contained in *The New York Times Index* and *The Encyclopedia Yearbook* for the period 1947-1981 (Thompson, 1982). Only natural disasters that met one or more of the following criteria were considered:

(a) at least US$2.8 million (1981 value) in damage;
(b) at least 100 persons killed; or
(c) at least 100 persons injured.

The similarity between these criteria and those noted in Table 1 is evident and the data set suffers from some of the difficulties already discussed. Clearly, "small" disasters were not considered. Furthermore droughts, the deaths that result from associated starvation and disease, and damage produced by soil erosion, perhaps the most deadly composite of disasters in the 20th century, were not included in the survey. Nonetheless, the results of the survey provide some clues as to the global significance of geological hazards. Table 2 provides a summary of part of Thompson's data.

Only 21% of the events included in this table are geological disasters. However, the mean number of deaths per geological event is 2066, whereas the equivalent figure for meteorological events is 992, suggesting that geological disasters, or at least those events included here, are twice as deadly. Earthquakes and tropical cyclones appear to be the most deadly events on a global scale.

Table 3 provides a regional perspective. Asia, as defined here, stretches from Turkey to Indonesia and includes 85.8% of the 1,208,044 deaths reported in the world in the period (remembering the provisos noted

Table 3 Loss of life by disaster type in Asia and Australasia, 1947-1981 (based on data in Thompson, 1982)

Event	Asia	Australasia
Geological disasters		
Landslides	3,576	0
Tsunamis	7,864	44
Volcanoes	2,805	4000
Earthquakes	333,623	133
Meteorological disasters		
Cyclones	476,816	289
Tornadoes	4,876	0
Floods	171,435	2
Heatwaves	4,155	100
Thunderstorms and gales	20,410	0
Total deaths	1,036,113	4,675

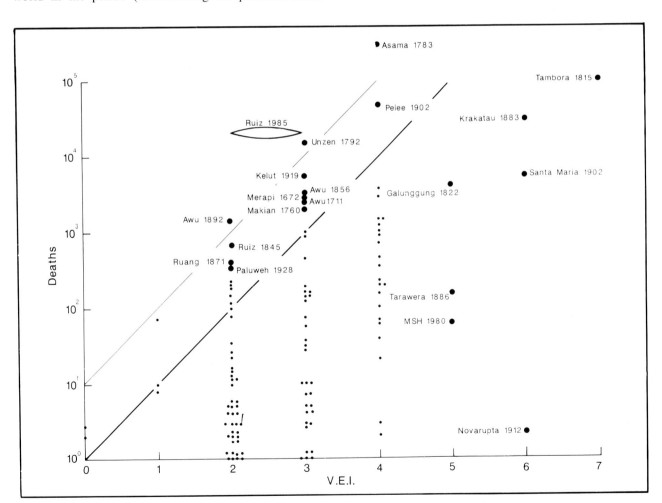

Figure 1 Relationship between Volcano Explosivity Index (see Newhall and Self, 1982) and death toll for a number of eruptions between 1600 and 1986. Note the apparent threshold, limiting death tolls in relatively small eruptions (after Blong, 1988).

Table 4 Average values for major losses, 1960-1983 (based on data in Berz, 1984)

Hazard	No. of events	Average loss (US$ m)	Average insured loss (US$ m)	Ratio insured/ total loss
Eruptions	1	2,700	27	1.0
Earthquakes	8	2,034	35	4.0
Bushfires	1	230	151	65.7
Cyclones	17	999	239	30.7
Storms	7	582	206	35.4
Surges	1	600	10	1.7
Tornadoes	1	1,000	430	43.0
Average		1,163	178	25.4

above). The Australasia region, which includes Australia, New Zealand, Papua New Guinea and Oceania, recorded only 0.4% of the global deaths. The deaths attributed to the listed events do not sum to the totals given, as hazards such as snowstorms, avalanches and coldwaves have been omitted from this table.

In terms of human deaths, Asia is the most disaster prone part of the globe. However, only 34% of the deaths recorded in the table resulted from geological disasters, slightly less than the global average of 37%. While geological hazards killed 89% of those who died

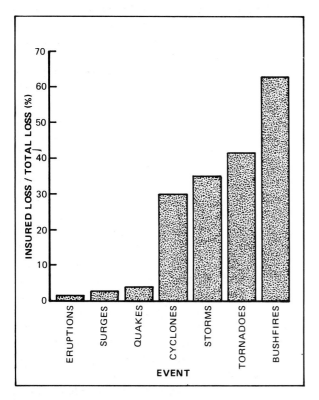

Figure 2 Variations in the average ratio (%) of total losses insured for a small number of hazards. Data from Berz (1984).

in hazards in the Oceania region, the total number of deaths are inconsequential on a global scale.

Despite the inadequacies of the data, it seems that geological disasters are less frequent than meteorological disasters but, on average, more people die per event.

While these data, despite their numerous difficulties, establish which natural disasters are the most deadly, they do not provide direct information about the natural hazards that are most damaging. While most human deaths occur in dwellings or, depending on the time of day, workplaces, there is not always a direct correspondence between deaths and damage. Furthermore, as Figure 1 (after Blong, 1988) illustrates there may be only a limited relationship between hazard magnitude and the death toll. Similarly, Kanamori (1980) has shown for earthquakes that there is often little correspondence between quake magnitude and the number of deaths. This lack of correspondence results from variations in population at risk, ground conditions, the quality and location of buildings and other shelter and, in the case of volcanic eruptions, the nature of the eruption.

Economic losses

While many would consider the available data on deaths in natural hazards to be inadequate, they are certainly more complete and more readily obtained than information on the costs of natural disasters. Not infrequently, estimates are made of first order costs or direct damages, but second order effects (replacement costs, clean-up expenses, homelessness) and third-order effects (changes in property values, unemployment, and the reduced socio-economic viability of affected communities) are rarely considered (cf. Petak and Atkisson, 1982). Moreover, even the first order costs do not apply equally to communities across the globe.

Nonetheless, data collected by Munich Reinsurance Company for the period 1960-1983 provide information on total losses and insured losses for a limited number of events. These data are summarised in Table 4 (after Berz, 1984).

Cyclones, earthquakes and storms dominate the list of events in Table 4 in terms of number of events included. The single eruption (Mount St Helens, USA, 1980) posted the highest individual loss at US$2.7 billion, but the eight earthquakes averaged losses of over $2 billion. These data suggest that geological disasters are amongst the more expensive in terms of costs (again, taking the earlier provisos into account). However, for all the data in Table 4 it must be emphasised that the sample sizes are small and that standard deviations and coefficients of variation are extremely high.

From the data in Table 4 the average proportion of total losses that were insured can be calculated. The variation in this ratio from hazard to hazard is summarised in Figure 2. The most significant point is that only

4% of losses in major earthquakes were insured (range 0.1–12.5%), while cyclones averaged 30.7% (range 3.6–73.3%) and storms averaged 35.4% (range 16.7–53.6%). The average for all the data summarised in Table 4 is 25.4%. Insurance can be viewed as a method of distributing the losses incurred as a result of an individual event to a wider community. In these terms, the geological disasters included in the sample affect the victims more severely than do meteorological disasters.

Hazard pairing

Although geological hazards often occur in combination with other hazards (both geological and meteorological), too often this is forgotten, either during hazard mapping, or by planners and disaster managers. Attempts at hazard prediction or disaster consequence analysis need to take into account the potential diversity of influences in the one geological hazard impact and/or the likelihood that the impact of one hazard will produce other hazards. Some examples illustrate the point.

Earthquakes can produce significant landslides (e.g. Yungay, Peru, 1970), soil liquefaction (Niigata, Japan, 1964), tsunami (Alaska, 1964), or fires (San Francisco, 1906). Large landslides (debris avalanches) from volcanic edifices can produce tsunamis (Ritter, Papua New Guinea, 1888; Unzendake, Japan, 1792). Landslides are also produced by extreme rainfall associated with typhoons (Hong Kong, 1972).

The multiplicity of hazards in one event can produce a variety of consequences; for example, the 1953 earthquake in Fiji killed only six people—two were killed by building collapse induced by ground shaking, three were drowned by a tsunami, and one died in a landslide (see Blong and Johnson, 1986).

Similarly, volcanic eruptions should not be regarded as a single hazard but, often, as a complex of interrelated hazards with differing consequences for both humans and structures at different locales at varying distances and directions from the vent. Table 5 (after Blong and Aislabie, 1988) illustrates the point with regards to the likely consequences for humans in the next eruption of Rabaul volcano (Papua New Guinea), assuming that the eruption is similar in character and magnitude to the 1937 eruption (which killed 441 people).

While the analysis summarised in Table 5 is based on limited data, it suggests relative magnitudes and indicates wide variations in the consequences of differing types of volcanic hazards. For example, the injury/death ratio estimates vary by two orders of magnitude.

Risk mapping

In collision margin zones of the world such as Japan, the Phillipines, Indonesia or Papua New Guinea, it is not uncommon for earthquake, volcanic, tsunami and landslide hazards to present problems for the one community

Table 5 Areas affected and hazard risks, Rabaul Volcano (after Blong and Aislabie, 1988)

Hazard	Areas affected (square km)	Death rate (per thousand)	Injury-to-death ratio	Most common injury types
Quakes > magnitude 6	30	1	4	head injuries, fractures, soft tissue injuries
Ballistic projectiles	23	10	2	head injuries
Tephra falls				head injuries, burns, fractures, respiratory problems
>183 cm	8	50	20	
183–61 cm	18	25	40	
61–15 cm	51	10	50	
15–2.5 cm	518+	1	100	
Pyroclastic flows and surges	21	400	1.5	burns, fractures
Tsunamis	15	50	3	fractures
Toxic gases	6	100	5	respiratory problems

or region. Excellent maps exist in numerous countries of such hazards at scales and with explanations that indicate practical use is intended.

Nonetheless, it is rare to find vulnerability, probability or risk maps that consider more than one type of geological hazard. Where multiple maps have been produced of the one area, they are often at different scales, based on different criteria, or use boundaries to categories which are incompatible from one map to another. Maps frequently rank areas as "low", "medium" and "high" hazard areas. In such situations it is almost impossible to compare a "low" earthquake risk with a "medium" volcanic risk. Often, it is difficult to determine what might be meant by a "high" risk.

Planners, politicians and others who might wish to consider questions such as "which area of jurisdiction has the greatest accumulated geological risk?" or "where would we receive the best return for money spent on hazard reduction?" are unlikely to find answers in studying the products of an uncoordinated mapping program.

If coordinated mapping programs were easy, they would certainly have been undertaken in many countries; however, it is difficult to visualise clearly a common basis for mapping. One possibility involves a common denominator such as the one in 50-year return period event, taking into account the proportions of administrative areas likely to be affected by events of this magnitude, and the expected levels of damage. The 1 in 50 year event has merit in that the time period approximates the anticipated life of many structures.

An example illustrates the benefits of such an

Table 6 Geological Risk Rating for selected towns in Papua New Guinea (after Blong, unpublished, 1988)

Town	Earthquake rating	Volcano rating	Tsunami rating	Risk rating
Arawa	6	2	0.1	8
Daru	2	0.1	0.03	2
Goroka	5	0.1	—	5
Kavieng	3	0.1	0.1	3
Kieta	6	2	0.1	8
Kimbe	10	4	0.1	14
Kokopo	10	4	0.2	14
Lae	5	0.1	0.2	5
Madang	5	1	0.2	6
Mount Hagen	3	0.1	—	3
Port Moresby	2	1	0.01	3
Rabaul	10	5	0.2	15
Wewak	6	2	0.2	8

approach. Table 6 sets out values which provide estimates of earthquake, volcanic, and tsunami risk for several of the major towns in Papua New Guinea. The table has been constructed in such a way that numbers can be added or divided, to assess relative earthquake and volcanic consequences in a town or to compare the total risk in one town with that in another (Blong, unpublished). Undoubtedly, the estimates are far from perfect and rely (as do most risk assessments!) too heavily on expert judgement, but at least they result from a practical assessment method which allows realistic comparisons to be made.

It is interesting to note that the framework outlined above and the results in Table 6 stemmed from a widespread view that the 1983-84 volcanic crisis in Rabaul indicated that the town site should be abandoned in favour of less hazard-prone areas such as the towns of Kokopo or Kimbe. The table indicates: (i) that earthquake risk in Rabaul is significantly greater than the volcanic hazard—at least in the terms of the present analysis; and (ii) the towns of Kokopo and Kimbe are no less at risk than Rabaul!

The analysis also indicated those towns in which further investigation of geological hazards and their potential consequences was warranted. Coordinated microzonation studies, one of the most important global tasks for those with a professional interest in geological hazards and disasters, are now underway in those towns.

Extreme events

The term "extreme event" is often used to refer to natural hazards with recurrence intervals of less than once in a hundred years. While 1 in 1000 year events are only rarely considered in planning, locating and designing everyday structures, such events have a better than 1 in 20 chance of occuring in the lifetime of many dwellings. In recent years notions of Probable Maximum Floods with return periods of perhaps 1 in 10,000 years (or even less frequently), have

been taken into account in the design or upgrading of large dams. Such events presumably have a probability of occurence in the life of a structure of about 2% or less. But what of geological extreme events with similar low frequencies of recurrence?

Although most notions of an extreme volcanic eruption probably equate with something like the 1883 eruption of Krakatau, this eruption rated as Volcanic Explosivity Index (VEI) = 6, indicating that 10-100 km^3 of eruptive products were released (Newhall and Self, 1982; Simkin and others, 1981). On the logarithmic scale, the maximum VEI = 8 indicates the production of more than 1000 km^3 of eruptive products. No eruptions of this magnitude have been recorded in the Holocene, but a good example of a VEI = 8 eruption is the 75,000 years BP eruption of Toba in Sumatra which produced an estimated 2800 km^3 (dense rock equivalent), probably in a period of 9-14 days (Rose and Chesner, 1987).

Even eruptions with a VEI = 7 have not been witnessed by literate humans. However, large eruptions have been recorded in the last 2000 years; the 1800 year BP eruption of Taupo in New Zealand, possibly dated to 186 AD by Chinese chronicles (Wilson and others, 1980), was believed to be the most powerful eruption ever known. Similarly, the 1400 BP eruption of Rabaul, possibly dated to 536 AD by Mediterranean chronicles (Stothers, 1984), was at least a VEI = 6.

While eruptions of such magnitude have global consequences, and recurrence intervals of 1 in 1000 to 1 in 2000 years, to my knowledge no one has even modelled the global consequences of such events; some aspects of nuclear winter scenarios may be an appropriate starting point.

Although of more limited consequences, cataclysmic tsunamis are also possibilities—that is, high probabilities on a Holocene timescale. Seismic sea waves produced by the Aleutian (1946), Chile (1960) and Alaska (1964) earthquakes produced extensive damage across the Pacific Basin, while tsunami generated by the 1883 eruption of Krakatau caused damage up to 30-40 m above sea level (Wharton, 1888).

In recent years evidence has emerged of the potential for widespread damage that could result from tsunamis similar in magnitude to those generated by prehistoric landslides. Dawson and others (1988) and Smith and Dawson (1990) have suggested that the earthquake-triggered submarine Storegga slides of about 7000 years BP produced widespread tsunamis along the east coast of Scotland. On a more cataclysmic scale, prodigious submarine landslides on the Hawaiian Ridge (Moore and others, 1989) have produced tsunamis large enough to cause damage around much of the Pacific margin. Lipman and others (1988) have suggested that the Alika slide from Hawaii Island, with a volume of 1500-2000 km^3, produced tsunamis with run-up to a height of as much as 325 m on the island of Lanai, more than 100 km away. Submarine landslides and associated tsunamis of this magnitude from the Hawaiian Ridge have recurrence intervals estimated at 1 in 25,000 to 1 in 100,000 years.

In some areas of the world (e.g. United States, Australia), considerable thought, design, and capital have gone into the modification of dams threatened by Probable Maximum Floods with return periods of 1 in 10,000 years or even less frequently, in order to protect valuable downstream assets. On the other hand, knowledge of geological hazards with similar recurrence intervals but with severe regional or even global consequences has produced little response and negligible action. At the least, as Lipman and others (1988) noted, such infrequent events need to be considered in worst-case scenarios.

Future disasters

Future disasters are likely to be larger than past disasters. In making this statement several factors have been taken into account.

Firstly, efforts at prediction of geological hazards have not been very successful. While notable successes have occurred in the prediction of some earthquakes (e.g. Haicheng, China, 1976), there have also been notable failures. An earthquake prediction system with a high probability of success and effective across much of the globe is unlikely to appear in the next 20 years. In the case of volcanic eruptions, there is some evidence that the biggest eruptions are likely to be produced by volcanoes that have long been quiescent. If this is the case, few of the right volcanoes are even being monitored.

Secondly, for a prediction to be useful, it must be coupled with an effective warning system that delivers clear, unambiguous information to a population sufficiently prepared (or trained?) in adaptive responses which enhance the chances of survival, perhaps within a matter of minutes. Predictions may be possible, even in real-time, but without warnings and responses by those at risk the consequences may be much the same. The elegant use of electronic technology and satellite communication in the THRUST program allows early warning of locally generated tsunamis to be sent to coastal cities in underdeveloped countries (Bernard and others, 1988). But how are the warnings disseminated to those at risk? The death of more than 20,000 people in the city of Armero during the 1985 eruption of Nevado del Ruiz, Colombia (see Hall, this volume) illustrates the tragic consequences that can occur when the chain "prediction→ warning→ response" is broken.

Thirdly, careful monitoring and successful prediction is most likely in industrialised countries where geological hazards produce relatively few deaths but result in large damage bills (see the model outlined by Burton and others, 1978). Recent, relatively small, earthquakes such as Loma Prieta (California, 1989) and Newcastle (Australia, 1989) emphasise the point—in the latter case economic losses averaged about A$100 million per death!

However, in underdeveloped countries geological disasters can produce moderate property losses and very high death tolls. Recent earthquakes in Iran and Armenia, and the Nevado del Ruiz eruption in Colombia, provide examples.

In underdeveloped countries death tolls will continue to rise because of larger populations and increasing marginalisation of the populace whereby greater proportions of citizens live on the peripheries of urban conglomerations on land more prone to flood, storm surge, ground shaking, or landsliding than the city centre areas. In industrialised countries, where populations have lower (even negative) growth rates, geological disasters will claim more lives because an ageing population is more vulnerable (see, for example, Baker and others, 1974).

The costs of major disasters will also escalate rapidly, partly as the world moves further into the global economy, partly because of increasing dependence on electronic communication. The next Richter magnitude 7.5+ earthquake in the vicinity of Los Angeles or Tokyo will easily produce damage in excess of the total US$42 billion listed in Table 4 for the period 1960–1983 (see for example, Ziony, 1985). What would be the global human death toll resulting from the devastation of the North American grain belt and the world's grain surplus by a VEI = 7 or 8 eruption of the Yellowstone caldera? Similarly, what would be the consequences for the world's financial markets, global communications and the global economy of a volcanic eruption that spread a few centimetres of wet, magnetic tephra over the whole of Tokyo—in effect a repeat of the 1707 eruption of Fujiyama?

Conclusions

The perspectives on geological hazards and their consequences presented here are somewhat cynical, perhaps over-dramatised and, some would argue, sensationalised. Undoubtedly, most investigations of geological hazards and disasters will remain focussed on small areas and single events in developed parts of the world. However, at the most simplistic level, the viewpoints presented here suggest:

(a) That global efforts at hazard and disaster investigation should be focussed on Asia, where 85% of the deaths in geological hazards occur;

(b) that attention should be directed toward earthquake-prone areas because these have historically proved to suffer most in terms of human life loss and damage;

(c) that, nonetheless, efforts should be made to produce hazard maps and risk assessments in which it is possible to compare the risks and consequences of a variety of geological (and meteorological?) hazards;

(d) that some attention should be directed at the risk of and consequences from extreme events with low probabilities of occurrence but very high magnitudes;

(e) recognising, despite valuable efforts at prediction of geological hazards, that burgeoning populations, economic marginalisation, the vulnerability of the more numerous elderly (and the young in less developed countries), and increasing reliance on expensive technology and rising standards of living, will result in greater life loss and damage in future disasters, and increased insurance loss;

(f) that geologists need to be involved not only in hazard identification and consequence assessment but also in integrating efforts at strengthening links in the chain "prediction→warning→response".

The 1990s, the International Decade for Natural Disaster Reduction, is an appropriate time to tackle these issues with determination.

References

Baker, S.P., B.O. Neill, W. Haddon and W.B. Long, 1974. The injury severity score: a method for describing patients with multiple injuries and evaluating emergency care. *Journal of Trauma*, vol 14(3), p 187–196.

Bernard, E.N., R.R. Behn, G.T. Hebenstreit, F.I. Gonzalez, P. Krumpe, J.F. Lander, E. Lorca, P.M. McManamon and H.B. Milburn, 1988. On mitigating rapid-onset natural disasters: Project THRUST (Tsunami Hazards Reduction Utilizing Systems Technology). *EOS*, June 14, 1988.

Berz, G., 1984. Research and statistics on natural disasters in insurance and reinsurance companies. *The Geneva Papers on Risk and Insurance*, vol 9(31), p 135–157.

Blong, R.J., 1988. Assessment of eruption consequences. *Kagoshima International Conference on Volcanoes, Proceedings*, p 569–72.

Blong, R.J., and C. Aislabie, 1988. The impact of volcanic hazards at Rabaul, Papua New Guinea. *Papua New Guinea Institute of National Affairs*, Discussion Paper no 33, p 1–201.

Blong, R.J., and R.W. Johnson, 1986. Geological hazards in the southwest Pacific and southeast Asian region: identification, assessment, and impact. *Journal of Australian Geology and Geophysics*, Bureau of Mineral Resources, vol 10, p 1–15.

Britton, N.R., 1987. Towards a reconceptualization of disaster for the enhancement of social preparedness. In: R.R. Dynes,

B. de Marchi and C. Pelanda (eds), *Sociology of Disasters—Contribution of Sociology to Disaster Research*, Franco Angeli, Milan, p 32–55.

Burton, I., R.W. Kates and G.F. White, 1978. *The Environment as Hazard*. Oxford University Press, New York.

Dawson, A.G., D. Long and D.E. Smith, 1988. The Storegga slides: evidence from eastern Scotland for a possible tsunami. *Marine Geology*, vol 82, p 271–76.

Heathcote R.L., 1979. The threat from natural hazards in Australia. In: R.L. Heathcote and B.G. Thom (eds), *Natural Hazards in Australia*, Australian Academy of Science, Canberra, p 3–12.

Hewitt, K., and I. Burton, 1975. *The Hazardousness of a Place: a Regional Ecology of Damaging Events*. University of Toronto Press, Toronto.

Johnson, R.W. and R.J. Blong, 1984. Vulnerability and the identification and assessment of geological hazards in the Southwest Pacific and Southeast Asia. In: M.B. Katz and E.J. Langevad, (eds), *The Geosciences in International Development*, AGID-ILP Workshop Proceedings, p 217–225.

Kanamori, H., 1980. The size of earthquakes. *Earthquake Information Bulletin*, vol 12, p 10–15.

Lipman, P.W., W.R. Normark, J.G. Moore, J.B. Wilson and S.E. Gutmacher, 1988. The giant submarine Alika debris slide, Mauna Loa, Hawaii. *Journal of Geophysical Research*, vol 93 B5, p 4279–4299.

Moore, J.G., D.A. Clague, R.T. Holcomb, P.W. Lipman, W.R. Normark and M.E. Torresan, 1989. Prodigious submarine landslides on the Hawaiian Ridge. *Journal of Geophysical Research*, vol 94 B13, p 17465–17484.

Newhall, C.G., and S. Self, 1982. The volcanic explosivity index (VEI): an estimate of explosive magnitude for historical volcanism. *Journal of Geophysical Research*, vol 87 C2, p 1231–1238.

Petak, W.J., and A.A. Atkisson, 1982. *Natural Hazard Risk Assessment and Public Policy: Anticipating the Unexpected*, Springer-Verlag, New York.

Rose, W.I., and C.A. Chesner, 1987. Dispersal of ash in the great Toba eruption, 75 ka. *Geology*, vol 15, p 913–917.

Simkin, T., L. Siebert, L. McClelland, D. Bridge, C. Newhall and J.H. Latter, 1981. *Volcanoes of the World*, Hutchinson Ross, Stroudsburg.

Smith, D., and A. Dawson, 1990. Tsunami waves in the North Sea. *New Scientist*, vol 127(1728), p 30–33.

Stothers, R.B., 1984. Mystery cloud of AD 536. *Nature*, vol 307, p 344–345.

Thompson, S.A., 1982. Trends and developments in global natural disasters, 1947 to 1981. *University of Colorado Institute of Behavioral Science Natural Hazards Research*, Working Paper 45.

Westgate, K., and P. O'Keefe, 1976. Some definitions of disaster. *Bradford University Disaster Research Unit*, Occasional Paper 4.

Wharton, W.J.L., 1888. On the seismic sea waves caused by the eruption of Krakatoa, August 26th and 27th, 1883. In: G. Symons (ed.), *The Eruption of Krakatoa and Subsequent Phenomena*, Royal Society of London, p 89–107.

Wilson, C., N. Ambraseys, J. Bradley and G. Walker, 1980. A new date for the Taupo eruption, New Zealand. *Nature*, vol 288, p 252–53.

Ziony, J.I. (ed.), 1985. Evaluation earthquake hazards in the Los Angeles Region—an earth-science perspective. *US Geological Survey Professional Paper*, vol 1360, p 1–505.

20 The International Decade for Natural Disaster Reduction and the Geohazards Unit at Polytechnic South West, Plymouth, UK

S.C. Scott

Abstract Many natural disasters have been caused by geological events. The current response of governments and commercial interests to geohazards tends to be *reactive* rather than *proactive* which, in developing countries, has led to disastrous and largely unnecessary loss of life and property. These problems are currently being addressed by UN resolutions 42/169 and 44/236, declaring the decade 1990–2000 an International Decade for Natural Disaster Reduction. The Department of Geological Sciences at Polytechnic South West responded to this initiative by consolidating its geohazard expertise in the *Geological Hazards Assessment, Mitigation and Information Unit*, which offers a geohazard assessment and mitigation consultancy service, a geohazard information and database facility, and geohazard training and awareness courses. It is committed to the geohazards education of politicians, administrators and the public, particularly from developing countries, in order to create the climate for a wider proactive response to geohazards.

Introduction

I tuned into British television news last year and saw graphic images of the Iranian earthquake of 21 June 1990, only hours after the event had taken place. This brought home the potency of modern communication technology as a rapid means of disseminating information about geohazard events, initiating an immediate international relief response. Efficient and fast worldwide satellite communication systems now ensure that we are made aware of major geohazard events and their damaging effect on life, property, production and communications, almost as they happen! Such rapid and graphic communication of disaster events arouses international public concern and was certainly instrumental in prompting the immediate international responses to the 1989 Loma Prieta earthquake in California; the 1988 Spitak, Armenia (USSR) earthquake, the 1985 Nevado del Ruiz (Colombia) lahar (see Hall, this volume), the 1985 Mexico City earthquake (see Degg, this volume),

and the 1980 Mount St Helens (Cascade Range, USA) pyroclastic flow eruption. Major geodisasters have now become a world problem rather than a local problem, calling for international co-operation in proactive and reactive responses.

The term "major geodisaster" used in the above context is normally defined using "western" criteria (Hewitt and Burton, 1975). For developing countries these criteria are in many cases totally inappropriate and need to be redefined on the basis of local, more realistic socio-economic and environmental conditions. For example, a geodisaster such as a landslide or mudflow, which destroys the crop fields and food supply of a small Indonesian island community would, in their view, be considered as a "major" geodisaster. However, "western" criteria applied to the same event would not recognise it as such. Criteria for defining the seriousness of geohazard events must be relevant to the region being considered, its peoples, their vulnerability and their way of life (see also Blong, this volume).

Another modern development affecting the problems that geodisasters pose has been the concentration of a growing world population into cities. The increased severity of geodisasters in terms of loss of life and property is in part due to this concentration trend. Degg (this volume) emphasises that by the year 2000 the weight of the world's most populated cities would have switched to the developing world, further increasing its vulnerability to geodisaster. Geohazard assessment and risk mitigation in developing countries currently receives a lower priority than economic advancement. Many of these countries also lack the specific skills to monitor geohazards effectively and formulate plans for effective hazard management. This lack of economic and scientific resources, plus the lack of political will, has severely influenced the manner in which, and the degree to which, the problem of geohazards in developing countries can be addressed. Unfortunately, the current response tends to be *reactive*—responding to the disaster—rather than *proactive*—responding to the threat. *There is an urgent need to make governments, development planners and other decision makers aware of the human and economic advantages of proactive geohazard assessment and mitigation measures.*

Although the developed nations have the relevant scientific knowledge, research and skills base, these need to be (i) integrated with the socio-economic aspects of geohazards, and (ii) more effectively applied to geohazard assessment and mitigation, particularly in the developing countries, for a more complete response to be achieved. It is precisely these problems which are currently being addressed by United Nations Resolution 42/169 declaring the decade 1990–2000 an International Decade for Natural Disaster Reduction.

The UN International Decade for Natural Disaster Reduction (IDNDR)

Events leading to the declaration of the IDNDR started in 1984 when Dr Frank Press, seismologist and President of the United States Academy of Sciences, delivered a speech to the 8th World Conference on Earthquake Engineering, arguing the case for an international programme to reduce natural hazards. In this, he proposed the establishment of an International Decade for Natural Hazard Reduction commencing in 1990. The proposal, which had unanimous support at the meeting, was followed by the appointment of a US National Research Council Advisory Committee on the IDNDR under the chairmanship of G.W. Housner. This committee was given the brief to evaluate the potential for such an effort and how it might be realized. The report of the committee, entitled *Confronting Natural Disasters; An International Decade for Natural Hazard Reduction*, was

published in 1987.

The proposal for such a decade was taken up by the United Nations General Assembly which unanimously passed resolution 42/169 on 11th December, 1987, declaring the decade beginning 1990 as an *International Decade for Natural Disaster Reduction* (see panel for full transcript of the resolution; also note the change from "Hazard" to "Disaster" in the title). The aim of the resolution was "*to reduce catastrophic life loss, property damage and social and economic disruption from natural hazards*" by specifically addressing rapid-onset (short duration) hazards associated with earthquakes, windstorms, floods, tsunamis, landslides, volcanic eruptions and wildfires. Long-term hazards such as drought, insect plagues and desertification were not specifically addressed. The aim of the IDNDR is to be achieved through a number of objectives:

1. To improve the capacity to mitigate the impact of natural disasters, particularly through the use of early warning systems.
2. To apply existing knowledge.
3. To foster scientific and engineering research.
4. To disseminate information.
5. To develop measures and programmes.
6. To train, educate and evaluate.

These objectives are amplified by Brook (this volume). The UN resolution requires that:

(a) each government establishes a National Committee to act as a focus for IDNDR activities;
(b) all nations should participate in the decade;
(c) the UN should promote and facilitate the IDNDR through UNDRO and UNESCO.

The full objectives for the IDNDR and the institutional framework for technical conduct of the programme are defined in the recent UN resolution 44/236 (see panel for the full transcript). UN bodies will provide the link between engineering capabilities, social strategies and scientific endeavours.

Further detailed discussions of the IDNDR and its implications can be found in Housner (1989), Holland (1989) and Brook (this volume).

The Geohazards Unit: the response of Polytechnic South West to the UN initiative

Many natural disasters, especially the spate of incidents during the past 10 years, have been caused by geological events. Non-implementation of a hazard assessment programme prior to many of the hazard events, and poorly co-ordinated post-disaster response mechanisms, particularly in developing countries, have led to disastrous and largely unnecessary loss of life and property.

The Department of Geological Sciences at Polytechnic South West (Plymouth, UK) responded to the UN initiative on disaster reduction by consolidating its wide range of geological hazard expertise into the *Geological Hazards Assessment, Mitigation and Information Unit*, which was established in October 1989. The aims of the unit are:

1. To increase the awareness of governments, government agencies, development planners, civil authorities and commercial and industrial companies to potential geological hazards, and
2. To assist them in strategic pre-disaster planning by providing a range of geohazard consultancy services:
   (a) geological hazard assessment;
   (b) mitigation advice;
   (c) geological hazards information and database facility;
   (d) training courses in geological hazard processes, assessment techniques and mitigation procedures.

The unit offers specific expertise in the following types of geological hazards:

Volcanic—eruption of pyroclastic flows; lava flows; lahars

Earthquakes

Glacial—lake overflow and collapse of ice or moraines;

Slope stability—engineering aspects of rock slopes and slopes in unconsolidated materials

Hydrogeological—groundwater fluctuations; water geochemical hazards; waste disposal hazards

Subsidence—mine workings collapse; natural cavern collapse

Coastal erosion—storm events; longshore drift; tidal processes; erosion rates

Coastal progradation—storm events; longshore drift; estuarine and spit advance.

The Unit is particularly committed to the geohazard education of politicians, administrators and the public, in order to create the climate for promoting a wider proactive response to geohazards.

The Geohazards team currently consists of Dr S.C. Scott (unit co-ordinator, volcanic hazards), Prof D. Tarling (earthquake hazards), Prof D. Huntley (coastal hazards), Dr J.M. Reynolds (glacial, slope stability and earthquake hazards), Dr K. Vines (hydrogeological hazards), Prof W. Dearman (subsidence and slope-

stability hazards) and Ms C. Brown (subsidence and slope stability hazards). The unit also makes use of the geological expertise offered by other colleagues within the Department of Geological Sciences, as and when appropriate. Negotiations are currently under way for the participation of a sociologist and financial analyst in the team, in order to identify and integrate the broader social/financial/political consequences of geohazard events and mitigation measures.

Since its inception in late 1989, members of the unit have been quick to develop a portfolio of geohazard activity in the following areas:

Geohazard assessment and mitigation

(a) Landslide, glacial and mudflow hazards in Peru (see Reynolds, this volume).
(b) Slope stability hazards on china clay/quartz waste tips in South West England.
(c) Volcanic hazards associated with the Kumo-Wakunai road project, Bougainville Island.
(d) Volcanic hazards associated with Recent volcanoes in Mexico.
(e) Subsidence hazards within the Plymouth Limestone.

Geohazards information

(a) Radio commentaries on *Geohazards* and *The International Decade for Natural Disaster Reduction*, on BBC (both national and Radio Devon) and on Australian Broadcasting Corporation.
(b) Informed replies to media, public service and commercial geohazard enquiries.
(c) Prominent participation in nationally convened geohazard meetings.
(d) Publication of geohazard research results.

Geohazards education

(a) Presentation of geohazards assessment techniques and mitigation procedures in our undergraduate *Applied Geology, Applied & Environmental Geology* and *Environmental Science* degree courses.
(b) Presentation of Geohazards public-awareness lectures to local and national audiences.

Further development of the Unit's activities and its covering of operating costs is to be achieved by (i) an expansion of its geohazards consultancy provision, (ii) marketing of its growing geohazard information and database facility, (iii) extending its national and international collaboration on geohazard research projects, and (iv) a continued commitment to the provision of geohazards education and training. To meet this latter commitment, it will shortly be embarking on a survey to

UN International Decade for Natural Disaster Reduction

The General Assembly,

Recalling its resolution 3345 (XXIX) of 17 December 1974, in which it requested the Secretary-General to take appropriate measures to provide facilities for co-ordinating multidisciplinary research also at regional level aimed at synthesizing, integrating and advancing existing knowledge on the interrelationships between population, resources, environment and development, in order to assist Member States, particularly the developing countries, and the organizations of the United Nations system in their efforts to cope with the complex and multidimensional problems related to this field in the context of social and economic development,

Noting with appreciation the important contribution made by the World Commission on Environment and Development, (20) which calls for new national and international approaches in dealing with the various factors affecting the environment, including natural disasters,

Considering that natural disasters, such as those caused by earthquakes, windstorms (cyclones, hurricanes, tornadoes, typhoons), tsunamis, floods, landslides, volcanic eruptions, wildfires and other calamities of natural origin, have claimed about 3 million lives worldwide in the past two decades, adversely affected the lives of at least 800 million more people and resulted in immediate damages exceeding $23 billion,

Considering also that, among disasters of natural origin, drought and desertification are resulting in enormous damage, particularly in Africa, where the recent drought threatened the lives of more than 20 million people and uprooted millions of others,

Recognizing that the effects of such disasters may damage very severely the fragile economic infrastructure of developing countries, especially the least developed, land-locked and island developing countries, and thus hamper their development process,

Recalling the report of the Secretary-General on the work of the Organization, particularly the section concerning natural disasters and the merits in proposals that have been made to stimulate international study, planning and preparations on this subject over the next decade under the auspices of the United Nations, (21)

Also taking note with appreciation of the report of the Secretary-General concerning the existing mechanisms and arrangements within the United Nations system for disaster and emergency assistance and co-ordination, (22)

Recognizing the responsibility of the United Nations system for promoting international co-operation in the study of natural disasters of geophysical origin and in the development of techniques to mitigate risks arising therefrom, as well as for co-ordinating disaster relief, preparedness and prevention, including prediction and early warning,

Convinced that concerted international action for the reduction of natural disasters over the course of the 1990s would give genuine impetus to a series of concrete measures at the national, regional and international levels,

Recognizing that the primary responsibility for defining the general goals and directions of efforts undertaken in the framework of an international decade for natural disaster reduction and for implementing the measures that would result from its activities lies with the Governments of the countries concerned,

Considering that the concept of a global programme for natural disaster reduction is predicated on collaborative efforts among culturally and economically diverse nations, together with relevant organizations of the United Nations system and other national and international non-governmental organizations, including the scientific and technological institutions concerned,

1. *Recognizes* the importance of reducing the impact of natural disasters for all people, and in particular for developing countries;

2. *Recognizes further* that scientific and technical understanding of the causes and impact of natural disasters and of ways to reduce both human and property losses has progressed to such an extent that a concerted effort to assemble, disseminate and apply this knowledge through national, regional and world-wide programmes could have very positive effects in this regard, particularly for developing countries;

3. *Decides* to designate the 1990s as a Decade in which the international community, under the auspices of the United Nations, will pay special attention to fostering international co-operation in the field of natural disaster reduction, and to take a decision at its forty-third session on the content and modalities of United Nations participation therein after having considered the report of the Secretary-General referred to in paragraph 9 of the present resolution;

4. *Decides* that the objective of this decade is to reduce through concerted international action especially in developing countries, loss of life, property damage and social and economic disruption caused by natural disasters, such as earthquakes, windstorms (cyclones, hurricanes, tornadoes, typhoons), tsunamis, floods, landslides, volcanic eruptions, wildfires and other calamities of natural origin, such as grasshopper and locust infestations, and that its goals are:

   (a) To improve the capacity of each country to mitigate the effects of natural disasters expeditiously and effectively, paying special attention to assisting developing countries in the establishment, when needed, of early warning systems;
   (b) To devise appropriate guidelines and strategies for applying existing knowledge, taking into account the cultural and economic diversity among nations;
   (c) To foster scientific and engineering endeavours aimed at closing critical gaps in knowledge in order to reduce loss of life and property;
   (d) To disseminate existing and new information related to measures for the assessment, prediction, prevention and mitigation of natural disasters;
   (e) To develop measures for the assessment, prediction, prevention and mitigation of natural disasters through programmes of technical assistance and technology transfer, demonstration projects, and education and training, tailored to specific hazards and locations, and to evaluate the effectiveness of those programmes;

5. *Requests* the Secretary-General, in co-operation with the appropriate organizations of the United Nations system and relevant scientific, technical, academic and other non-governmental organizations, to develop an appropriate framework to attain the objective and goals referred to in paragraphs 3 and 4 of the present resolution and to submit a report thereon to the General Assembly at its forty-fourth session through the Economic and Social Council;

6. *Recommends* that, if necessary, extrabudgetary resources be provided for the preparation of the above-mentioned report and considers that, for this purpose, voluntary contributions from countries, international organizations and other organizations are highly desirable;

7. *Calls upon* all Governments to participate during the decade for concerted international action for the reduction of natural disasters and, as appropriate, to establish national committees, in co-operation with the relevant scientific and technological communities, with a view to surveying available mechanisms and facilities for the reduction of natural hazards, assessing the particular requirements of their respective countries or regions in order to add to, improve or update existing mechanisms and facilities and develop a strategy to attain the desired goals;

8. *Further calls upon* Governments to keep the Secretary-General informed of their countries' plans and of assistance that can be provided so that the United Nations may become an international centre for the exchange of information, the storing of documents and the co-ordination of international efforts concerning the activities in support of the objective and goals referred to in paragraphs 3 and 4 above, thus enabling each Member State to benefit from the experience of other countries;

9. *Requests* the Secretary-General to report to the General Assembly at its forty-third session on progress made in the preparations outlined above with particular emphasis on defining the catalytic and facilitating role envisaged for the United Nations system.

(20) A/42/427, annex.
(21) *Official Records of the General Assembly, Forty-second Session, Supplement No 1 (A/42/1), section II.*
(22) A/42/657.

determine the national and international market needs for short geohazards awareness courses aimed at the non-specialist. The results of this survey will dictate course content, marketing strategy and targeted client groups, particularly those in developing countries. These non-specialist awareness courses, designed to create a climate for promoting a wider proactive response to geohazards, will form an additional component of a training course portfolio covering specific geohazard processes, assessment techniques and mitigation procedures. All courses are tailor-made to cover specific client needs—be they local or international.

Our geohazard information files are constantly kept up to date and we are further developing our computerised geohazards database facility. This facility integrates an events database, a techniques database and a literature database, all of which will be accessible to client enquiry.

The IDNDR presents a timely opportunity to develop and apply our geohazard assessment and mitigation knowledge by collaborating with scientific colleagues and decision makers/governments, particularly in the developing countries. As a scientific community, let's not miss the opportunity presented to us!

References

Hewitt, K., and I. Burton, 1975. *The hazardousness of a place: a regional ecology of damaging events.* University of Toronto Press, Toronto.

Holland, G.L., 1989. Observations on the International Decade for Natural Disaster Reduction. *Natural Hazards*, vol 2, p 77–82.

Housner, G.W., 1989. An International Decade for Natural Disaster Reduction: 1990–2000. *Natural Hazards*, vol 2, p 45–75.

Subject index

223

Place name index